EXPERIENTIA SUPPLEMENTUM 15

Comparative Physiology of the Heart: Current Trends

Proceedings of a Symposium
held at Hanover, New Hampshire (USA) on 2 to 3 September 1968

Edited by
F. V. McCann, Dartmouth Medical School, Hanover, New Hampshire, USA

A satellite symposium
of the XXIV International Union of Physiological Sciences

Supported by
the Council on Basic Science of the American Heart Association

1969   Springer Basel AG

ISBN 978-3-0348-6788-7   ISBN 978-3-0348-6800-6 (eBook)
DOI 10.1007/978-3-0348-6800-6

Library of Congress Catalog Card Number
© Springer Basel AG 1969
Originally published by Birkhäuser Verlag Basel in 1969.
Softcover reprint of the hardcover 1st edition 1969

Foreword

As a discipline, Comparative Physiology has been subjected to numerous attempts to prescribe its content, outline its boundaries and define its objectives. In view of diverse interpretations, it has thus eluded categorization into the more definitive areas of mammalian, invertebrate, plant and bacterial physiology. Comparative Physiology should be considered not as a distinct and separate discipline, but rather as a method of approach that allows the study of broad physiological problems of a fundamental nature. Extrapolation of these studies can then define the extent to which common mechanistic principles can be applied to explain function in living cells. Thus, the selection of the problem for investigation and the astuteness of the questions posed supersede in immediate importance the detailed behavior of the particular system itself. How biological systems of simpler design and/or specialized function can be utilized or adapted as models in the search for common denominators of function, and the manner in which physiological mechanisms have evolved to ensure survival in a hostile or changing environment constitute the substrate of Comparative Physiology.

Modern concepts of fundamental physiological mechanisms have evolved from earlier studies initiated on biological systems uniquely suited to the problem at hand. The squid axon has served as a model for membrane excitability studies, insect flight muscle provided material for classic descriptions of ultrastructure and biochemical processes of striated muscle and invertebrate eyes have been extensively used to elucidate mechanisms of retinal activity.

The formulation of basic concepts of cellular activity with regard to the heart has not generally followed this line of development. By far the major research effort in the field of cardiac physiology has centered on vertebrate and, in particular, mammalian hearts. Despite the complexities encountered in these hearts, there has been little interest in searching for hearts of biological forms that, because of cellular specialization, might afford a simpler system with which to work. On the other hand, those working with less specialized hearts have not extensively applied current techniques that have been so well developed and applied to the more advanced myocardia. It appeared that those working on the hearts of various animals might benefit from an exchange of information and that the research perspectives of all could be broadened. It was to this end that this meeting was organized.

Frances V. McCann
Dartmouth Medical School

Preface

This two-day meeting convened at the Dartmouth Medical School on the campus of Dartmouth College in Hanover, New Hampshire, USA. Invited papers pertaining to fundamental mechanisms common to myocardial activity in a variety of animals representing various levels of phylogenetic evolution and physiological adaptation were presented. The presentation of conflicting interpretations of physiological data was by intent and invitation, so that we might be exposed to both sides of issues of major concern.

This meeting was recognized as a satellite symposium of the 24th International Union of Physiological Sciences through the auspices of the Comparative Physiology Division of the American Society of Zoologists. I wish to thank the Council on Basic Science of the American Heart Association for their support and interest.

I wish to express great appreciation to Dr. BERNARD C. ABBOTT, Director of the Hancock Foundation and Chairman of the Department of Biological Sciences, University of Southern California, for his encouragement and advice in initiating this symposium and for his invaluable assistance in the preparation of the manuscripts for publication.

F. V. M.

Contents

I
Structural and Functional Correlates of the Myocardium

Introduction
by E. H. Sonnenblick

At the outset of this meeting, we may briefly explore the purpose of a conference on Comparative Physiology of the Heart. Indeed, the present state of our knowledge of cardiac physiology is replete with comparative physiology. Thus, the frog heart has yielded fundamental information about ion movements and the primary role of calcium in excitation-contraction, while the frog sartorius has provided the basis of mechanics and energetics in skeletal muscle. These later concepts have recently been extended to the mammalian myocardium, and while the details may differ from skeletal muscle, the model remains useful and informative, and recent studies have been directed toward refining the model. It is with this background in mind that one turns to comparative physiology of the heart in invertebrates as well as lower forms of vertebrates in order to reveal useful models and basic mechanisms.

High resolution microscopy, made possible by the development of the electron microscope, has renewed interest in structure-function studies in many tissues and particularly in myocardial cells. Renewed attention is being focused on the ultrastructural characteristics of excitable cells in an attempt to correlate structural details with information derived from electrical and chemical studies. Investigations on a number of animal species continue to seek structural features that may be associated with specialized cell functions, e.g. pacemaker activity, intercellular communication, conduction pathways, and electromechanical coupling. Observations on animals representing various levels of functional complexity will be presented and discussed and attention will be focused on the general area of myocardial structure as it relates to function.

As will become apparent in the conference, structure, excitation-contraction coupling, and contractile systems are often revealed with greater clarity and simplicity in lower animal and insect forms. Nature's experiments and alternative pathways to reach a functional end may well provide insight into mechanisms, as well as a readily available source of material with which to work. Perhaps the mutual enlightenment of a conference such as this will extend all our models and approaches to mutual problems.

The Structure and Function of the Intercalated Disc in Vertebrate Cardiac Muscle

by Maynard M. Dewey
Department of Anatomy, The Woman's Medical College of Pennsylvania, Philadelphia

A Brief Historical Review
of the Cellular Versus Syncytial Nature of Cardiac Muscle

The cellular nature of mammalian cardiac muscle was demonstrated in the 1850's. Kolliker[1]* described this cellularity from teased human embryonic preparations and Eberth[2], in the 1860's, corroborated Kolliker's findings by describing the end-to-end apposition of adult cardiac muscle cells. Eberth's interpretation was based on the finding that silver nitrate would impregnate regions between cells and delineate cell junctions. It was assumed that this procedure stained intercellular cement. Following these earlier descriptions, a number of subsequent investigators demonstrated intercalated discs in preparations of adult cardiac muscle and attributed these structures to cell boundaries (e.g. see Schweigger-Seidel[3], Zimmermann[4] and Palczewska[5]).

In the same decade as Eberth's work, however, doubt was cast on the proposition that the stainable darklines, which traversed normal to the axis of the fiber, were cell junctions. This doubt was based on Aeby's[6] and Wagener's[7] observations of embryonic cardiac tissues. Since these stainable lines did not appear in embryonic tissues, they interpreted this as evidence that the muscle was syncytial rather than composed of discrete cells. Further, a consensus developed during the latter half of the 19th century that embryonic mesenchyme consisted of a reticular syncytium. Nuclei occurred at nodal points in the reticulum and protoplasmic processes continued from one nodal point to another. Because cardiac muscle arose by differentiation of mesenchyme, it was argued that it retained the syncytial nature of the original tissue. Since the intercalated discs appeared later in ontogeny, this was taken as evidence for their being intracellular structures.

Thus by the turn of the century there was general agreement that cardiac muscle was a syncytium; the major exponents of this view were Heidenhain[8], Von Ebner[9], and Jordan et al.[10]. The intercalated discs were considered to be: (1) contraction waves at rigor mortis[7], (2) non-contractile zones hindering contraction of the fibers[9], (3) growth centers for new muscle segments[8], (4) tendinous continuities into the cell interiors[11], (5) supportive structures for the

---

* Numbers refer to References, p. 27

maintenance of myofibrillar alignment during contraction[12], or (6) isotropic bands evident only during contraction, 'irreversible contraction bands'[10]. The observations of JORDAN and his collaborators included a wide spectrum of vertebrate and some invertebrate types[10,13]. Consideration of his drawings of cardiac muscle from vertebrates below mammals suggests that, in these forms, he was looking at contraction bands or irregularly spaced regions of insertion of myofibrils into the sarcolemma, since he does not illustrate continuous discs across fibers. In fact, it was this latter finding that convinced him that cardiac muscle was syncytial.

The view that cardiac muscle was a syncytium and that the intercalated discs were some kind of intracellular structure was held through to the 1950's. The most recent major work employing the light microscope to analyze the structure of cardiac muscle was that of AURELL[14]. Assuming the syncytial nature of cardiac muscle, he concluded that the intercalated disc was primarily a supporting structure which maintained myofibrillar alignment.

The question as to whether cardiac muscle is cellular was not raised until the early descriptions of the intercalated disc as seen with the electron microscope were reported. The intercalated disc was first described in electron micrographs as a region 'produced by collagenous invasion at cell wall junctions'[15]. With better resolution the intercalated disc was shown to consist of the plasma membranes of adjacent cells apposed to within 150–200 Å of each other[16]. These observations were confirmed in adult cardiac muscle by a number of workers[17–20]. Similar but less well-developed structures were described in embryonic mammalian tissue[21].

Early Descriptions of the Intercalated Disc as Seen with the Electron Microscope

In a study of the structure of cardiac muscle from frog, mouse and guinea-pig, SJÖSTRAND, ANDERSSON-CEDERGREN and DEWEY[22] described three different regions of the intercalated disc: (1) interfibrillar region, (2) intersarcoplasmic region and (3) longitudinal connecting surface. The interfibrillar portion was defined as the region of the intercalated disc where myofilaments of the I-band of the myofibril inserted into cell membranes. This region consisted of an expansion of an interlacing network of myofilaments which adhered to the cytoplasmic surface of the apposing cell membranes. In addition, even in this early study, a dense material was observed to bridge across the gap between the two plasma membranes at this region of the intercalated disc. The intersarcoplasmic region was defined as that region of the intercalated disc where adjacent cell membranes separated sarcoplasmic columns of the apposing cells. These sarcoplasmic columns contained elements of the sarcoplasmic reticulum, mitochondria, ribosomes and ground cytoplasm. The space between membranes in this region was estimated to be 130 Å and was designated the L-space. Along these intersarcoplasmic regions of the intercalated disc, specialized areas of the opposed membranes were described which we now identify as desmosomes[23]. At the third component of the intercalated disc, which they called the longitudinal connecting surface, the two plasma membranes were interpreted as coming together in a close contact. Using osmic acid as a fixing agent, these workers were able to see only a single layer (one osmiophilic line) of each cell membrane and thus they interpreted this region as being a very narrow intercellular cleft about 50 Å in width. Under optimal conditions they could see a central line between the membranes (see Fig. 7[22]). Subsequently, DEWEY and BARR[24], using permanganate as the fixing agent (which gives a triple-layered image of the plasma membrane), demonstrated that these regions were areas of fusion of the adjacent cell membranes and termed them nexuses. These areas of fusion have been identified by a number of investigators[25-27].

The term intercalated disc has been used variously by different investigators. The range of usage has extended, on the one hand, from calling the composite of structures along the cell junctions (i.e. the interfibrillar, intersarcoplasmic and longitudinal connecting surfaces or nexuses) an intercalated disc, to singling out, on the other hand, individual regions of specialization along the membranes (i.e. desmosomes and regions of myofibrillar insertion) as intercalated discs[28, 29]. The latter usages are confusing and do not relate the junction to the overall geometry of the cell.

The Intercalated Disc of Mammalian Cardiac Muscle and the Geometry of the Cardiac Muscle Cells

Let us consider the shape of cardiac muscle cells and their intercellular contacts from a number of vertebrates. It will be most convenient to begin with

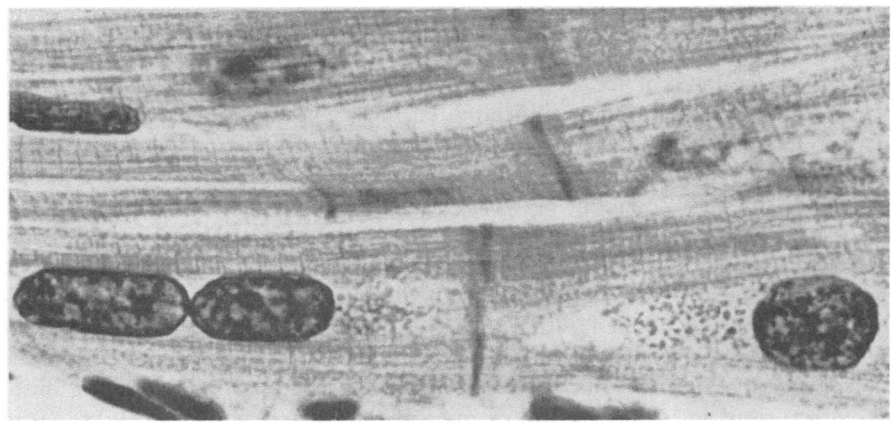

Fig. 1.
Light micrograph of human ventricular cardiac muscle. × 1700

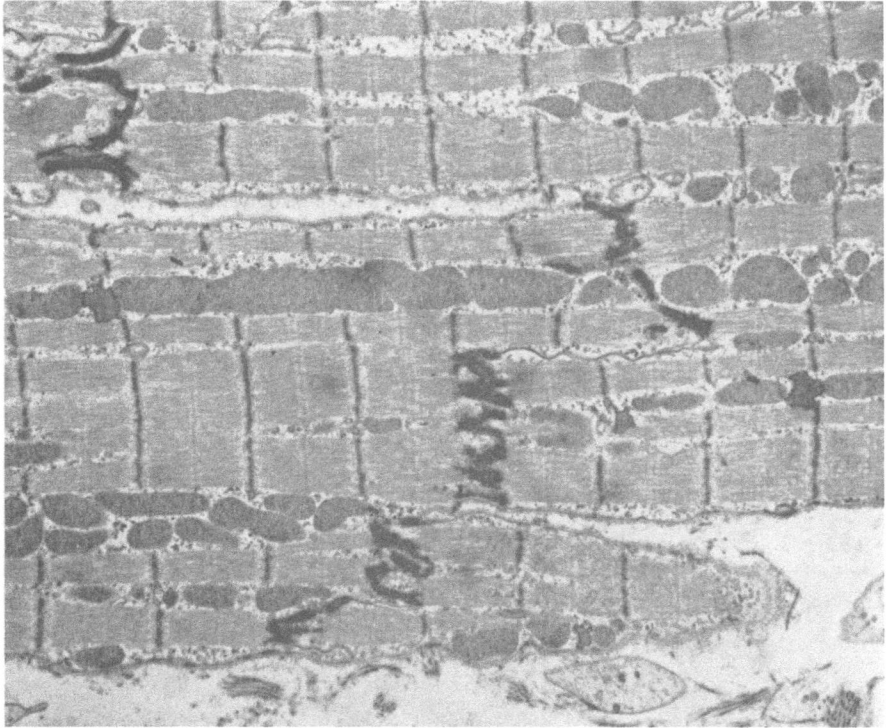

Fig. 2.
Electron micrograph of ventricular fiber of guinea-pig heart. Note the increased
density at the intercalated disc due to regions of myofibrillar insertion. Osmic acid.
× 13,000

mammals and define the intercalated disc for this group of animals. We can
then consider the applicability of the term to cardiac muscle of animals of the
other vertebrate classes.

The intercalated disc of mammalian cardiac muscle, as observed with the
light microscope, appears as a dark line of variable thickness which occurs
across the muscle fiber, at the level of one or more Z-bands (Figs. 1 and 2). It
may pass directly across the fiber at one sarcomeric level or it may pass some
distance into the fiber, jump one or more sarcomeric levels, pass some distance,
and then change to another sarcomeric level at the other side of the fiber. Such
structures are readily apparent in adult atrial and ventricular mammalian cardiac muscle. Electron microscopy has shown that the apparent planar shifts of
the intercalated disc are due to rather broad interdigitating processes of the
adjacent cells. Three-dimensional reconstructions have not been made of the
intercalated disc. Thus the extent, number and size of the processes which occur
in the plane of the intercalated disc, normal to the axis of the fiber, are not

accurately known but have been suggested in a drawing constructed by SJÖ-STRAND and CEDERGREN[30]. Because some of these undulations or interdigitating processes are 1 μ or more in length, optical summation from their refractive surfaces results in the appearance of a broad line. For descriptive purposes, based on its ultrastructure this line or disc, which constitutes the cellular junction is subdivided, most reasonably, into two regions. The first is the interfibrillar region as described by SJÖSTRAND and coworkers[22]. Here myofibrils insert into plasma membranes of adjacent cells at what would be the Z-band level of each myofibril (Fig. 3). In high-resolution electron micrographs, it can

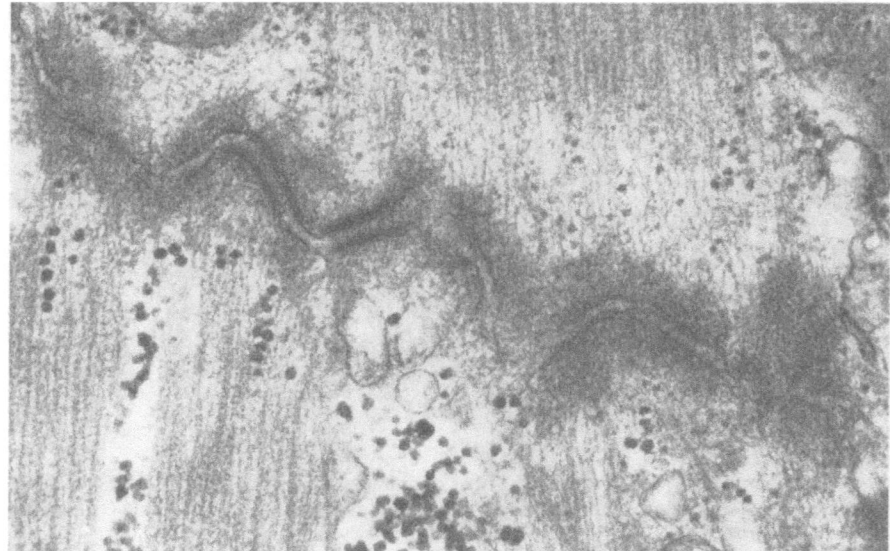

Fig. 3.
Regions of myofibrillar insertion along an intercalated disc. Note the density in the gap between plasma membranes at these regions. Occasionally myofibrils appear to insert into desmosomes as seen here. Glutaraldehyde-osmic acid. ×75,000

be seen that the filaments of the I-band branch and form a network. This causes an increased density near the membrane, which extends from the membrane as much as 1 μ or more into the I-band of the myofibril of each cell. This density and the interdigitating processes described above are the major factors which result in the intercalated disc being observed with the light microscope.

In optimal preparations filamentous material is seen to pass across the gap between the outer leaflets of the opposing plasma membranes at regions of myofibrillar insertion. The fine filaments insert into the outer leaflet of each membrane. Thus a proteinaceous material bridges the gap between the membranes of adjacent cells at this region. We will see later that this is a site of mechanical adherence between adjacent cardiac muscle cells.

The second subdivision of the intercalated disc is the intersarcoplasmic region. This region of the disc occurs between opposing sarcoplasmic columns of adjacent cardiac muscle cells. Three morphologically distinct areas can be identified along this region of the cell membranes.

The first is an area of gap that varies in width from 100 to 250 Å (Fig. 4). The space in this gap is assumed to be in continuity with the extracellular space along the lateral sides of the cardiac muscle fiber. No estimate has been made of the area of the intercalated disc which is occupied by this gap. The basement membrane or external lamina along the lateral surfaces of the fiber does not extend into this space. The space in the gap appears to have the same electron density as the extracellular space along the sides of the muscle fiber.

The second region that occurs along the intersarcoplasmic portion of the intercalated disc is a region of modified cell membrane, the desmosome (Fig. 4).

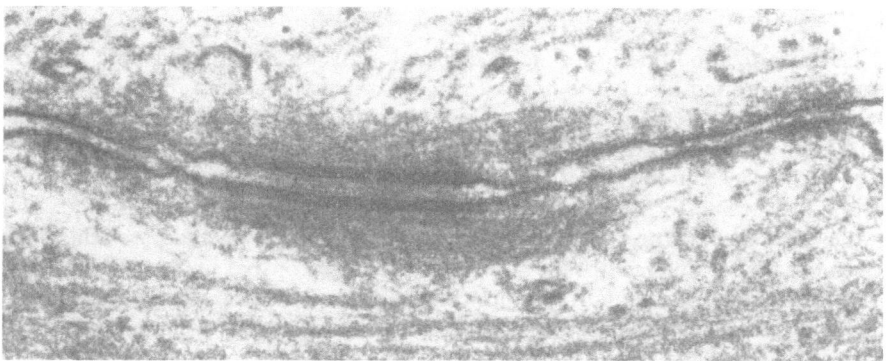

Fig. 4.
Intersarcoplasmic region of an intercalated disc. Three regions can be identified
in this portion of the disc: A. unspecialized gap, B: desmosome and C. an area in which
the plasma membranes come close together with an increased density of the cytoplasm
immediately subjacent to each membrane. Glutaraldehyde-osmic acid. ×150,000

Structurally, the desmosome is bipartite and consists of modifications of each membrane of the two opposing cells. The cytoplasmic leaflet of each membrane is usually denser even though the total thickness of the membrane is unchanged. A dense region occurs immediately subadjacent to the plasma membrane. Inserting into this region are tonofibrils from the cytoplasm of the cell. Bridging between the membranes of the adjacent cells are discrete filamentous structures. In certain preparations the region of gap between the cell membranes at the desmosome seems not only to contain bridging filaments but a central dark line through which the filaments pass. The desmosomes in mammalian cardiac muscle vary in size and while they have not been analyzed in serial sections, it is assumed that they are plaque-like in shape and thus similar to the desmosomes described in epithelia[32].

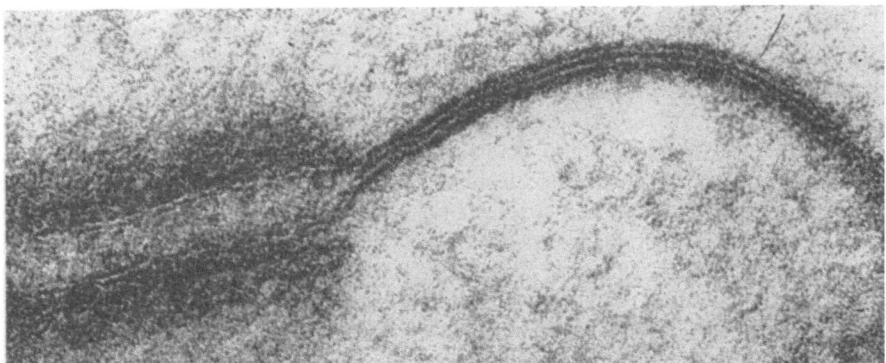

Fig. 5.
Intersarcoplasmic portion of an intercalated disc showing a desmosome (left) and
nexus (right). Permanganate. × 320,000

The third modification which occurs along the intersarcoplasmic portion
of the intercalated disc is the nexus (Fig. 5). In permanganate-fixed specimens,
the nexus is identified as a region in which the apposing cell membranes approximate each other and form a structure which has a thickness of approximately
150 Å when measured from the cytoplasmic surface of the cytoplasmic leaflet
of one membrane to include the cytoplasmic surface of the cytoplasmic leaflet
of the other membrane. The nexus appears as a five-layered structure: the cytoplasmic leaflet of one membrane; the light interspace of that membrane; a single
dark line which represents the region of overlap of the apposing outer leaflets
of the adjacent membranes; the light interspace of the second membrane; and
finally, the dark line of the cytoplasmic surface of that membrane.

The extent of the nexus along the intercalated disc of mammalian cardiac
muscle is variable. The nexus is restricted in its distribution to the intersarcoplasmic portions of the intercalated disc and most commonly occurs in that
region of the intersarcoplasmic portion of the disc which lies parallel to the
myofibrils. These are regions where the disc passes from one sarcomeric level
to another or where, in certain cases, the disc membranes simply extend up
into the sarcoplasmic column and return to the same sarcomeric level before
passing on to the next myofibril.

In certain cardiac muscles, such as that of the atrium of the guinea-pig,
an additional region may occasionally be seen along the intersarcoplasmic portion of the intercalated disc. This is a region of close approximation of the
plasma membranes with a slight increase in the density of the sarcoplasm on
either side of the two membranes, as well as an increased density in the gap
(100 Å) between the adjacent membranes. These regions are usually very small
and are similar in appearance to the membrane specializations at the zonula
adherens of epithelia[31] (Fig. 4).

From the above discussion it is clear that the intercalated disc, as we have
defined it, constitutes the entire junctional region between neighboring cardiac

muscle cells which is observable with the light microscope as a stainable line of varying thickness. Specifically, it is most probably the region of myofibrillar insertion along the disc which renders it visible with the light microscope. Implicit in this definition is the recognition that the disc occurs where cells are essentially cylindrical and myofibrils terminate in that portion of the plasma membrane which covers each end of the cell (Fig. 6).

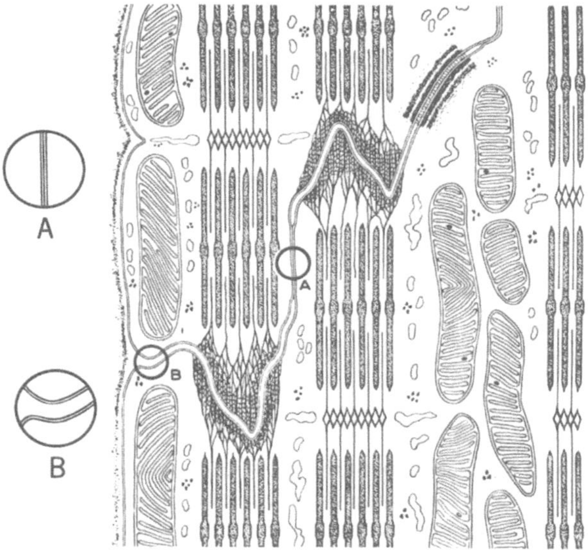

Fig. 6.
Schematic drawing of portion of intercalated disc of mammalian cardiac muscle. A shows nexus along intersarcoplasmic portion of the intercalated disc and B shows the region of gap in continuity with the extracellular space along the lateral margin of the fiber. In the upper right a desmosome is illustrated in an intersarcoplasmic portion of the disc. Modified from SJÖSTRAND, ANDERSSON-CEDERGREN, and DEWEY[22].

At least two other possibilities might exist for junctions to occur between cardiac muscle cells which would not be identifiable as intercalated discs with the light microscope. If cardiac muscle cells were fusiform, then myofibrils would insert into the sarcolemma all along the cell surface and thus be so widely spaced as to not appear as a continuous structure with the light microscope. Regions where cardiac muscle cells contact each other along their lateral surfaces could also occur. Such regions would not be readily apparent with the light microscope since no areas of myofibrillar attachment would occur here. Both of these situations are found not only in mammalian cardiac muscle but in the cardiac muscle of other vertebrate classes. For these reasons, intercalated discs have been described as missing in specific regions of the mammalian heart and absent in all vertebrate classes below the mammals.

From the light microscope studies of the early part of this century and the limited number of electron microscope studies of the past decade, certain generalizations about the shape of cardiac muscle cells in the various parts of the mammalian heart are apparent. In the case of most mammalian cardiac muscle, a muscle fiber is composed of a series of cells linked together by intercalated discs which, in general, pass normal to the long axis of the fiber. This is true even

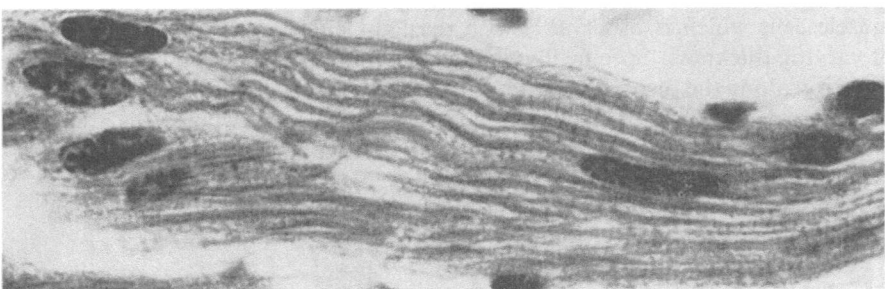

Fig. 7.
Light micrograph of atrial muscle bundle from the frog heart. The cells are fusiform
and several fibers are surrounded by a single basement membrane. No intercalated
discs are visible. ×1500

though some cells may be relatively stellate in shape. The chain of cardiac
muscle cells is ensheathed by a continuous basement membrane or external
lamina except at regions of lateral contact (see below). This morphology is in
contrast to the structure of cardiac muscle of the other vertebrate classes.
Cardiac muscle cells of all vertebrate classes except mammals are essentially
fusiform and occur with junctional complexes along their entire cell surfaces
(see for example electron microscope observations on the frog heart[40]). Thus a
bundle of cardiac muscle contains overlapping fibers which are ensheathed in a
single basement membrane. In this case, then nearly the entire surface of each
muscle cell is available for contacts with neighboring cells. These contacts are
formed by nexuses and desmosomes which occur alternatingly along the lateral
surface of the cells and regions of myofibrillar insertion which occur primarily
at the tapered ends of the cells. Since the cells do not have interdigitating
processes and the regions of myofibrillar insertion are not spaced closely together, intercalated discs are not seen with the light microscope in these types of
cardiac muscle (Figs. 7 and 8). There are mammalian cardiac muscles which
have a similar cellular geometry. Purkinje fibers[34], which constitute the terminal
bundles of the conduction system of the ungulate heart, are composed of
cylindrical to fusiform cells and nearly the entire surface of each cell is involved
in contact with adjacent cells. These contacts include desmosomes, nexuses,
and regions of insertion of myofibrils[25, 33, 35]. Several overlapping cells are surrounded by a single basement membrane. Purkinje fibers are structurally different from the terminal conducting fibers of other mammalian hearts. In the
terminal branches of the conduction tissue of man[36], dog[37] and guinea-pig
(DEWEY and BARR, unpublished) the fibers while they are similar to those of the
ungulate heart in that they are large in diameter, have few myofibrils and contain large quantities of glycogen, are dissimilar because they do have typical
intercalated discs. In this regard these 'conduction' fibers are similar to the
cardiac muscle of other regions of the heart, i.e. the cells are cylindrical, aligned
end-to-end, and occur as a single file of cells surrounded by one continuous

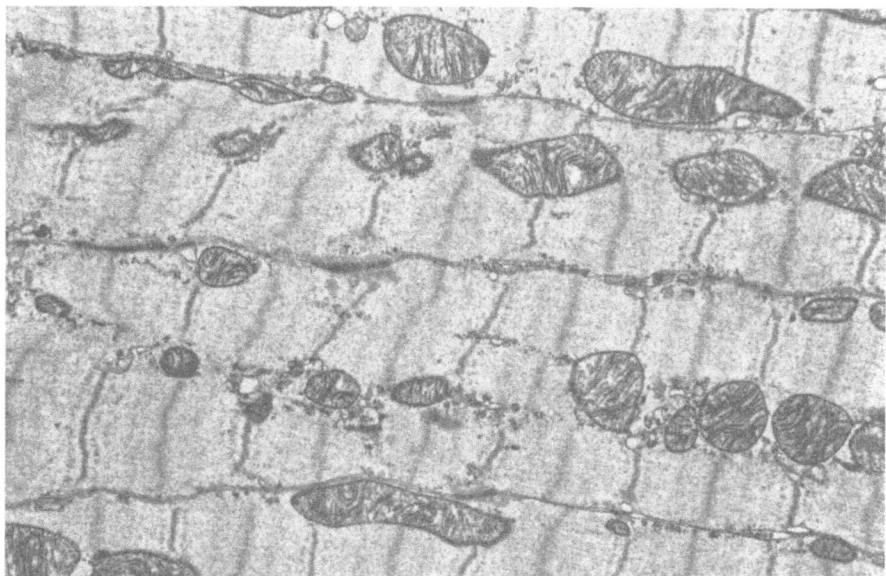

Fig. 8.
Electron micrograph of several cardiac muscle fibers from the atrium of the
frog heart. No basement membrane intervenes between fibers; the entire surface of
each fiber is in contact with a neighboring fiber. Desmosomes, nexuses and gap
regions occur along these junctions. Regions of myofibrillar insertion occur separated
by several sarcomere lengths along the tapered ends of the cells. Permanganate. × 9000

basement membrane. For these reasons it is probably worthwhile to restrict
the use of the term 'Purkinje fibers' to the conducting fibers of the ungulate heart.

Morphologically, the situation is even more striking when comparisons are
made between the cardiac muscles of lower vertebrates and that of the sinoatrial and atrioventricular nodes of mammalian hearts. In mammalian heart,
the nodal fibers are nearly fusiform and several muscle cells are enclosed in a
single basement membrane forming a bundle of cells[38,39]. Junctional contacts
between cells, including myofibrillar attachments, desmosomes and nexuses
occur at nearly any point along the surface of the cell. Thus, while the whole
cellular surface may be involved in contact with adjacent cells, these contacts
are not discernible as intercalated discs with the light microscope. The structural
modifications along the membranes are not spaced sufficiently close together
to appear as a single structure. Thus, there are geometric differences in the
contact relationships between cardiac muscle cells in various regions of the
mammalian heart and also differences in these relationships between mammalian
atrial and ventricular muscle and the cardiac muscle of all other classes of
vertebrates. At least two major groups are identifiable: (1) cardiac muscle
composed of cylindrical or nearly cylindrical cells joined together in a single
file by intercalated discs and enclosed in a basement membrane, and (2) cardiac

muscle composed of overlapping fusiform cells enclosed in a single basement membrane with intercellular contacts anywhere on the surface of the cells.

That other morphologic differences occur between various cardiac muscles of the vertebrates is intuitively obvious. The differences just described, however, are particularly significant electrophysiologically. The spread of excitation along a single fiber, composed of a row of cells arranged end-to-end, will be different from the spread of activity along a bundle of overlapping cells enclosed in a common basement membrane.

Lateral Contact between Mammalian Cardiac Muscle Cells

In addition to the intercalated disc which serves as a junction between mammalian cardiac muscle cells, lateral contacts between cells occur. In the early drawings of cardiac muscle observed with the light microscope, it would appear that workers were able to see lateral contacts between muscle fibers[13]. However, these did not stain in such a way as to be identifiable as intercalated discs. From ultrastructural studies, it is apparent that lateral contacts between cardiac muscle cells do occur in mammalian cardiac muscle and that they have structural components identical with those found along the intercalated disc except they lack regions of myofibrillar insertion. Thus a lateral contact between cardiac muscle constitutes a region in which the membranes of adjacent cells come close together along the lateral surfaces of the fibers and are mechanically adherent to each other because of desmosomes which occur intermittently along the lateral contact (Fig. 9). Between desmosomes there are the regions similar

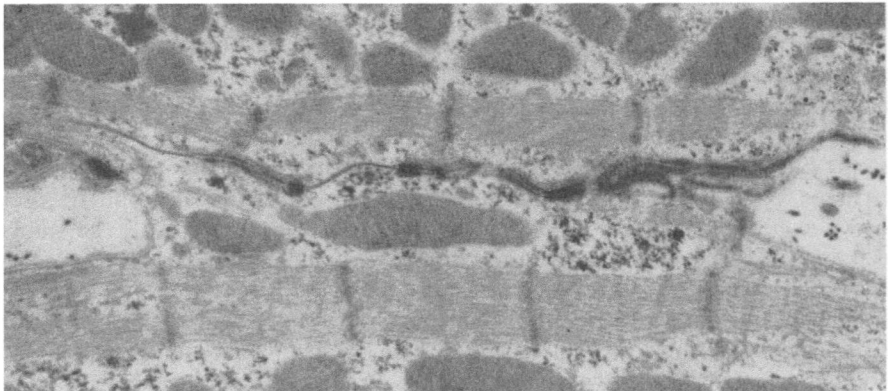

Fig. 9.
Lateral contact between two fibers of guinea-pig ventricular muscle.
Desmosomes, nexuses and unspecialized gap regions occur along such junctions.
The geometry of these contacts is similar to that which occurs along most of the surface of cardiac muscle fibers of lower vertebrates. Osmic acid. ×20,000

to those seen at the intercalated disc at which the apposing cell membranes are separated by a space of some 100–400 Å and in other regions, along the contact, the adjacent plasma membranes fuse to form nexuses. The nexuses are similar in structure to those observed at the intercalated disc. No basement membrane occurs within these contact regions.

The occurrence and extent of these lateral contacts have not been studied in a systematic way in any mammalian heart. However, SPIRA[37] in an analysis of the ultrastructure of canine cardiac muscle reports that lateral contacts are more common in auricular muscle (trabeculae carnae or pectinate muscle) than in ventricular muscle (papillary muscle). He indicates that they are fairly common in cardiac fibers of the false tendon of the canine heart. Observations in our laboratory on cardiac muscle of the guinea-pig also indicate that lateral contacts are most numerous in auricular muscles and relatively rare in ventricular muscle. In any case a quantitative estimate of the occurrence of these contacts has not been made. It is interesting to note that SPIRA did find a greater percentage of the surface area of lateral contacts involved in nexuses than he found in the intercalated discs of canine atrial and ventricular musculature. A comparison made between the percentage of length of lateral contact containing nexus in auricular muscle and muscle of the false tendon showed that in conduction fibers nearly 21% of the contact occurred as nexus while only 12% of the contact occurred in auricular muscle. The significance of the lateral contact between cardiac muscle cells is apparent from electrophysiological considerations, since these lateral contacts should affect the spread of current in the muscle bundle transverse to the long axis of the muscle fibers. The need for a quantitative estimate of the number of lateral contacts in various types of mammalian cardiac muscle is thus apparent. Unfortunately, this can only be done with the electron microscope since lateral contacts are not identifiable with the light microscope.

The Structure of the Nexus

The nexus is defined operationally as the structure underlying the image of three dark lines of uniform thickness separated by two light lines which results from the fusion of the outer surfaces of the plasma membranes of two neighboring cells. This structure is observed most clearly in preparations of cardiac muscle fixed with permanganate. Unfortunately, the chemical moieties that are stabilized and stained by permanganate are not yet known. However, it seems reasonable to assume that the two dark lines in the image of the plasma membrane fixed with permanganate are due to the non-hydrocarbon portions of the membrane. Further, it is of interest that the thickness of the individual plasma membranes at the nexus is remarkably constant from one muscle preparation to another and from preparations from different species. In permanganate-fixed cardiac muscle the image of the plasma membrane is approximately 75–85 Å wide. It comprises two dark lines separated by a lighter one, all of about equal thickness. At the nexus the distance across both membranes is in the range of 120–150 Å. The difference between the thickness at the nexal region and the sum of the thicknesses of the constituent membranes is about the thickness of one dark line. The width of the inner dark line and the lighter inner zones are unchanged. The loss of thickness occurs because, in the region of the nexus, the two outer dark lines are usually no thicker than one alone. This is the reason for considering the nexal relationship as a fusion of the plasma membranes. If there were no loss of thickness and the membranes simply abutted, the central line should appear twice as thick. These interpretations are based on a number of assumptions derived by comparison with other more thoroughly studied biological membranes. This is necessary since no structural information on the plasma membrane of cardiac muscle is available other than that obtained by electron microscopy. We have no independent measure of the thickness of the plasma membrane of cardiac muscle. In fact, this holds for all biological membranes with the exception of myelin. In the latter case structural data is available from the hydrated membrane. There is a good correspondence between the thickness of the membrane of myelin fixed with permanganate and that obtained from interpretations of X-ray diffraction data from living myelin[41]. Thus assuming that the thickness of the plasma membrane of cardiac muscle is close to that measured in permanganate-fixed specimens and assuming a lipid bilayer model of the plasma membrane, we can make some tentative hypotheses about the molecular structures involved in formation of the nexus. First, to judge from the loss of thickness of one dark line, as well as the experimental reversibility of the separation of nexal membranes (see effects of hypertonicity, below), it seems possible that the membrane fusion at the nexus is due to interdigitation of non-hydrocarbon moieties of the membrane. Second, since the light area of each plasma membrane image is unchanged in the nexal region it appears that the hydrocarbon layers of the membrane do not take part in the formation of the nexus. Third, since the central dark line is only 25–30 Å across and bounded by hydrocarbon any water in this region is probably highly crystallized.

The nexus as observed following osmic acid fixation and staining with heavy metals appears somewhat different from that seen following permanganate fixation[24]. The total thickness of the nexus is usually greater, being as much as 200–250 Å. The inner line or cytoplasmic leaflet of each membrane is thicker than when observed following permanganate fixation and measures 50–60 Å. The central line of the fused membranes is usually thinner and may appear beaded. With osmic acid fixation the outer leaflet and perhaps some portion of the central light region of each membrane lateral to the nexus are not observed. After fixation with glutaraldehyde followed by osmic acid and staining with heavy metals, the nexus appears similar to that seen following permanganate fixation. Occasionally in this circumstance, the central dark line appears less thick than that observed in fixation with permanganate. Since the images of the nexus obtained with various fixatives cannot be superimposed, it must be concluded that we do not have as yet a clear understanding of its structure.

Recently, REVEL and KARNOVSKY[42] have shown that treatment of cardiac muscle with a colloidal suspension of lanthanum following fixation in glutaraldehyde results in the penetration of lanthanum into the extracellular compartments of the tissue. Lanthanum under these conditions also penetrates into the nexus itself. Following lanthanum penetration, these workers were able to see a substructure along the membranes involved in the nexal fusion. This substructure had a long-range ordering in the form of a hexagonal array. It consisted of light unstained globular regions with a dark core surrounded by regions penetrated by lanthanum. The period, from center of light region to center of the adjacent light region, was about 90 Å. Even in these preparations, the thickness of the nexus itself, from cytoplasmic side to cytoplasmic side, is in the order of 200 Å (DEWEY and BARR, unpublished). From these observations KARNOVSKY and REVEL[42] argued that an extracellular space exists along the nexus into which lanthanum is able to penetrate. As yet, we are unsure whether such treatment is truly a function only of diffusion of the lanthanum into the aqueous phase or whether there is binding to specific components within the membrane.

An additional bit of information which tells us something of the structure of the nexus is the observation of KRIEBEL[43]. He has shown in the tunicate heart that the nexus forms a functionally tight junction around the muscle cells thus excluding any pathway for shunting of extracellular current between the lumen of the heart and the extramural space. Thus while we are able to say little about the molecular structure of the membranes at the nexus, it would appear that it does represent some kind of fusion between adjacent cellular membranes.

These observations of a substructure in the nexuses of cardiac muscle seem to distinguish them from the nexus which forms a tight junction or zonula occludens around the apical ends of epithelial cells. In the latter case, it would appear that lanthanum is unable to penetrate the nexus which forms the zonula occludens. The ease with which nexuses could be demonstrated in the various tissues is interesting in this regard. It has been easy to demonstrate the nexus in various epithelia following osmic acid fixation alone, while in cardiac muscle and particularly in smooth muscle the demonstration of the nexus as representing a fusion of the outer leaflets of the outer cell membranes was not clearly demonstrated until permanganate was used as a fixative[24]. Further, there is a difference in stability of the nexus of the involuntary muscles and those of epithelia. The nexuses of involuntary muscles are ruptured by hypertonic treatment. As yet such a mechanical disruption of the nexus has not been possible where it forms the zonula occludens in epithelia.

Calcium plays a significant role in the maintenance of structures involved in cell adherence. It has been shown that calcium removal causes rupture of desmosomes of the intercalated disc but apparently does not affect the structure of the nexus[44]. Specifically, calcium removal followed by fixation causes the rupture of desmosomes and results in the nexus being torn from the surface of the one cell leaving a hole in that cell and the nexus itself intact on the adjoining cell[26]. So far, the only effective mechanism in rupturing the nexus in cardiac muscle has been that of hypertonic treatment[40].

Estimates of the Surface Area Involved in Nexus between Cardiac Muscle Cells

No systematic attempt has been undertaken from a comparative point of view to estimate the surface area which is nexus between cardiac muscle cells. Perhaps the only effort along these lines has recently been reported by SPIRA[37]. He estimated the surface area involved in nexuses along the intercalated disc of atrial, ventricular and false tendon of the canine heart. By planimetry he measured the length of the nexuses and the length of the intercalated disc in montages of these muscles. His most consistent data came from auricular muscle. The average length of intercalated discs was 18 μ. The average total length of nexus was 1 μ giving approximately 6% of the length of the disc involved in nexus. While his data from ventricular muscle and false tendon showed great variation, the percentage of the length of the disc involved in nexus ranged between 10 and 40%. Further work along these lines is needed. It is possible that a correlation exists between the amount of nexus per disc and the conduction velocity of the various cardiac muscles.

Evidence for the Nexus as a Site of Electrical Transmission between Cardiac Muscle Cells

There is compelling evidence that the transmission of action potentials from one cardiac muscle cell to another is due to electrotonic spread of current. There is extensive electrotonus along frog atrial fibers[45], Purkinje fibers[46] and rat atrium[47]. The most direct evidence for electrical transmission comes from proof that there is sufficient current flow across a segment of cardiac muscle isolated in a sucrose gap to excite a post-gap segment of muscle when an appropriate shunt-resistor allows current flow between the extracellular electrolyte solutions[48]. These experiments (see article by BARR, this volume) have been performed on both frog atrial muscle[40] and mammalian cardiac muscle (atrium of the guinea-pig; BARR, BERGER and DEWEY, unpublished). An obvious corollary of this electrical coupling between cardiac muscle cells is the morphologic structure responsible for such coupling. It has been suggested that the nexus allows current flow from the interior of one cell to the interior of the adjacent cell since it presumably excludes extracellular space between adjacent cells and thereby eliminates an extracellular pathway which would shunt current[24,49]. One of the immediate problems faced in attempting to establish whether it is the nexus that is the low-resistance coupling, is the question of whether the nexus along the intercalated disc forms a seal around the periphery of adjacent fibers trapping extracellular space within the interior of the disc (a type of zonula occludens). If such were the case, it would be impossible to distinguish whether it was current flow across the nexus or any region of the disc enclosed by this seal that was responsible for the electrotonic coupling (Fig. 10). Several attempts

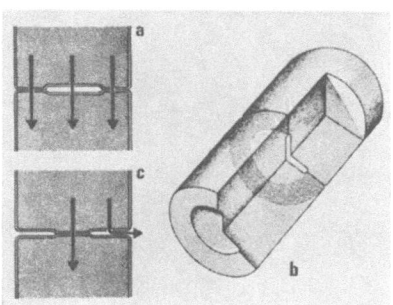

Fig. 10.
Schematic diagram illustrating two alternatives for the arrangement of nexuses along the intercalated disc. In A and B the nexus forms a seal around the perimeter of apposition of two cells, a type of zonula occludens. In this case, it would be difficult to determine whether current flow between cells is solely across the nexus or across the gap region and nexuses. The gap region would not act as a current shunt. In C the nexus does not form a seal and any gap region along the disc can act as a current shunt since it communicates with the extracellular space. The latter case holds for the intercalated disc of mammalian cardiac muscle. At electrical synapses, the former case must be excluded before we can be sure that the nexus is the site at which current passes between electrically interacting neurons.

have been made to demonstrate that the 200 Å space between apposing membranes along the intercalated disc is continuous with the extracellular space at the lateral side of the fiber, i.e. that the nexus does not form a zonula occludens surrounding the perimeter of opposing cells. BARR, DEWEY and BERGER[40] have demonstrated that ferrocyanide apparently can penetrate the space along the intercalated disc. Ultrastructural studies have shown that ferritin diffuses into this space[50] and that perioxidase also diffuses into this space (DEWEY, BARR and BERGER, unpublished). Thus it would appear that there is continuity between the extracellular compartment and a space that extends along the intercalated disc through regions of myofibrillar attachment and desmosomes.

Evidence that the nexus is the structure involved in electrotonic coupling between cardiac muscle cells comes from studies in which a correlation between the blockade of propagation and the rupture nexuses was demonstrated. When small strands of cardiac muscle are incubated in electrolyte solutions made hypertonic with the addition of sucrose to three times the normal tonicity, pronounced shrinkage of the cells occurs and a greatly increased space develops along the intercalated disc (Fig. 11). From inspection of the various structures along the intercalated disc, it is apparent that there is no rupture at the region of the myofibrillar insertion since the small filaments can be seen to bridge across the gap of the disc in this region. Likewise, there seems to be little disruption of the desmosomes along the intercalated disc. In these experiments, the nexuses are ruptured. It is difficult to ascertain quantitatively that there is a complete loss of nexuses along the intercalated disc of all fibers since the effect

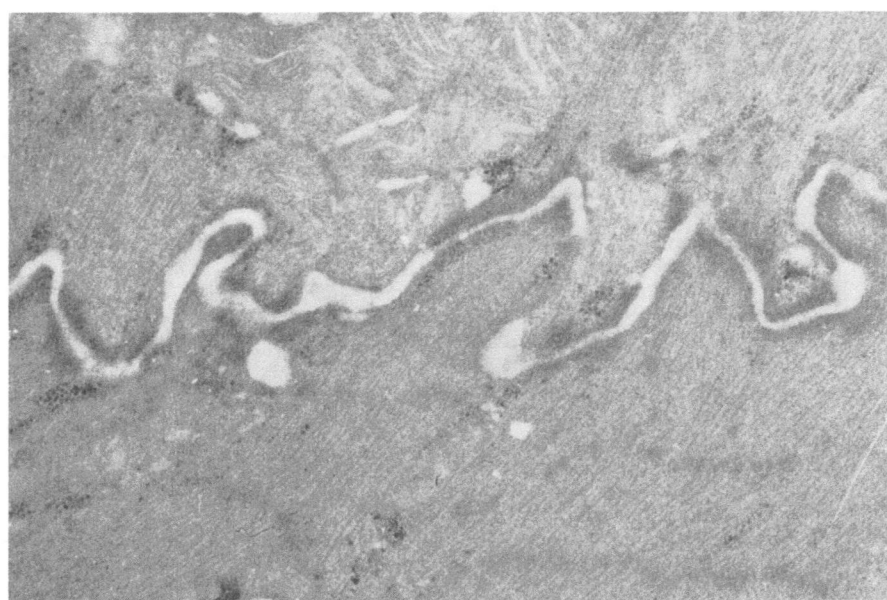

Fig. 11.
Ventricular muscle of guinea-pig heart incubated in hypertonic sucrose solution prior to fixation. Note the widened gap between cells at the intercalated disc. Nexuses are ruptured by desmosomes and regions of myofibrillar insertion are intact. Osmic acid. × 29,000

of tonicity is greatly dependent on the size of the muscle bundle. However, in small bundles it would appear that there is a complete or nearly complete rupture of all nexuses along the disc. This structural modification correlates well with the block of propagation which occurs in these muscle preparations treated with hypertonic medium (see article by BARR in this volume). Further, it can be shown when muscle preparations are first soaked in hypertonic media, then returned to isotonic media, they again demonstrate propagation. There is almost a complete reconstitution of the morphology of the intercalated disc (DEWEY, BARR and BERGER, unpublished). The gap between adjacent membranes decreases to the value of control specimens and all of the nexuses appear restored. Thus it would appear that there is good evidence that it is the nexus which is involved in the electrotonic coupling between adjacent cardiac muscle cells. These experiments seem to eliminate the other morphologic continuities between muscle cells (regions of myofibrillar insertion and desmosomes) as structures responsible for electrotonic coupling. In fact, in complimentary studies on different cardiac muscle it was shown that cells were electrotonically decoupled when nexuses were broken and not electrotonically decoupled when only the desmosomes were destroyed by calcium removal[44].

Similar experiments have been performed on smooth muscle, and it is clear that the electrotonic coupling which occurs in this muscle is due to the nexuses between adjacent smooth muscle cells[51]. In fact, in this situation the morphological involvement of the nexus is more clear-cut since it is the only structural continuity between smooth muscle cells. That this structural continuity may be formed and ruptured under physiologic conditions has been suggested for smooth muscle. Uterine muscle of the rat becomes electrically inexcitable or shows only non-propagating responses in the absence of adequate amounts of estrogen. Estrogen therapy to castrated rats restores excitability and propagation[52]. This is correlated morphologically with the lack of nexuses between smooth muscle cells in the castrate rat and their occurrence following estrogen treatment[53].

Nexuses have been implicated in the electrotonic coupling at a number of synapses and between epithelial cells. In these cases, however, morphologic rupture of the nexuses correlated with decoupling of these cells has not been demonstrated. Further at these junctions there is a considerable problem associated with the demonstration that the nexus does not isolate islands of extracellular space between the cells (Fig. 10). Such islands themselves might serve as regions of electrotonic coupling.

References

[1]   A. KOLLIKER, *Handbuch der Gewebelehre des Menschen*, 560 (Wilhelm Engelmann, Leipzig 1859).
[2]   C. J. EBERTH, Virchows Arch. path. Anat. *37*, 100 (1866).
[3]   SCHWEIGGER-SEIDEL, in *Strickers Handbuch: Das Herz*, 1 (Leipzig 1871).
[4]   K. W. ZIMMERMANN, Arch. Mikr. Anat. *75*, 40 (1910).
[5]   I. VON PALCZEWSKA, Arch. Mikr. Anat. *75*, 41 (1910).
[6]   C. AEBY, Z. ration. Med. *17*, 195 (1863).
[7]   G. K. WAGENER, Sber. Beförd. ges. Naturw. Marburg *141* (1872).
[8]   M. HEIDENHAIN, Anat. Anz. *20*, 33 (1901).
[9]   V. EBNER, Sber. Akad. Wiss. Wien, Math.-naturwiss. Klasse, Abt. III, *109*, 700 (1900).
[10]   H. E. JORDAN and J. B. BANKS, Am. J. Anat. *22*, 285 (1917).
[11]   F. MARCEAU, Annls. Sci. nat., Zool. *19*, 191 (1904).
[12]   A. DIETRICH, *Die Elemente des Herzmuskels* (Gustav Fischer, Jena 1910).
[13]   H. E. JORDAN and K. B. STEELE, Am. J. Anat. *13*, 151 (1912).
[14]   G. AURELL, *Die Glanzscheiben des Herzmuskelgewebes und ihre Verbindungen* (Almgvist and Wiksells Boktryckeri, Uppsala 1945).
[15]   V. L. VAN BREEMAN, Anat. Rec. *117*, 49 (1953).
[16]   F. S. SJÖSTRAND and E. ANDERSSON, Experientia *10*, 369 (1954).
[17]   R. POCHE and E. LINDNER, Z. Zellforsch. mikrosk. Anat. *43*, 104 (1955).
[18]   K. C. PRICE, J. M. WEISS, H. DAIKICHI and J. R. SMITH, J. expl. Med. *101*, 687 (1955).
[19]   E. LINDNER, Z. Zellforsch. mikrosk. Anat. *45*, 702 (1957).
[20]   D. H. MOORE and H. J. RUSKA, J. biophys. biochem. Cytol. *3*, 261 (1957).
[21]   A. R. MUIR, J. biophys. biochem. Cytol. *3*, 193 (1957).
[22]   F. S. SJÖSTRAND, E. ANDERSSON-CEDERGREN and M. M. DEWEY, J. Ultrastruct. Res. *1*, 271 (1958).
[23]   D. W. FAWCETT, *Structural Specializations of the Cell Surface. Frontiers in Cytology* (Yale University Press, New Haven, Conn. 1958).
[24]   M. M. DEWEY and L. BARR, J. Cell Biol. *23*, 553 (1964).

[25]   H. E. KARRER, J. biophys. biochem. Cytol. *8*, 135 (1960).
[26]   A. R. MUIR, J. Anat. *99*, 27 (1965).
[27]   D. W. FAWCETT, *An Atlas of Fine Structure: The Cell, Its Organelles and Inclusions* (Saunders, Philadelphia 1966).
[28]   D. W. FAWCETT and C. SELBY, J. biophys. biochem. Cytol. *4*, 63 (1958).
[29]   J. R. SOMMER and E. A. JOHNSON, J. Cell Biol. *36*, 497 (1968).
[30]   F. S. SJÖSTRAND and E. ANDERSSON-CEDERGREN, *Intercalated Discs of Heart Muscle*, in: *Structure and Function of Muscle. I* (Ed. BOURNE; Academic Press, New York 1960).
[31]   M. G. FARQUHAR and G. E. PALLADE, J. Cell Biol. *17*, 375 (1963).
[32]   D. E. KELLY, J. Cell Biol. *28*, 51 (1966).
[33]   J. A. G. RHODIN, P. DEL MISSIER and L. C. REID, Circulation *24*, 349 (1961).
[34]   J. E. PURKINJE, Arch. Anat. Physiol. wiss. Med. *12*, 281 (1845).
[35]   R. CAESAR, G. A. EDWARDS and H. RUSKA, Z. Zellforsch. *48*, 698 (1958).
[36]   K. KAWAMURA, Jap. Circ. J. *25*, 594 (1961).
[37]   A. SPIRA, *The Ultrastructure of the Intercalated Disc in Auricular, Ventricular and Conduction Tissues of the Canine Heart* (Thesis: University of Michigan, Ann Arbor 1967).
[38]   K. KAWAMURA, Jap. Circ. J. *25*, 973 (1961).
[39]   W. TRAUTWEIN and K. UCHIZONO, Z. Zellforsch. *61*, 96 (1963).
[40]   L. BARR, M.M. DEWEY and W. BERGER, J. gen. Physiol. *48*, 797 (1965).
[41]   C. R. WORTHINGTON and A. E. BLAUROCK, Nature *218*, 87 (1968).
[42]   J. P. REVEL and M. J. KARNOVSKY, J. Cell Biol. *33*, C7 (1967).
[43]   M. E. KRIEBEL, J. gen. Physiol. *50*, 2097 (1967).
[44]   J. J. DREIFUSS, L. GIRARDIER and W. G. FORSSMAN, Arch. ges. Physiol. *292*, 13 (1966).
[45]   W. TRAUTWEIN, S. W. KUFFLER and C. EDWARDS, J. gen. Physiol. *40*, 135 (1956).
[46]   S. WEIDMAN, J. Physiol. *118*, 348 (1952).
[47]   J. W. WOODBURY and W. E. CRILL, in *Nervous Inhibition* (Ed. E. FLOREY; Pergamon Press, London 1961), p. 124.
[48]   L. BARR and N. BERGER, Pflügers Arch. ges. Physiol. *279*, 192 (1964).
[49]   L. BARR, J. theor. Biol. *4*, 73 (1963).
[50]   W. G. FORSSMAN and L. GIRARDIER, Z. Zellforsch. *72*, 249 (1966).
[51]   M. M. DEWEY and L. BARR, Alimentary Canal, Section 6 in *Handbook of Physiology* (Ed. C. F. COLE, Am. Physiol. Soc., Washington, D.C. 1968), p. 1629.
[52]   C. E. MELTON JR. and J. T. SALDIVAR JR., Am. J. Physiol. *207*, 279 (1964).
[53]   R. A. BERGMAN, J. Cell Biol. *36*, 639 (1968).

    This work was supported in part by Public Health Research Grants AM 05197, AM 11327, NB 07199 and a grant from the Michigan Heart Association. The author wishes to acknowledge the technical assistance of Mrs. JULIA TERNAK and Mrs. VIVIAN WILSON.

Ultrastructure and Function in an Insect Heart

by FRANCES V. MCCANN and JOSEPH W. SANGER
Department of Physiology, Dartmouth Medical School, Hanover,
New Hampshire

The heart of the adult moth *Hyalophora cecropia* fulfils in several ways the requirements of a model myocardium on which to study fundamental mechanisms of cardiac physiology. The gross structure of the tubular heart appears less complex than that of the vertebrates and thereby allows an evaluation of the role of morphologically specialized tissues. The unusual ionic milieu that constitutes the cellular environment presents unique problems in itself and further serves to test widely held concepts of the ionic requirements for bioelectrogenesis. The phase of diastole is an active period in the sense that reexpansion of the heart following systole is mediated by the contraction of alary muscle fibers, an extra-cardiac device that may provide a convenient system for investigating cell-to-cell communication. The phenomenon of beat reversal, the sporadic shift in location of impulse origination, with a concomitant change in direction of wave propagation, provokes questions basic to our understanding of the factors that initiate and control pacemaker activity and that provide for preferred directions of conduction.

Studies on the moth heart in this laboratory have been directed to these unique features as they pertain to cardiac cellular physiology. Since current quantitative information regarding cellular physiology of the heart derives almost exclusively from vertebrate hearts in general, and mammalian myocardia in particular, these data serve as our frame of reference.

Various aspects of cellular activity have been studied in single cells of the moth heart with electrophysiological techniques in conjunction with the electron microscope in this laboratory. This paper will briefly summarize some of the general features of these cells and present further evidence of unusual permeability qualities of the membrane. Ultrastructure and function in heart and alary muscles will be presented and discussed.

The heart of the adult moth *H. cecropia* is a tubular structure that lies suspended in the mid-abdominal cavity by long (5 mm), cylindrical, small diameter (5–15 μ) alary muscle fibers. Muscle fibers project at intervals from the dorsal surface of the heart and attach directly to the dorsal integument. There are no chambers, and only rudimentary valve-like structures serve the paired, segmentally arranged ostia that perforate the tubular heart and allow the open circulatory system to function. Pericardial cells and fat bodies cluster around the pericardial area. A cross section through the heart (Fig. 1) shows some of these gross anatomical features.

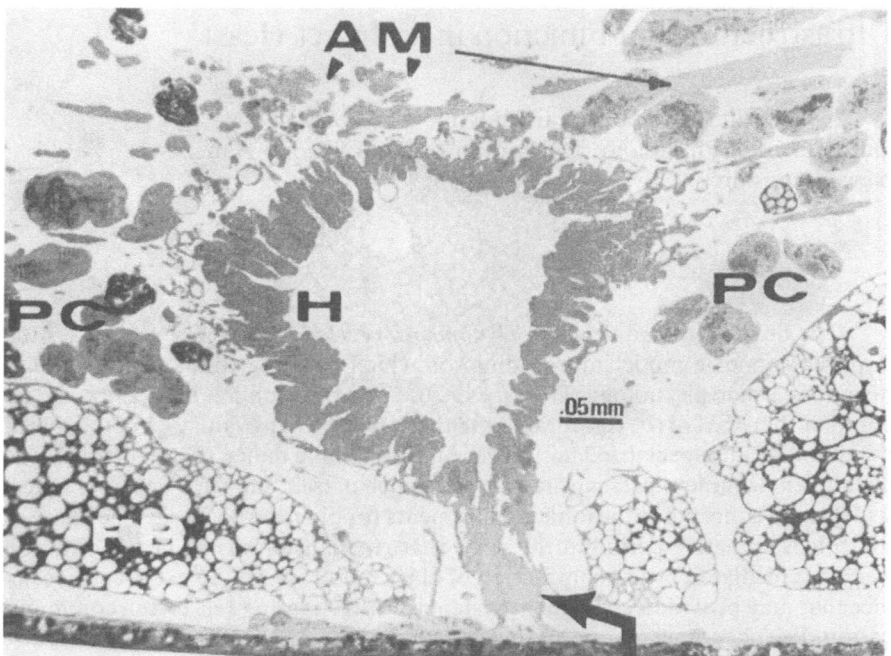

Fig. 1.
Cross section of the heart demonstrating the irregular shape of the wall (H)
and its attachment to the dorsal integument (arrow). The position of the alary
muscles (AM), the distribution of pericardial cells (PC) and the presence of fat bodies
(FB) are also indicated.

A number of functional characteristics of the heart have recently been
summarized[1]*. The heart beats at a rate of 60–100 beats/min, but this is a very
labile measure and undoubtedly many factors, both intrinsic and extrinsic,
contribute to its fluctuations. The contractile wave is peristaltic in nature and
usually proceeds from the caudal region forward toward the cephalic at low
(about 25 mm/sec) velocity. Conduction is believed to proceed faster in the
forward direction than after reversal[2,3]. This observation, however, has not
been confirmed when studied with microelectrodes[4]. The heart cannot be tetanized; it is protected by a refractory period. Hearts driven by externally applied
square pulse stimuli will follow up to 3.2/sec; the contractile wave does not
summate[1]. The heart is accelerated by epinephrine and slowed by acetylcholine[1].

The blood, or hemolymph, does not serve a respiratory function, and its
composition provides, by mammalian standards[5], a most bizarre internal milieu
(Fig. 2). Some may object to this comparison, but it is not our purpose here to
consider general features that characterize insect hearts. Rather, we are con-

* Numbers refer to References, p. 46

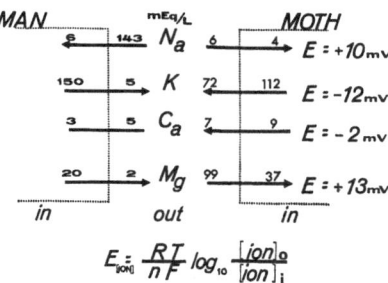

Fig. 2.
A comparison of the transmembrane ionic distribution between Man and a moth. Membrane potential values, E, predicted from Nernst Equation (lower figure) for moth are shown on far right. R, universal gas constant; T, absolute temperature; N, valency of ion; F, Faraday; ion concentration outside cell and inside, respectively.

cerned with a comparison of physiological mechanisms that govern the activity of cardiac cells. Fig. 2 serves to dramatize the transmembrane ionic distribution that, in conventional terms, must serve to generate and maintain electrical activity in the moth cardiac cell[6].

That this is one of Nature's practically sodium-free systems requires that bioelectric activity must prevail in its absence. Examination of the predicted transmembrane values in the far right column (Fig. 2) gives no immediate clue as to the identity of the major current-carrying ions.

The measured resting potential of the moth cardiac cell is about −65 mv[7]. Despite the lack of correlation between the resting potential value predicted from the transmembrane distribution of potassium (−11 mv) and the measured (−60 mv), the fundamental mechanism appears to share certain basic features common to other excitable tissues. A fraction (30%) of the original resting potential persists after poisoning with DNP. This suggests that part of the mechanism is dependent on metabolic energy and that at least a portion is independent of oxidative phosphorylation[8]. This observation has been reported for lepidopteran skeletal muscle, also[9]. Metabolic poisons affect action potentials and mechanical responses of moth cardiac cells in a manner similar to other vertebrate cardiac cells studied[10-13].

Action potentials overshoot as much as + 20 mv. Some action potentials repolarize slowly, resulting in a plateau. As in other heart cells, the plateau is the most labile phase of the action potential[1,14]. Total duration, depending on heart rate, averages 350–400 msec. Experiments in which pulses of constant current were introduced across the single cell membrane via a bridge circuit demonstrated that conductance does indeed increase rapidly during the upstroke and falls during the plateau as in mammalian hearts. This is evidence that the membrane resistance changes as depolarizing and repolarizing currents flow[1].

Action potentials in the moth heart display a variety of contours similar to those associated with topographically localized sites in mammalian myocardia[14]. A slow initial rise, indicative of pacemaker activity, may be a permanent or transitory component of any type seen. This observation has led us to the conclusion that pacemaker activity in this heart is not localized, but rather is a latent or dormant property of many if not all cells.

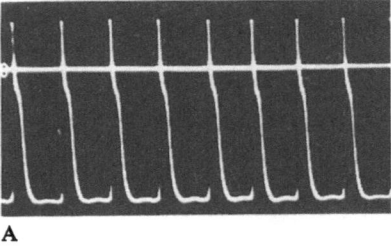

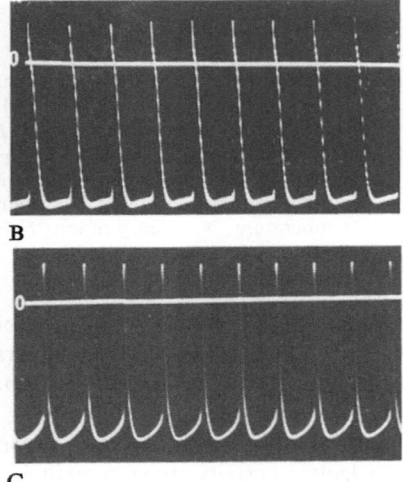

A

B

Fig. 3.
Types of action potentials recorded
from single cells of the adult moth myocardium. Note plateau in A, slight prepotential in B and pronounced pacemaker in C. Marker indicates 20 mv,
0.5 sec.

C

Some of the action-potential 'types' that are recorded from different cells
are shown in Fig. 3. The spontaneous heart rate in each case is nearly the same,
thus the different potential shapes do not reflect differing rates. It is also particularly significant that a single cell can generate action potentials of different
'types' as has been demonstrated by earlier studies in which the membrane
voltage was manipulated (depolarized) by the passage of pulses of constant
current. The repolarization phase of the action potential was delayed so that a
plateau phase resulted[15]. This maneuver effectively converted one 'type' into
another. Some action potentials are much smaller than these, and whether they
represent injured cells or some other specialization has not been methodically
determined.

Due largely to the development of the electron microscope, interest has
been renewed in the search for a correlation between structure and function in
single cells. That a distinct type of action potential can be associated with an
anatomical location in the mammalian heart has been extensively documented[14].
Action potentials exhibiting distinguishing characteristics are thus recorded
from the sinoatrial node, atrial fibers, A–V nodal and conducting pathways,
Purkinje cells and ventricular muscle fibers. If there exists an ultrastructural
basis for this electrophysiological specialization, it remains obscure. This is not
to say that the cells in the S–A node cannot be distinguished from ventricular
muscle; it is, rather, that the distinguishing characteristics such as cell size,
fibrillar composition, cell geometry, and interfiber boundaries, cannot as yet be
correlated with action potential magnitude and contour.

The search for structural features that might correlate with the generation
of more than one type of action potential by a single cell and other physiological
events herein described was implemented with the light and electron micro-
scopes. Since the ultrastructure of the myocardium has recently been detailed[16],

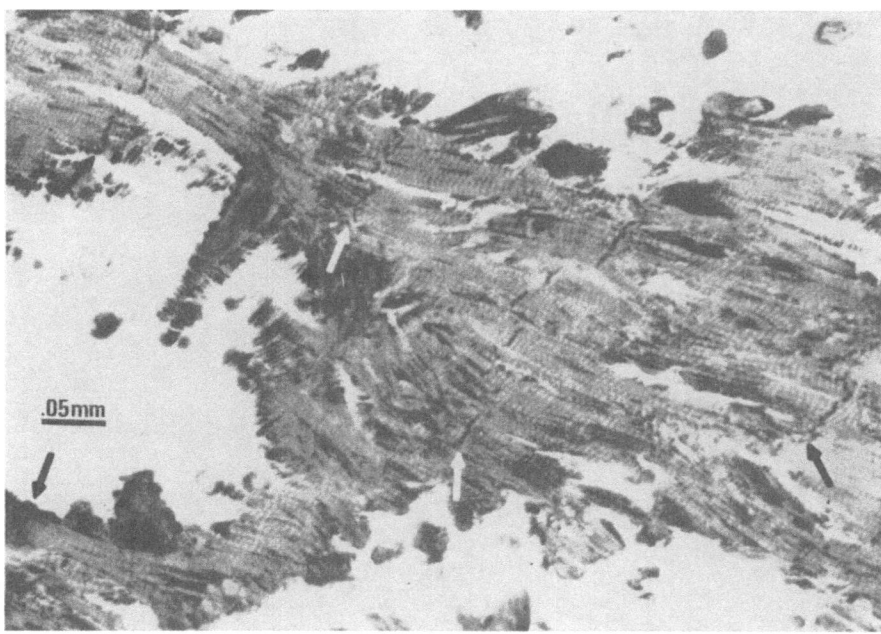

Fig. 4.
Longitudinal section through the ventral side of the heart showing a series of intercalated discs (arrows) with the light microscope. The long axis of the heart runs horizontally in the micrograph and the cells course in a helical fashion.

only the more salient features will be discussed. A sarcomere consists of an A-band (1.8 μ), reduced I-bands and no M-bands. The beaded appearance of the Z-bands results from an alignment of disconnected dense bodies[16]. The heart is composed of a single layer of striated muscle fibers averaging 25 μ in diameter and 100 μ in length that course in a helical fashion around the long axis of the tubular structure (Fig. 4). Interfiber boundaries demonstrated with the light

Fig. 5A.   Page 34 ▷
Transverse section of the heart indicating a septate desmosome (arrows) connecting two different cells (× 70,000).
Fig. 5B.
Transverse section of one cardiac cell demonstrating the extensive infoldings of the cell surface. Many trachea (T) are distributed throughout the wall. Basement membrane (bm) is demonstrated in the extracellular spaces (× 12,000).

Fig. 6A.   Page 35 ▷▷
Transverse section of the heart showing a portion of one of the many pockets of mitochondria which project into the lumen and pericardial spaces of the heart. Thin channels (arrows) can be traced from the lumen (L) into the interior of the cell (× 12,600).
Fig. 6B.
A higher magnification of another section of the same cell shows the attachment of sarcoplasmic vesicles (v) to the channels forming dyadic arrangements (× 70,000).

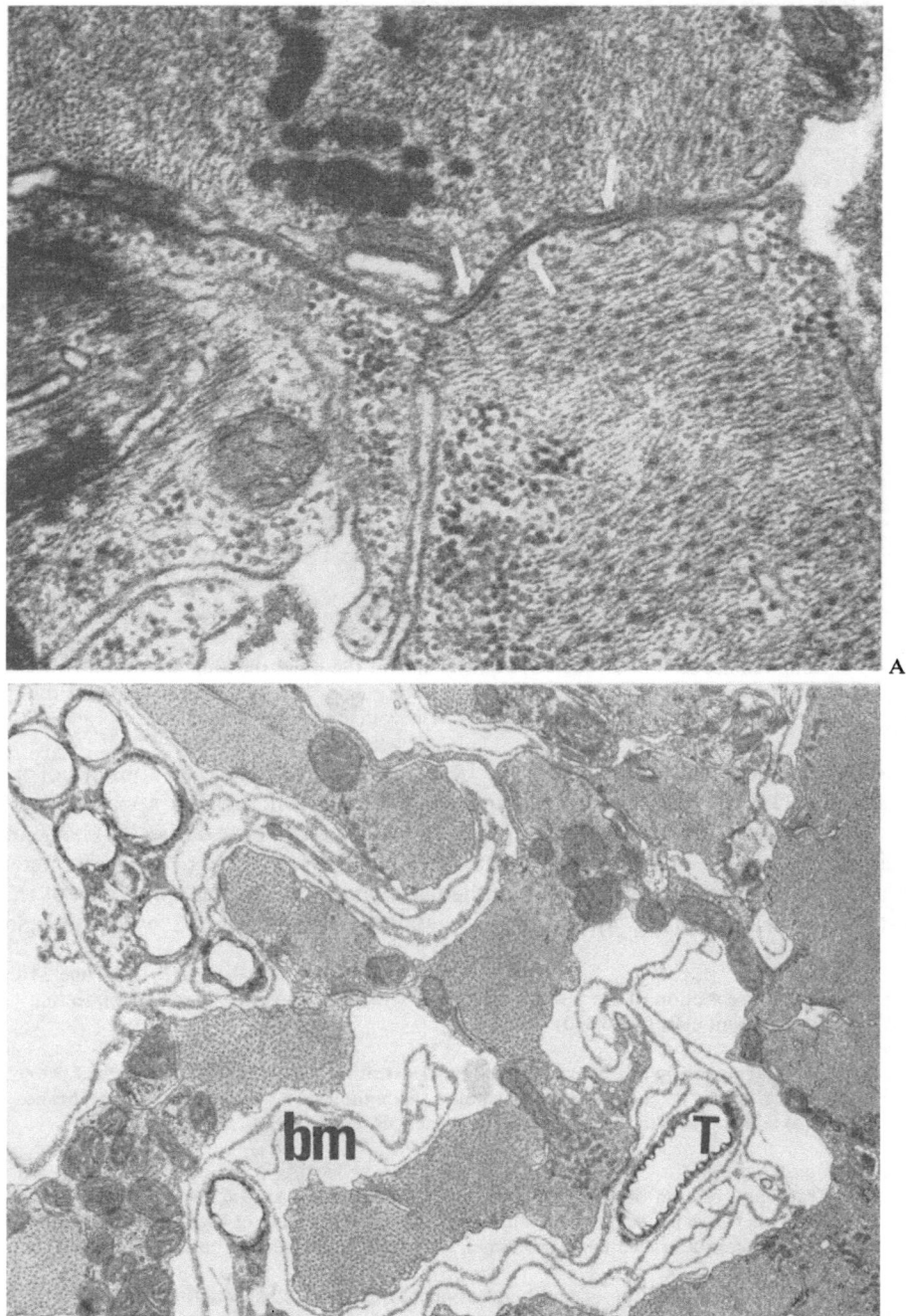

A

B

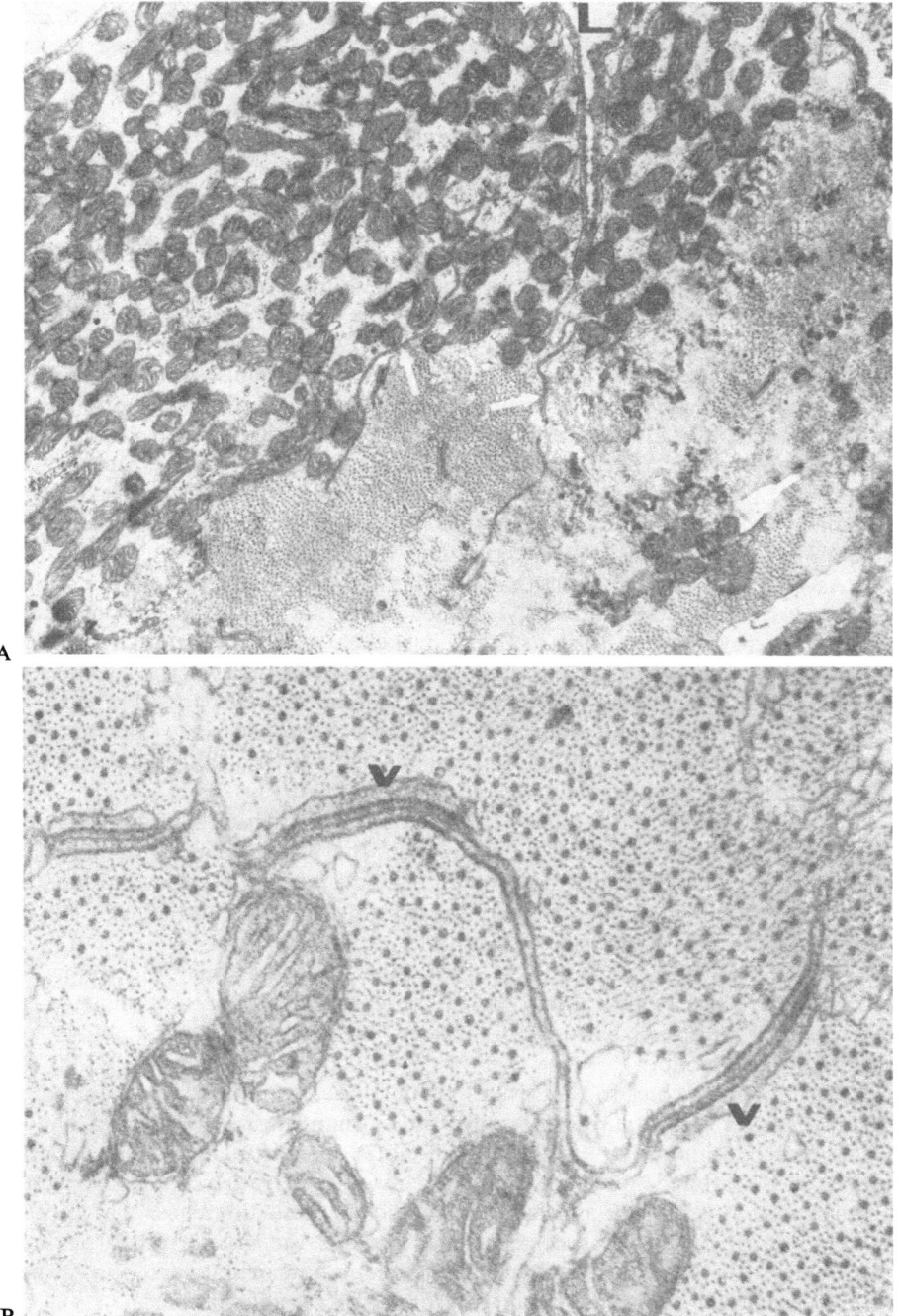

microscope appear similar to the intercalated disc structures observed in mammalian hearts as shown in Fig. 4. Examination of the intercalated disc region in the electron microscope revealed that two components could be identified[16], viz. the interfibrillar junction and septate desmosomes (Fig. 5A).

Multiple infoldings of the cell membrane contribute to a greatly augmented total cellular surface area. Fig. 5B illustrates the extensive invaginations into a single cell. At low magnification (Fig. 6A) these infoldings of the cells sometimes appear to be nexuses, however, at high magnification (Fig. 6B) they are found to be membranes of the same cell that are in close apposition. One can observe, with the electron microscope, places where these invaginations terminate and are surrounded by a common cytoplasm. This pattern results from infolding of the plasmalemma and would thus appear to serve a function similar to a T-system, that of providing a passageway for electrical impulses into the interior of the fiber. Observations of structural features that distinguish specialized cells in several vertebrate myocardia have led to speculation that pacemaker activity of a cell may be directly related to the relative amount of surface area of the plasma membrane to cell volume[17]. The large amount of surface area associated with each moth myocardial cell coupled with the physiological evidence of ubiquitous pacemaker activity is consistent with this suggestion.

As is typical of most cells that are rhythmically active and thus require ready sources of metabolic energy, mitochondria are abundant. Rather than an orderly array between fibrils as found in skeletal and cardiac muscle of some vertebrates[18], however, they are arranged in outpocketings of the membrane that project into the lumen of the heart and the pericardial spaces (Fig. 6A).

The sarcoplasmic reticulum appears poorly developed when compared to skeletal muscle[19] and is more like typical cardiac muscle in this respect. Both dyadic and triadic arrangements have been identified (Fig. 7).

The cells that comprise the myocardium appear structurally homogeneous. There is no apparent geometrical organization that would place tapering fibers in a converging or diverging pattern that would correlate with variations in segmental conduction velocities or preferred directions of conduction[3]. Populations of fiber sizes were also indiscernible.

We have concluded from these studies that we are at this time unable to recognize evidence of cell 'types', or of any distinctive structures that would identify cells as specialized to carry out functions such as initiating activity or facilitating conduction.

Perhaps some clues to the structural-functional correlates will be found in the appearance of septate desmosomes rather than nexuses, beaded Z-bands rather than continuous structures, mitochondria compartmentalized in packets rather than scattered, or a randomly organized S–R rather than an orderly arrangement. It is difficult at this time to associate these differences that appear so minor with functional specializations. Perhaps the unique properties are associated with the plasmalemma. The accumulation of data from a variety of myocardia will aid in making an interpretation feasible.

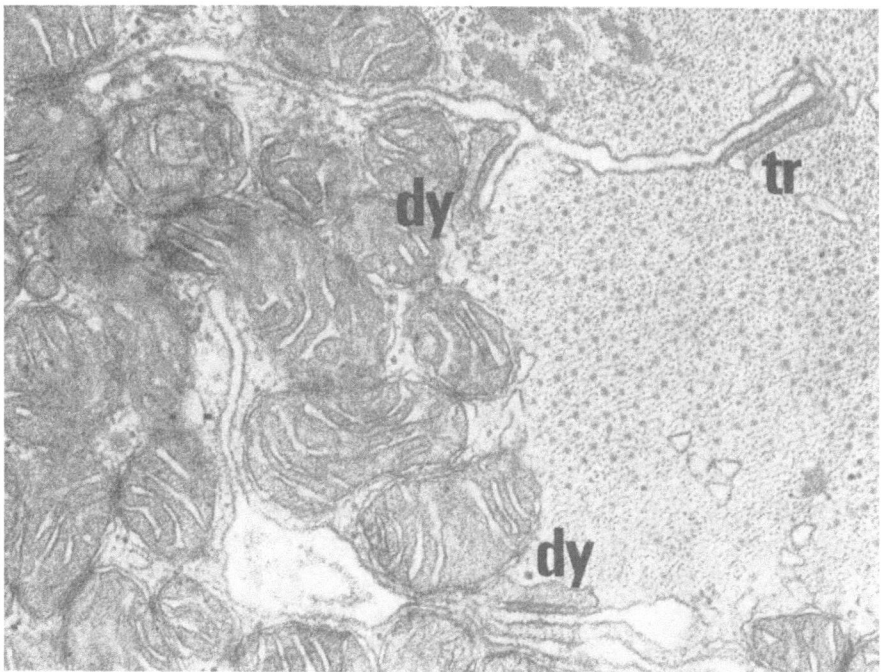

Fig. 7.
Vesicles of cardiac dyads (dy) and triads (tr) are filled with densely stained material
(×52,500).

In an attempt to mark the sarcotubular system with externally applied ferritin as has been done with other tissues[20,21], we found that the ferritin particles apparently penetrate the cell membrane so that they become distributed throughout the interior of the whole cell. The moth heart was bathed in situ with a 10% solution of cadmium-free ferritin in physiological saline. Experiments were carried out for durations of 5, 10, 15 or 45 min. At the end of each period, the heart was washed with physiological saline and plunged into glutaraldehyde fixative. The detailed procedure for fixation has been published[16].

Examination of these tissues in the electron microscope reveals that ferritin particles advance in a timed sequence across the membrane into the cytoplasm. Ferritin particles are distributed throughout the basement membrane bordering the plasmalemma of the cells after the 5-min exposure period. After 10 min, the ferritin particles had passed through the basement membrane and had concentrated at the outer surface of the plasma membrane. After 15 min, the particles were well inside the cell membrane and had migrated into the interior regions of the cytoplasm. After 45 min, ferritin was distributed throughout the cytoplasm as shown in Fig. 8.

Fig. 8.
Electron micrograph showing the distribution of ferritin particles throughout
the cytoplasm of a heart cell (H) and an alary muscle cell (A). The myomuscular
junction is indicated (M) (× 13,800).

Pursuant to our observation that ferritin was unsatisfactory as a sarcotubular marker since it invaded the whole cell, we conducted the same experiment while recording the electrical and mechanical activity of the heart to see
if the integrity of the membrane had been interrupted by toxic effects. Visual
observation of the heart during the timed ferritin experiments suggested that
the heart was not affected by its presence. Electrical activity coincident with the
spontaneous rhythm of the heart was recorded with a conventional microelectrode. A second measurement involved the simultaneous recording of the
maximum rate of rise of the action potential (dV/dt) as an indicator of the inward depolarizing current. A third trace recorded the contractile wave iso-
metrically. The technical details of this recording arrangement have been published[1]. Fig. 9 illustrates a recording taken at 0 time (A) and 1 h later (B). The
electrode remained in the same cell for the entire period. None of the parameters
changed during the experimental period. The heart was then prepared as described and examined in the electron microscope. The cells appeared as shown
in Fig. 8. Ferritin invades the cytoplasm of the cell, traversing the membrane

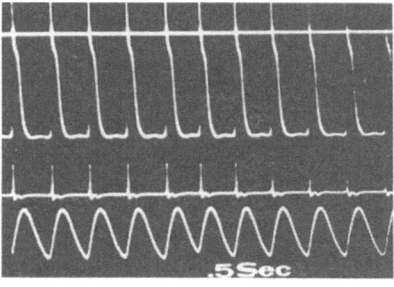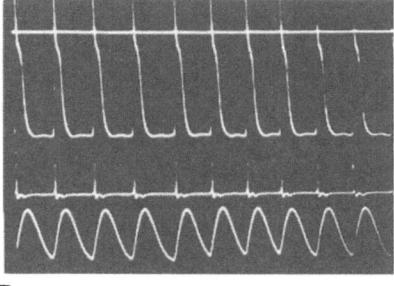

A   B

Fig. 9

Action potentials, top trace; dV/dt, second trace; and isometric tension, third trace during exposure of heart preparation to a solution of Cd-free ferritin. A shows control, B, one hour later. Marker shows 0.5 sec, 20 mv.

with ease and in so doing, does not affect the electrical or mechanical properties of the cell.

A widely held concept of the generalized cell membrane envisions water-filled pore structures about 7 Å in diameter[22] that differentiate between sodium (5 Å) and potassium (4 Å) ions. Orders of permeabilities for various anions have been proposed on the basis of the size of their hydrated radius relative to $K+$[23]. Evidence of this nature has been interpreted as support for the size-limiting function of membrane pore openings. It is indeed disconcerting to find that here ferritin particles, with a diameter of 100 Å, can easily penetrate the membrane and not disturb the inherent rate, rhythm, electrical or mechanical activity of the cell.

We have anticipated several questions that may be raised, and have considered several possible explanations of this phenomenon. It should be emphasized that we used cadmium-free ferritin to rule out the possible toxic effect of that ion. Since osmium fixation is known to break up membranes[24], glutaraldehyde fixation was employed to better preserve continuity of the membranes. No coated vesicles, as occur in moth pericardial cells[25], or pinocytotic vesicles were ever observed.

That the transmembrane entrance of ferritin results from a fixation artefact seems untenable, since we can document different degrees of ferritin distribution with time of exposure.

The movement of ferritin as well as saccharated iron oxide particles[26] into the cytoplasm of cells has been reported, but the validity of these observations has been questioned as resulting from either an artefact of fixation or an injury to the integrity of the membrane. Our time-studies coupled with electrical and mechanical recordings suggest that the penetration of ferritin is due neither to fixation artefact nor injury. The absence of vesicles or other evidence of pinocytosis leads us to conclude that either the membrane 'opens up' and allows the particles to enter or that the membrane is unselectively permeable because of exaggerated pore diameters.

That the permeability qualities of the membrane may indeed be unusual was suggested by the results from earlier experiments in which various anions, selected on the basis of the diameter of their hydrated radius, relative to K^+, were substituted for chloride. Marked hyperpolarization of the membrane resulted from the substitution of acetate (1.80) for chloride (0.96)[27], the increased negativity of the membrane being interpreted as resulting from the inward movement of acetate ion. The relatively small differences between the ratios of inside versus outside concentrations of ions also indicate that the membrane may not be a highly selective barrier to the passage of ions or particles of larger dimensions.

The presence of ferritin particles in the cytoplasm not only of the cardiac muscle cell but also of alary muscle fibers (Fig. 8) directed our attention to these muscle fibers and their point of contact with the heart.

Alary muscle fibers are long, filamentous, translucent, branching, cylindrical, striated muscle fibers. The diameter ranges from 5 to 15 µ and the length is uncertain, since the fibers may extend from their point of attachment at the lateral body wall to the surface of the heart, and continue to the other side of

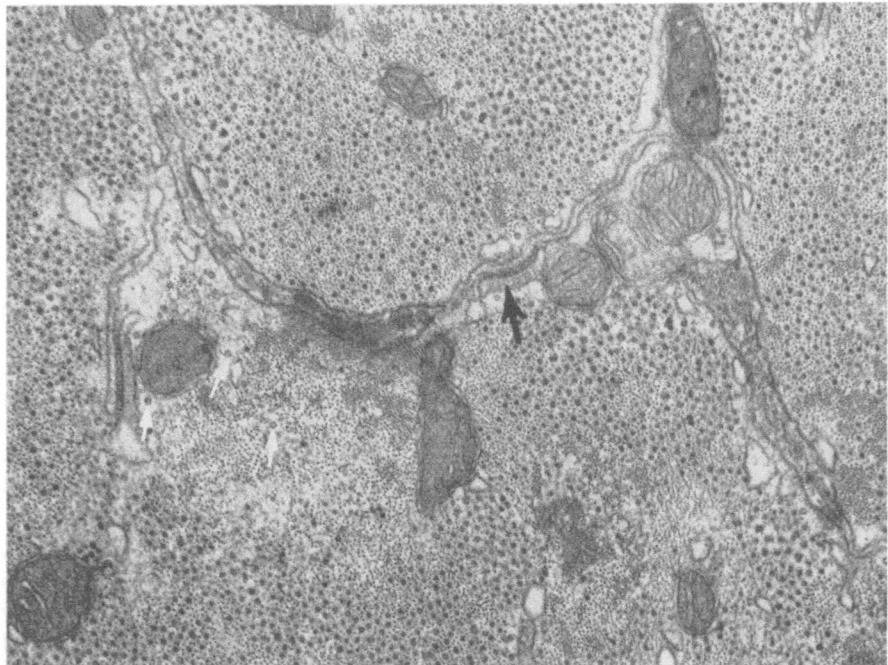

Fig. 10 A.
Transverse section through two alary muscle cells showing the presence of thin and thick filaments and a sarcoplasmic reticulum vesicle in dyadic (black arrow) relationship with the plasmalemma. Microtubules are indicated by white arrows (×35,000).

the body wall without the interruption of a cellular boundary. Since a detailed description of the ultrastructure of alary muscles has been published[28], only the most distinctive features will be presented here. As can be seen in the transverse section in Fig. 10 A, the fibers are characterized by an orbit of 10–12 thin filaments around each thick filament, a poorly developed sarcoplasmic reticulum and a profusion of microtubules.

Microtubules have been variously implicated in the intracellular transport of water and ions[29], cytoplasmic streaming[30] and especially development and maintenance of the skeleton or shape of the cell[31,32]. Why these should be present, and in such large quantity in a muscle cell, is, at present, unexplained. It may be that since the microtubules are always found with their long axes parallel to the long axis of the muscle fiber[28], they may have been involved in the initial extension of these long fibers and remained even after the muscle was completely formed.

While the electrophysiological work reported here on the alary fibers is somewhat preliminary, we have used the data collected to date to speculate on their role and propose a possible mechanism for their role as an extracardiac

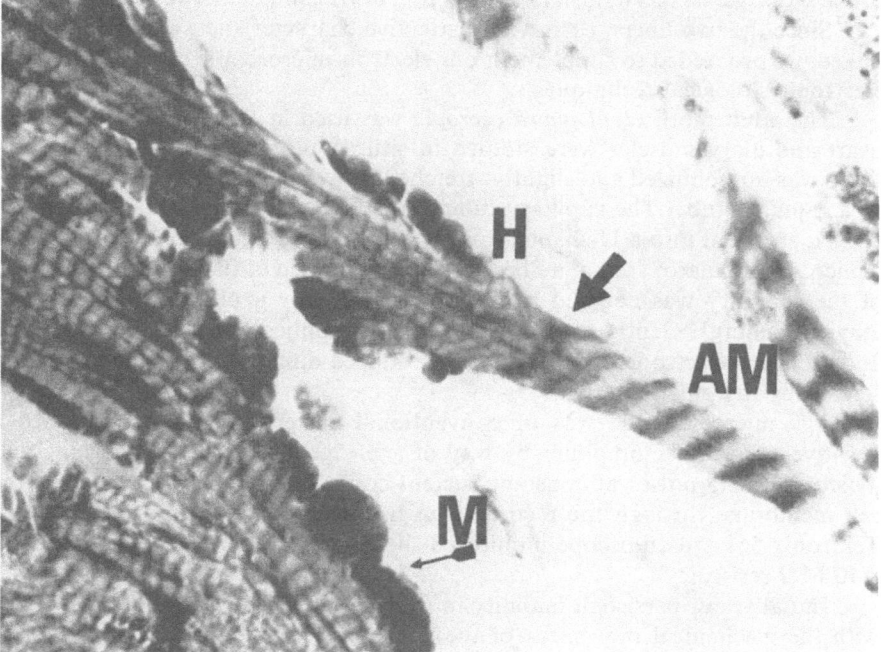

Fig. 10B.
Light micrograph of the attachment of an alary muscle (AM) to the heart wall (H) by means of a myo-muscular junction (arrow). The large sarcomeres of the alary muscle are easily distinguished from the smaller ones of the heart. The densely stained protrusions (m) along the heart are concentrations of mitochondria (see also Fig. 6A) (× 1000).

regulatory device. It must be emphasized that we are in no manner proposing that all alary muscle fibers of insects function in this manner. Our primary interest is to describe this particular system, not as a device peculiar to an insect, but rather as a physiological preparation in which two muscle cells, with distinctively different structural features, make direct contact through a structure that we have termed a myomuscular junction[28]. The junction between the two strikingly dissimilar striated muscles appears similar to the intercalated discs that serve as interfiber boundaries in cardiac cells and is shown in Fig. 10B. Examination of the intercalated disc in the electron microscope revealed that here, as in the heart, two component structures can be identified, viz. the interfibrillar region and a septate desmosome. A detailed picture of this structure as seen in the electron microscope is shown also in Fig. 8. At present, however, we have not observed any areas that can be identified as tight-junctions, nor are there any localized vesicles, granules or other particles that might be associated with chemical transmitter substances. Our interest is now directed to the functional significance of this myo-muscular junction, and to a study of the mechanisms whereby activity may be coordinated in the two cells. The nature of the message, its mechanism and direction of transmission awaits exploration.

Since the two fibers, each with distinctive characteristics, are anatomically linked, we proceeded to supplement our electron-microscopic observations with electrophysiological techniques.

The adult moth *Hyalophora cecropia* was used in all of these studies; the heart and alary muscles were studied in situ. A discrete section of the alary fibers was immobilized and slightly stretched by a gentle suction applied by way of a capillary tube. The capillary tubing was drawn to a final inside diameter of 25 μ and bent into a U-shape. This could be placed under an alary fiber with a micro-manipulator so that a fiber spanned the lumen of the pipette. The shank of the capillary was coupled to a syringe device by a piece of No. 50 polyethylene tubing[1]. Gentle suction could then be applied to immobilize the fiber and the microelectrode could then be introduced directly into the area circumscribed by the pipette lumen.

The microelectrode was of conventional design and was coupled to a negative-capacitance amplifier by way of a bridge circuit[15]. Depolarizing and hyperpolarizing pulses of constant current could thereby be passed across the cell membrane through the recording microelectrode. A second trace on the Tektronix 561A oscilloscope monitored the current as the voltage drop across a 10 MΩ resistor.

Initially, our persistent inability to measure any electrical signal associated with the mechanical movement of the alary muscle was indeed puzzling. The electrode could be observed with a microscope as it penetrated what appeared to be a single muscle fiber, and yet only very small resting potentials (-15 mv ± 4 mv) and small (10 mv ± 3 mv), slow-type action potentials could be recorded. They occurred rhythmically with a constant phase displacement with the heart cycle. One, obviously, would be reluctant to state unequivocally that these phasic changes recorded from the alary muscles were true intracellular electrical responses, for such a record might be reflecting only mechanical deformation of the electrode. It was not until inward pulses of constant current were passed across the membrane that the electrical responses, similar to cardiac type action potentials, were revealed. Hyperpolarization produced graded, rhythmic action potentials. The graded nature of these depolarizing responses is shown in Fig. 11 A–G as is the separation into fast and slow components that occurs as membrane voltage is increased. In this instance, the fiber initially exhibited practically no discernible spontaneous electrical activity, i.e. the cell appeared inactivated. The final coalescence of the fast and slow components into cardiac-type action potentials occurred at a membrane voltage of about − 60 mv. The frequency of firing was not altered by the manipulation of the membrane voltage but a slow initial rise indicative of pacemaker-type potentials was observed. Arrhythmias, independent of the heart rhythm, have been noted to develop in alary fibers. These observations suggest that the alary fibers may be independently rhythmic, i.e. myogenic.

At present we have been unable to demonstrate, either with conventional nerve stains or with the electron microscope, a nervous innervation, although a nervous innervation to alary fibers and the lepidopteran heart as well has been reported[33]. We have repeated classical staining techniques and serially sectioned the alary muscles for examination in the electron microscope, but have been unable to demonstrate a nervous innervation along the fibers. In view of technical difficulties we would be unwilling to state unequivocally that there is not a nervous innervation, for we have not examined the terminal area at the point of attachment to the lateral wall. However, we are certain that there is not a multiterminal innervation along the entire length of the fiber. This implies that the muscle fiber, in view of its length, embodies a mechanism that provides for a near-synchronous depolarization throughout its length. Whereas heart muscle contracts as a peristaltic wave, the alary muscle contracts uniformly throughout its length. A self-propagating, all-or-nothing action potential could serve this need. However, the graded potentials recorded appear more like non-propagated, post-synaptic potentials.

It is interesting that the myo-muscular junction, the area of direct anatomical linkage between two heterogeneous striated muscle cells, should have the same structural features that prevail between the homogeneous cells that comprise the myocardium. If one assumes that cell-to-cell activation as a result of current flow proceeds across this region, as across intercalated discs in other hearts[34], then the mechanism of transmission may be similar if not identical. In view of the direct anatomical linkage with myocardial cells, we suggest that the signal is of a mechanical nature and originates as stretch, produced by contraction of the heart. Stretch, serving as the mechanism to depolarize the entire fiber, could effect the apparent rapid conduction that allows the fiber to contract simultaneously throughout its length, a mechanism described for some non-striated muscles such as those found in *Golfingia*[35]. A mechanism such as

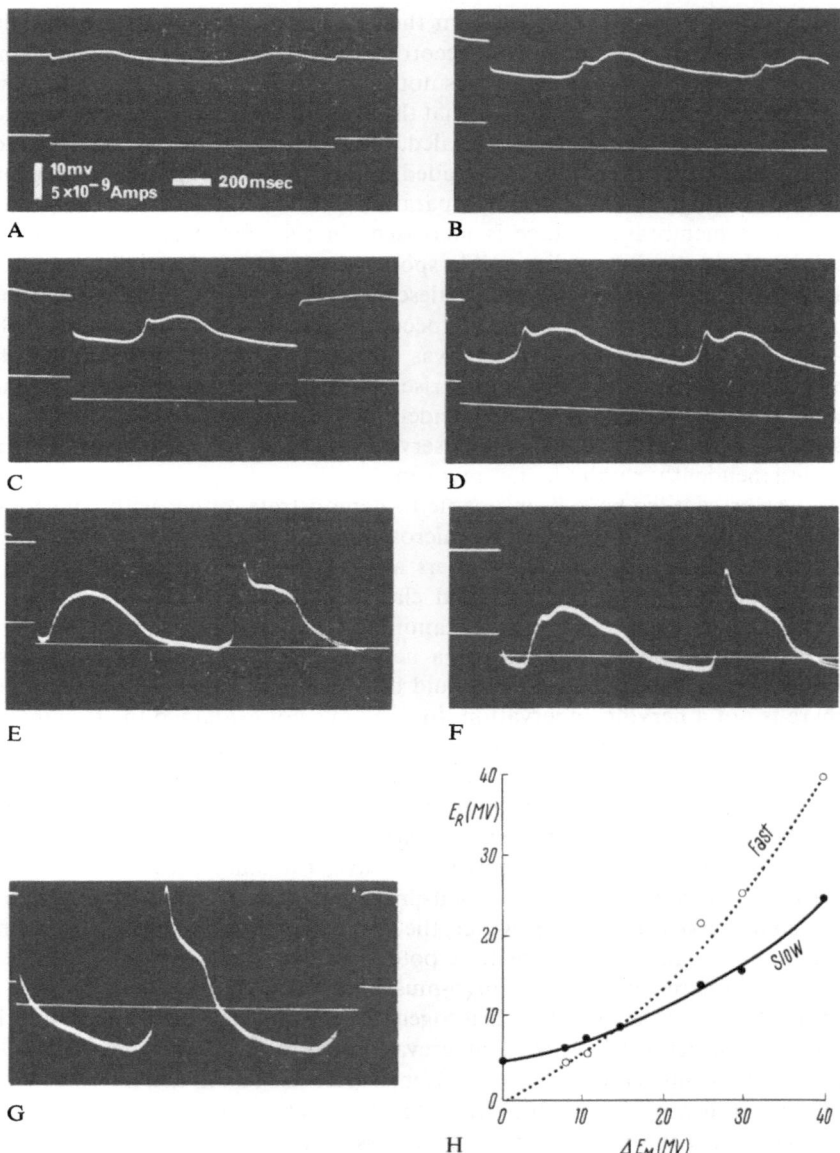

Fig. 11.
Intracellular electrical signals recorded from alary muscle fibers of the adult moth
H. cecropia. A–G. The step-wise separation of fast and slow components of the
cardiac-type action potential in an alary fiber. The graph (H) shows the relation between the change in membrane voltage (abscissa) and the voltage of the response
E_R (ordinate).

this would aid in explaining the curious inactivation of the membrane that appears to prevail during attempts at intracellular and extracellular recordings. This apparent inactivation may be the reason why all previous attempts to stimulate and record directly from these fibers have failed.

The records of repetitive cardiac-type action potentials that occur when the membrane is held at a hyperpolarized level, demonstrate that the cell can generate electrical signals coincident with mechanical activity. We propose, therefore, that inactivation of the membrane results from immobilization. Since the alary fibers essentially 'beat' in rhythm with the heart, it is necessary to restrain them in some fashion in order to impale them with a microelectrode. Most fibers are of extremely small diameter (5–10 μ), translucent and fragile, thereby further complicating the technical problems associated with the use of flexibly mounted electrodes. The gentle suction used here, as well as any type of hitherto employed external hook electrode or suction electrodes, would effectively immobilize the fiber and also impose a slight degree of stretch, thereby producing a depolarization with a resultant inactivation of the cell membrane. It would then seem a logical consequence that the myo-muscular junction need not be a highly specialized junction, i.e. similar to a synapse or tight junction, but rather that the contraction of the heart itself imposes a stretch on the directly attached alary muscle that effectively depolarizes the cell along its entire length and thereby activates it. The graded nature of the electrical signals and the curious separation of fast and slow responses may be related to gradations in the amount of imposed stretch, i. e. stimulus, and thereby serve as a control system to regulate the amount of pull needed to re-expand the heart. These proposals, while admittedly speculative, will now be rigorously examined in our laboratory, for this hypothesis would assign to the myo-muscular junction region the role of a structural reinforcement to insure integrity of the interfiber boundaries as they are subjected to mechanical pull (stretch).

Conclusion

Using the heart of an insect as a model myocardium, our studies on ultra-structure as related to function have been directed to fundamental processes of excitable tissues. The remarkable permeability characteristics of this membrane and the generation and transmission of electrical signals in single cells emphasize that we are still unable to recognize and associate the architectural modifications that may provide for physiological specialization.

Acknowledgments

This work was carried out during the tenure of an Established Investigatorship of the American Heart Association (FVM) and supported in part by the

Vermont and New Hampshire Heart Associations and the National Heart Institute.

J. W. SANGER, a Postdoctoral Fellow in Physiology and Trainee, was supported by National Institutes of Health Grant 5T1 HE 5322(05). His present address is Department of Anatomy, University of Pennsylvania, School of Medicine, Philadelphia, Pennsylvania.

References

1   F. V. McCANN, *Experiments in Physiology and Biochemistry* (Ed. G. KERKUT; Academic Press, New York 1969), vol. 2.
2   J. H. GEROULD, J. morph. Physiol. *48*, 385 (1929).
3   S. M. TENNEY, Physiologia comp. Oecol. *3*, 286 (1953).
4   F. V. McCANN, J. comp. biochem. Physiol. *11*, 45 (1964).
5   *Documenta Geigy Scientific Tables*, 6th ed. (Ed. K. DIEM; Geigy Pharmaceuticals, New York 1962).
6   F. V. McCANN and C. R. WIRA, Comp. Biochem. Physiol. *22*, 611 (1967).
7   F. V. McCANN, J. gen. Physiol. *46*, 803 (1963).
8   F. V. McCANN, Comp. Biochem. Physiol. *20*, 339 (1967).
9   H. HUDDART, Arch. Int. Physiol. Biochim. *76*, 519 (1968).
10   G. LING and R. GERARD, J. cell. comp. Physiol. *34*, 413 (1949).
11   J. M. MARSHALL, Am. J. Physiol. *180*, 350 (1955).
12   W. V. MAC FARLANE, Nature *178*, 1050 (1952).
13   C. DE MELLO, Am. J. Physiol. *196*, 377 (1959).
14   B. HOFFMAN and P. CRANEFIELD, *Electrophysiology of the Heart* (McGraw-Hill, New York 1960).
15   F. V. McCANN, Comp. Biochem. Physiol. *17*, 599 (1966).
16   J. W. SANGER and F. V. McCANN, J. Insect Physiol. *14*, 1105 (1968).
17   H. RUSKA, *Electrophysiology of the Heart* (Ed. B. TACCARDI and G. MARCHETTI; Pergamon Press, New York 1965), p. 9.
18   D. W. FAWCETT, Circulation *24*, 336 (1961).
19   K. R. PORTER and C. FRANZINI-ARMSTRONG, Sci. Am. *212*, 73 (1965).
20   H. E. HUXLEY, Nature *202*, 1067 (1964).
21   S. PAGE, J. Physiol. *175*, 10P (1964).
22   A. K. SOLOMON, J. gen. Physiol. *43*, 1 (1960).
23   T. ARAKI, M. ITO and O. OSCARSSON, J. Physiol. *159*, 410 (1961).
24   C. FRANZINI-ARMSTRONG and K. R. PORTER, Nature *202*, 355 (1964).
25   J. W. SANGER and F. V. McCANN, J. Insect Physiol. *14*, 1839 (1968).
26   L. FLOREY, Proc. R. Soc. [B] *166*, 375 (1966).
27   F. V. McCANN, J. comp. biochem. Physiol. *13*, 179 (1964).
28   J. W. SANGER and F. V. McCANN, J. Insect Physiol. *14*, 1539 (1968).
29   D. B. SLAUTTERBACK, J. Cell Biol. *18*, 367 (1963).
30   M. C. LEDBETTER and K. R. PORTER, J. Cell Biol. *19*, 239 (1963).
31   L. G. TILNEY and K. R. PORTER, Protoplasma *50*, 317 (1965).
32   L. G. TILNEY, Y. HIRAMOTO and D. MARSLAND, J. Cell Biol. *29*, 77 (1966).
33   K. DAVEY, Adv. Insect Physiol. *2*, 220 (1964).
34   L. BARR, M. M. DEWEY and W. BERGER, J. gen. Physiol. *48*, 796 (1965).
35   C. L. PROSSER, C. L. RALPH and W. W. STEINBERGER, J. cell. comp. Physiol. *54*, 135 (1959).

Functional Correlates of Fine Structure in the Heart of Achatinidae

by R. H. NISBET and JENIFER M. PLUMMER
Department of Physiology (E.M. Unit), Royal Veterinary College (University of London), London, N.W. 1

Introduction

Earlier studies on the nervous system and heart of Achatinidae have led to questions on the relationship of structure to function in the latter organ[1-4]*. There is a large and rapidly growing literature on molluscan muscle, much of which relates to the anterior byssal retractor of *Mytilus* and to adductor muscles in Bivalvia[5-9]. A significant contribution in the context of the present work is that of TWAROG[10]. Similarly, extensive pharmacological studies have been made on the hearts of many bivalves and of some gastropods[11-17], although these studies, as with those on the bioelectrical characteristics and the bioenergetics[18-22], have frequently been confined to the ventricle.

Important reviews are those of KRIJGSMAN and DIVARIS[23] and of HILL and WELSH[24]. The paper by DIVARIS and KRIJGSMAN[25] on heart function in *Cochlitoma zebra* is of interest because of the close relationship of *Cochlitoma* to the animals in the present work.

Functional Morphology of the Heart

Although the rhythmic contractility of the molluscan heart is said to be myogenic in origin[23-25] there are marked differences from the vertebrate heart in the behaviour of the organ[21]. It is desirable, therefore, to consider cardiac function in relation to some aspects of the blood system and to the behaviour of the gastropod.

In animals as phylogenetically diverse as the Trochacean genera *Monodonta* and *Gibbula*[26], the basommatophoran pulmonate *Lymnaea*[27,28] and the stylommatophora *Helix* and *Archachatina*[29] there is a basic similarity in the morphology of the blood system and the body musculature. The anterior aorta enters the cephalic haemocoel by passing through a cervical septum (Fig. 1). At this level there is a physiological valve (or an anatomical valve in the aorta). The effect of this valve can readily be seen when an attempt is made to perfuse a retracting animal. The septum and associated muscles isolate the hydrostatic

* Numbers refer to References, p. 67

skeleton of the head and foot from that of the rest of the body and simultaneously block forward movement of blood in the aorta.

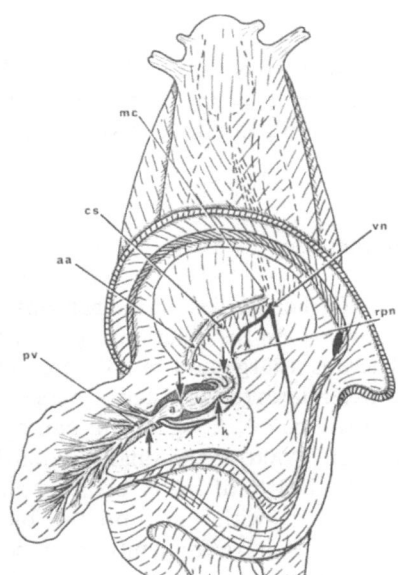

Fig. 1.
Archachatina marginata. Drawing to show heart, pallial vein, anterior aorta on floor of mantle cavity and point of muscular constriction at passage through cervical septum (physiological valve) where contraction blocks forward and backward passage of blood. Heart and its innervation also shown. Arrows indicate levels of insertion of cannulae for investigation of activity in auricular, ventricular, ventriculo-aortic and whole heart preparations (see p. 60).
a, auricle; *aa*, anterior aorta; *cs*, cervical septum; *k*, kidney; *mc*, muscular constriction (physiological valve) at cervical septum; *pv*, pallial vein; *rpn*, reno-pericardial nerve; *v*, ventricle; *vn*, visceral nerve.

Examination of actively crawling animals, in which the body whorl of the shell has been removed, shows that the heart beat is related not only to the general level of activity, but also to the type of activity. As soon as retraction occurs, the heart ceases to beat (Fig. 2). Even when the heart is beating, there

Fig. 2.
A. marginata. Tracing to show relative times of contraction of auricle and ventricle in an active snail with body whorl of shell removed. Signal markers indicate contractions of auricle (*A*) and ventricle (*V*). Note cessation of ventricular systole during contraction of the columellar muscle (*con. col.*). Time marker (*T*) = 1 sec.

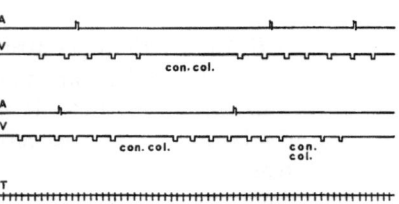

are independently variable frequencies of auricular and ventricular systole, the two chambers behaving as though they were in some degree subject to independent control.

The gross morphology and innervation of the Achatinid heart is similar to that of *Helix pomatia*[30]. There are no valves in the pallial vein at or near to the auricle, nor is there a pallial-auricular valve. A pair of auriculo-ventricular valves guards the junction of the two heart chambers and there is a single valve

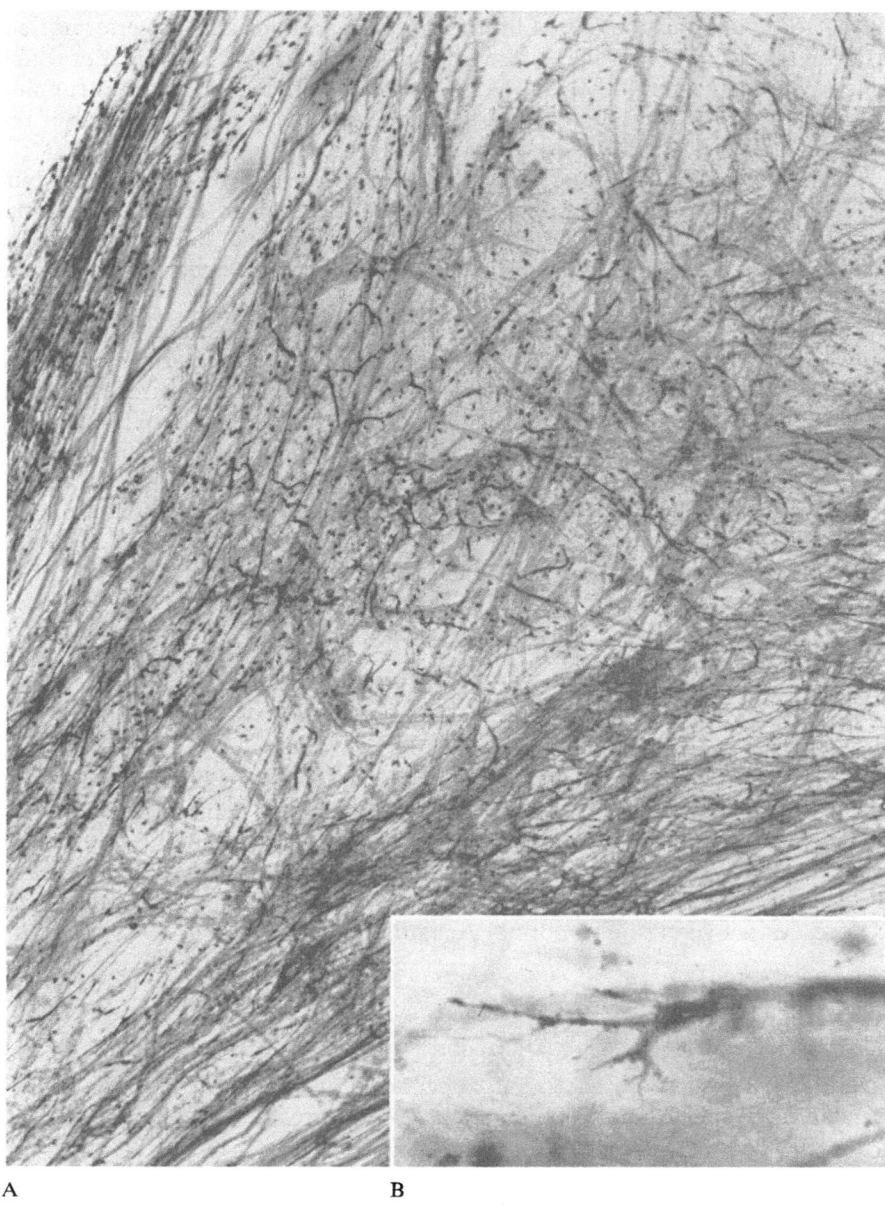

A   B

Fig. 3.
A. marginata. (A) Rongalit white preparation showing radiation of nerve fibres on inner surface of the auricle. × 50.
(B) High power print showing one of the possible nerve cells in the area of 3(A). × 600.

in the aortic bulb just before its bifurcation into anterior and posterior aortae. The reno-pericardial nerve gives rise to three cardiac branches, the first of which enters (or partly enters) the root of the aorta, the other two branches running in the ventral pericardium adjacent to the kidney and entering the wall of the auricle close to the entry of the main pallial vein.

Examination of the interior of the opened heart, with the aid of a binocular dissecting microscope, indicates a wide radiation of branches from the two auricular nerves over the inner surface of the auricle, but at this level of magnification there is little evidence of any extension of innervation from the aorta into the ventricle.

Rongalit white preparations[31] of the auricle demonstrate an extensive network of nerve fibres on its inner surface (Fig. 3A) and suggest the possibility that neurone somata may also be present (Fig. 3B). Similar preparations of the ventricle have not shown a corresponding plexus, but the evidence is equivocal because of the thickness of the ventricular muscle and the capricious nature of methylene-blue techniques. A systematic study of the fine structure of heart regions from the pallial vein to the aortic bulb has therefore been initiated, in an attempt to compare muscle structure and its innervation at each level. Some of this work is reported in the present paper.

Cardiac Fine Structure

Method

The hearts were fixed and embedded, sections were cut, mounted and stained in the ways already described[4], except that for some of the material, as indicated in the following legends, the physiological saline of CARRIKER[27] was replaced by that of DIVARIS and KRIJGSMAN[25]. The fine structural evidence agrees with the evidence from living hearts, viz. that the saline of DIVARIS and KRIJGSMAN is hypertonic to normal Achatinid blood.

Analysis of Fine Structure

Fig. 4 shows the characteristic appearance of auricular fibres, with fairly regular, in-register A- and I-bands, short (2.0 μ) 'sarcomeres', columns of mito-

Fig. 4.   ▷
A. marginata. Longitudinal section through two auricular muscle fibres, and a vesicular connective tissue cell. Note the regular (in register) arrangement of dense bodies and the presence of nerve fibres containing electron dense granules (one close to a myo-neural junction). Fixed in 3% formaldehyde and 2% glutaraldehyde, both in Carriker's (1946) solution followed by 1% osmium tetroxide in veronal acetate buffer. *db*, dense body; *m*, mitochondria; *m-n*, nerve fibre near myo-neural junction; *vct*, vesicular connective tissue cell. ×9,000. Bar represents 1 μ.

chondria and many nerve fibres containing electron dense granules. One nerve fibre (*m-n*) is close to the junction with the adjacent muscle fibre. The short sarcomeres are suggestive of a fairly fast contracting, phasic type of muscle, as are the thin (ca. 180 Å) paramyosin filaments and regular orbital arrangements seen in later prints in this paper. However, the appearance of an essentially

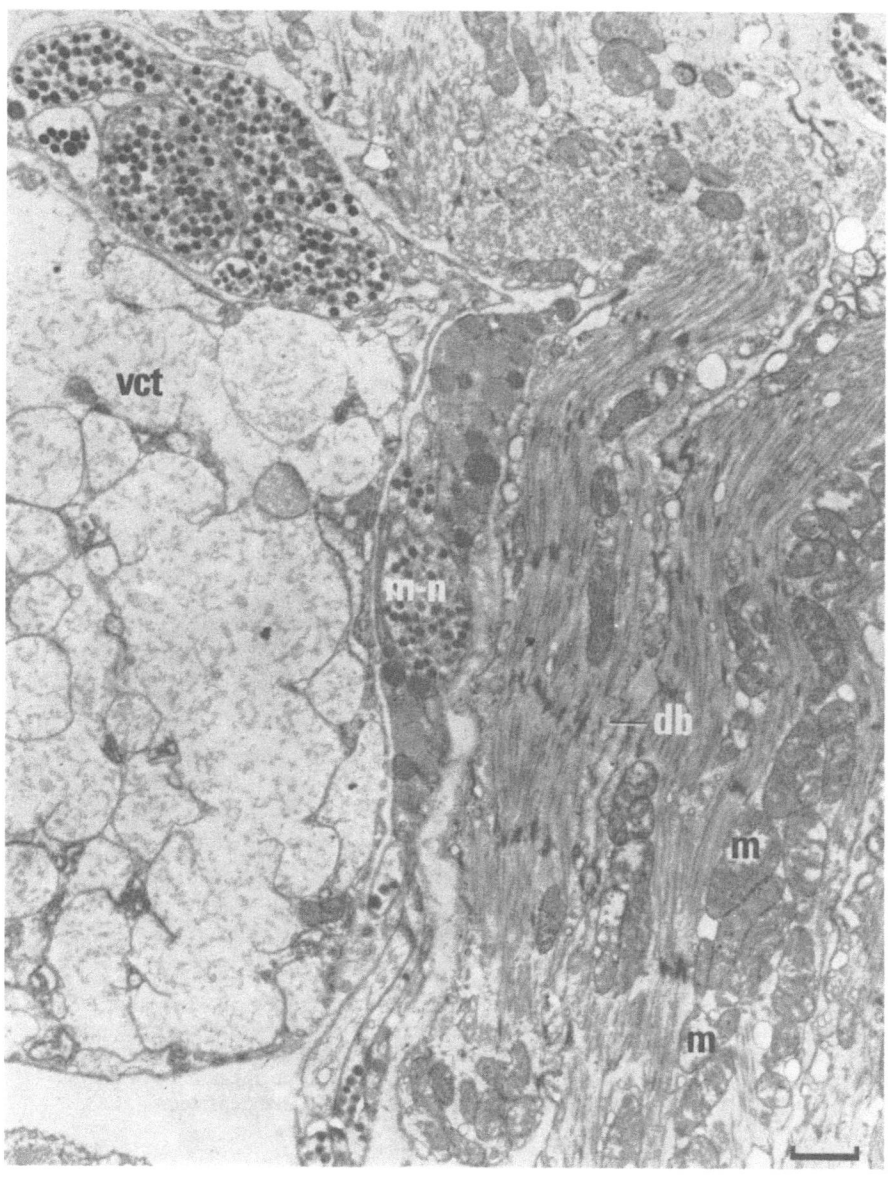

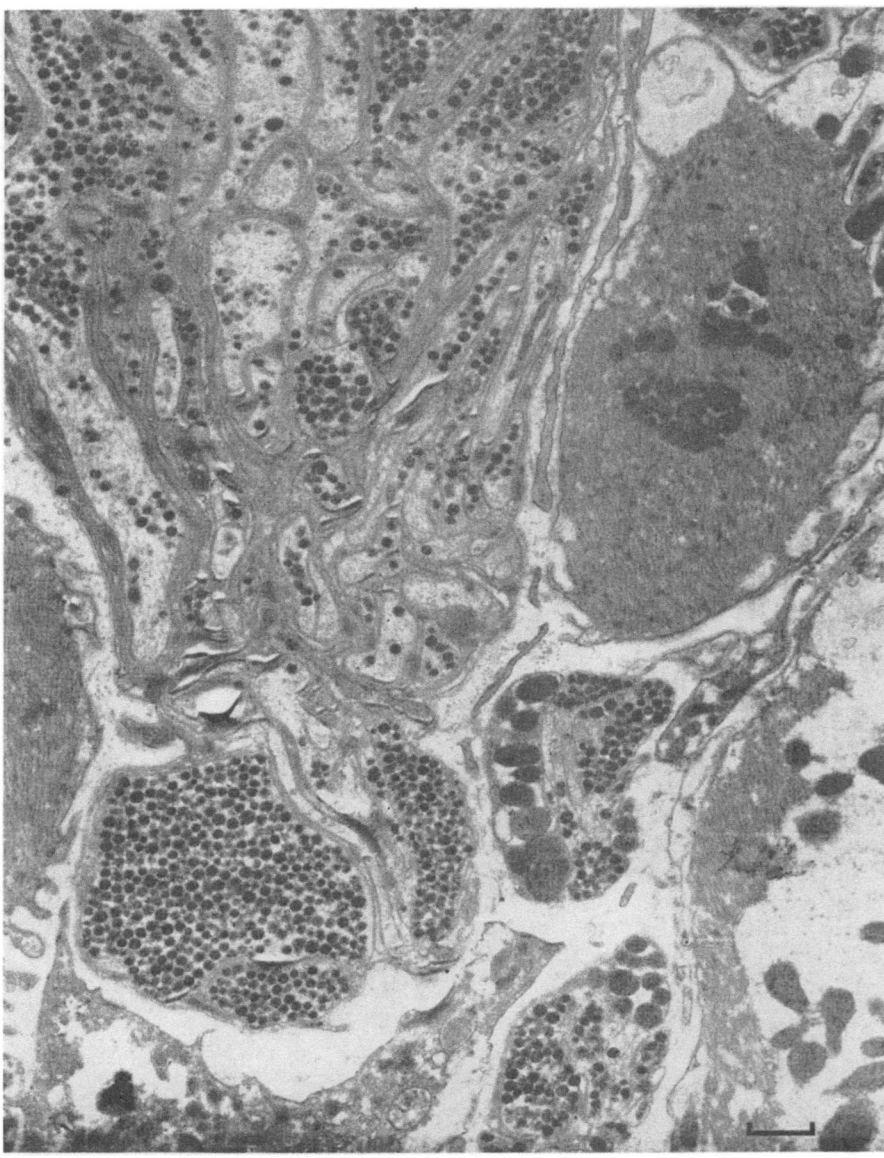

Fig. 5.
A. marginata. Section similar to that of Fig. 4, but physiological solution of DIVARIS
and KRIJGSMAN (1954, *Cochlitoma* solution) substituted for Carriker's solution. The
density of the mitochondria and evident shrinkage of muscle fibres reflect the higher
tonicity of the *Cochlitoma* solution. Note the characteristic appearance of the auricular muscle with a multiplicity of nerve fibres. ×9,000. Bar represents 1 μ.

spiral arrangement of the dense bodies[4] implies that each 'sarcomere', and therefore each muscle fibre, is capable of extensive increases in length. Cholinergic nerve endings are rarely seen and this may be significant (p. 61–62). Fig. 5 shows a similar area in auricular muscle fixed in *Cochlitoma* solution. Shrinkage

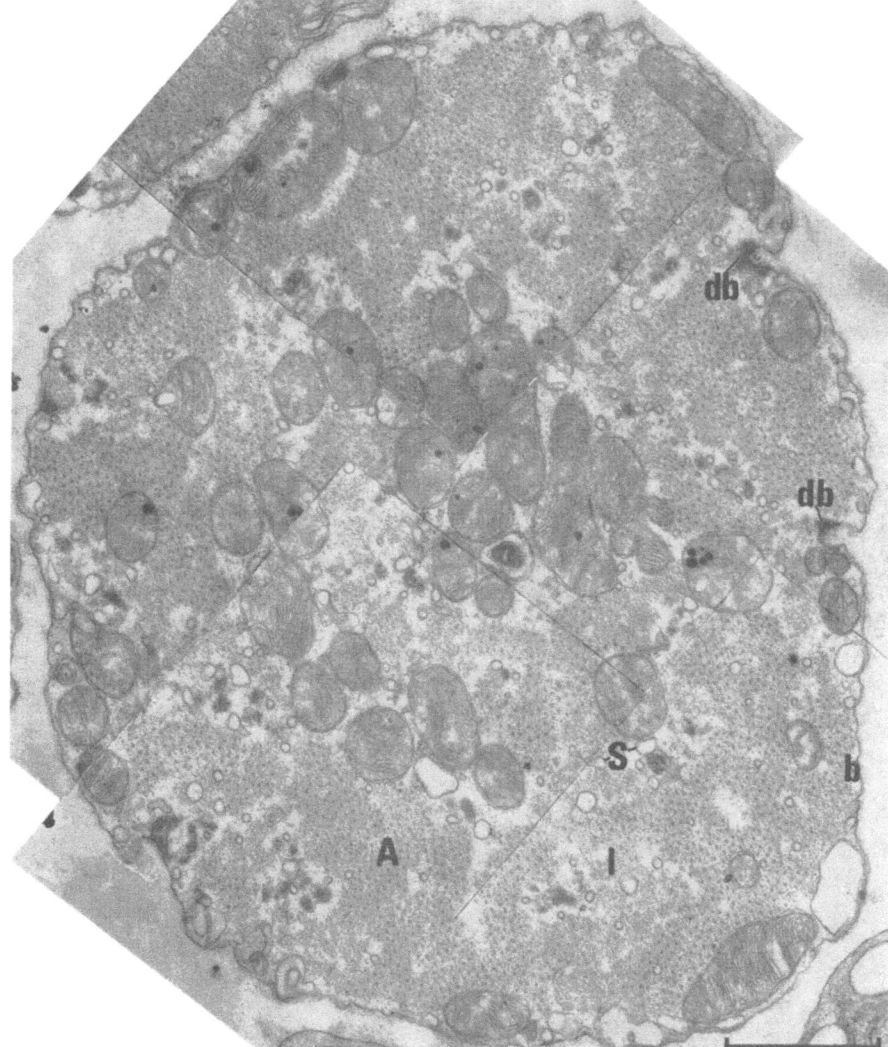

Fig. 6.
Achatina fulica. Transverse section through a typical muscle fibre from the body of the ventricle showing peripheral invaginations associated with dense bodies (*db*), peripheral and central zones of mitochondria, sarcotubules (*S*), A-band (*A*) and I-band (*I*) zones. Fixative in Carriker's solution. ×22,000. Bar represents 1 μ.

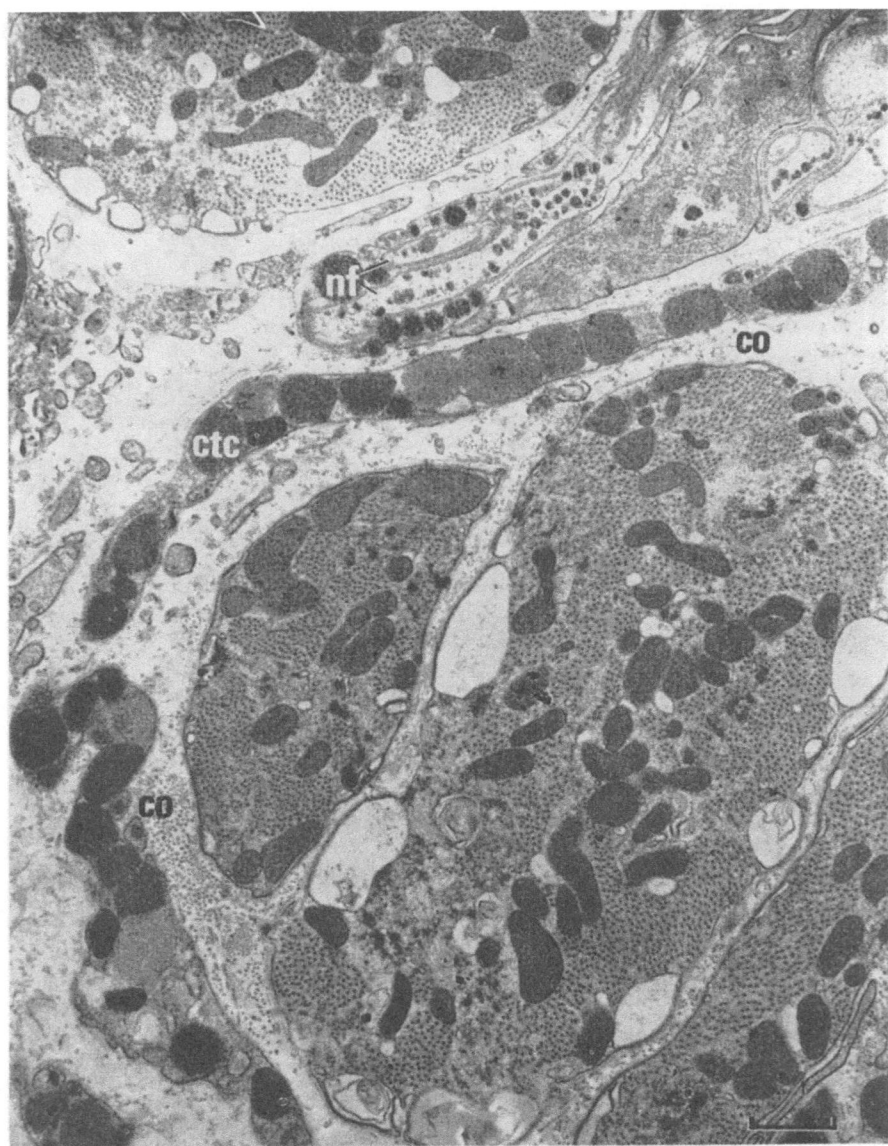

Fig. 7.
A. marginata. Transverse section through muscle fibres at the apex of the ventricle
showing progressive increase in the quantity of collagen and mucopolysaccharide (*co*)
and the entry of nerve fibres (*nf*) with associated Schwann cell elements into the
material. Note also connective tissue cells (*ctc*) with large pigment granules. ×11,000.
Bar represents 1 μ.

This and all following electron micrographs are from material fixed in 'Cochlitoma' solution. Note some shrinkage of contractile areas and swelling of sub-surface cisternae in muscle fibres.

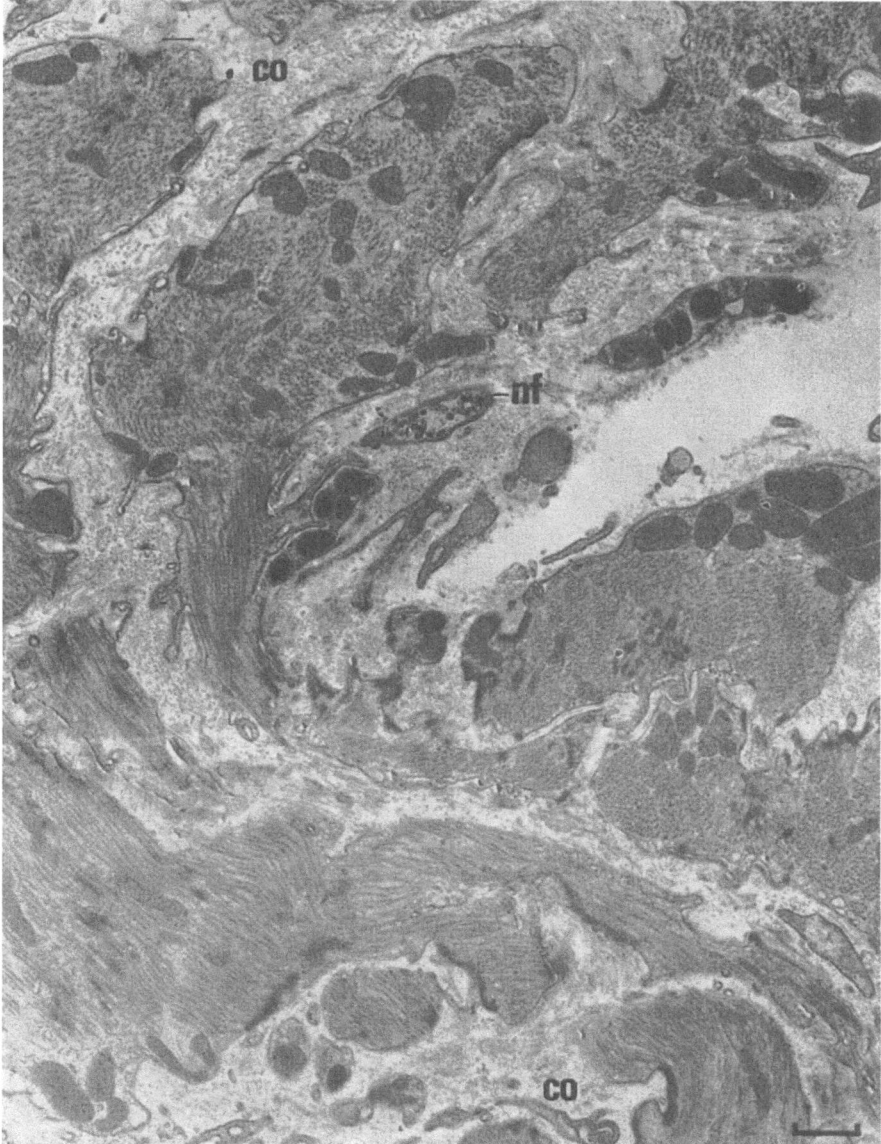

Fig. 8.
A. marginata. Muscle fibres round lumen at apex of ventricle. Note collagen (*co*), branching of muscle fibres with large hemi-desmosomes and one nerve fibre (*nf*) near the lumenal surface. ×9,000. Bar represents 1 μ.

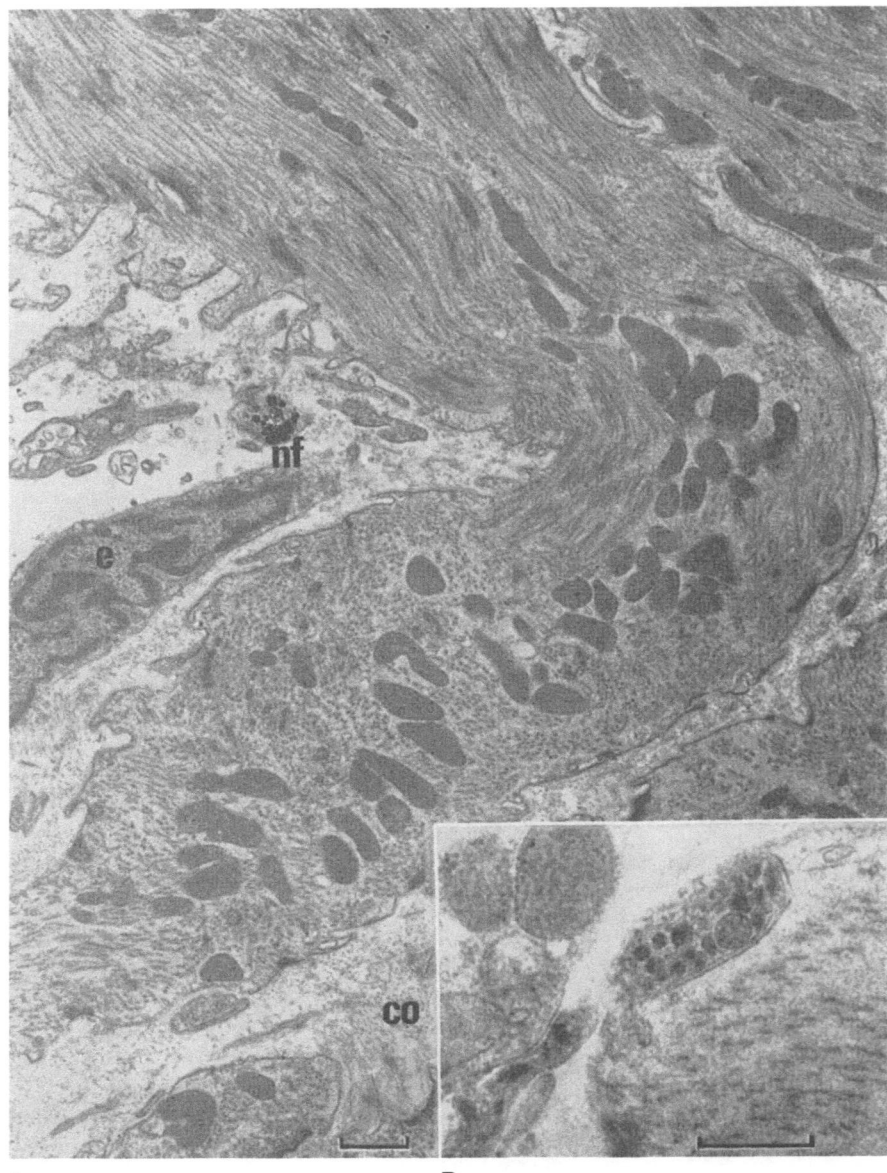

A   B

Fig. 9.
A. marginata. (A) Area adjacent to that of Fig. 8. Note processes on muscle fibres
and first appearance of an endothelial cell (*e*). (*nf*) nerve fibre, (*co*) collagen. ×9,000.
Bar represents 1 μ.
(B) Small portion of a muscle fibre showing the first appearance of a ventricular
myo-neural junction. ×29,000. Bar represents 0.5 μ.

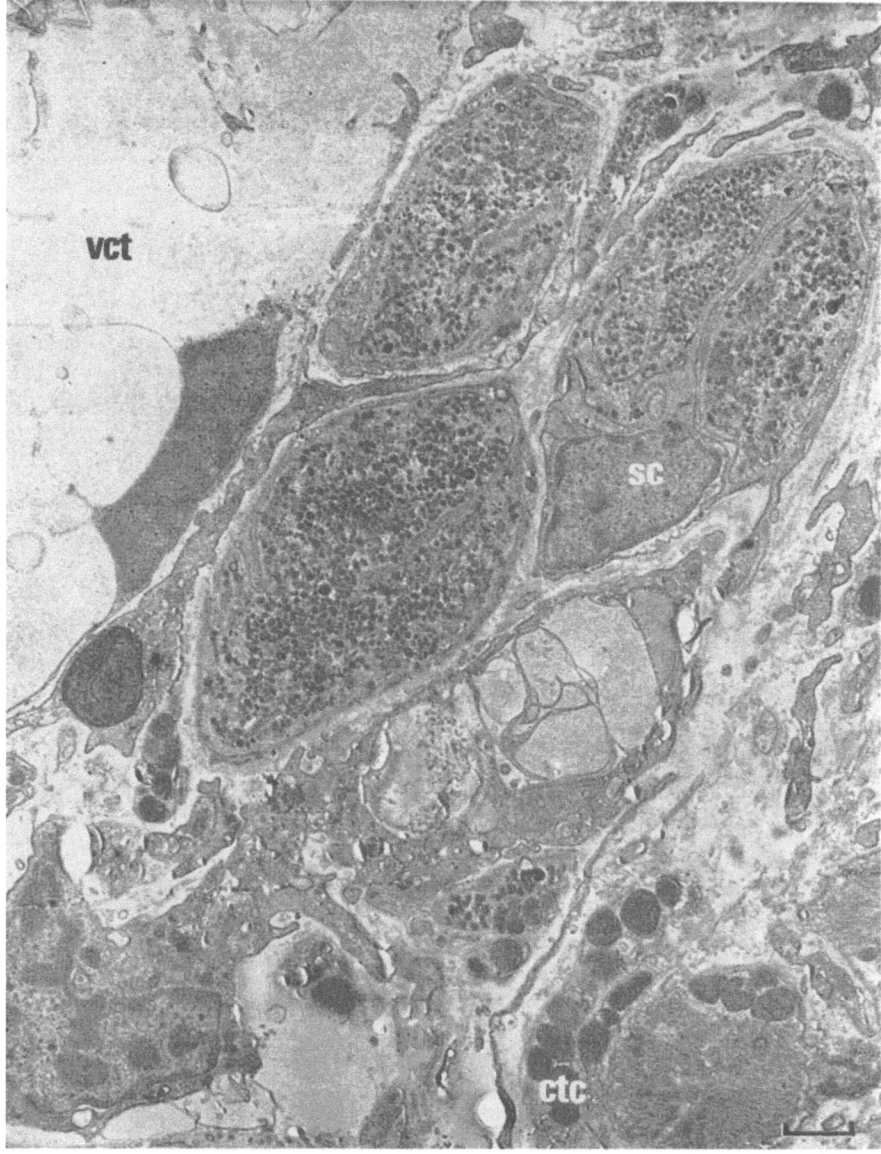

Fig. 10.
A. marginata. Junction of ventricle and aortic bulb. Note presence of many nerve
fibres, sheath cells, collagen and vesicular connective tissue cells. The electron dense
granules of the nerve fibres show qualitative differences from those in auricular nerve
fibres. *ctc*, connective tissue cell; *vct*, vesicular connective tissue cell; *sc*, Schwann
cells; ×9,000. Bar represents 1 μ.

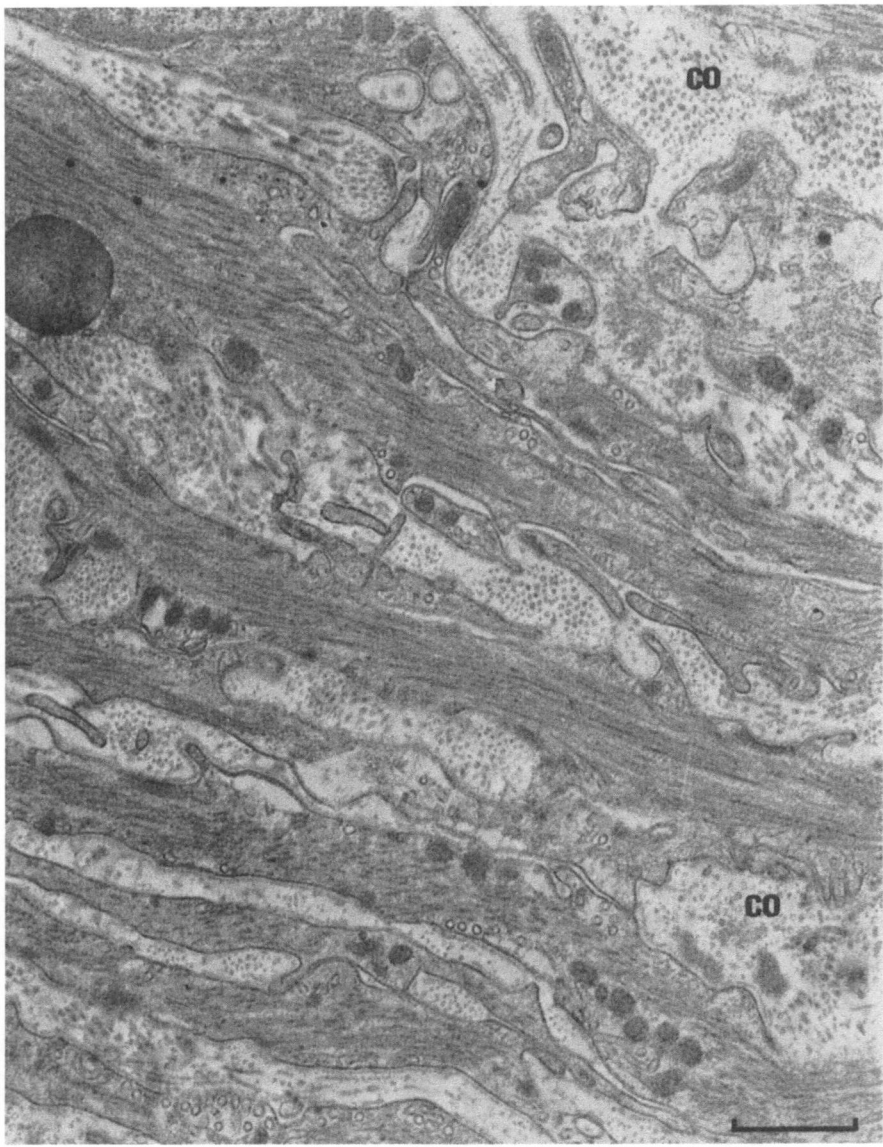

Fig. 11.
A. marginata. Longitudinal section through muscle fibres in the aorta. Note the extensive branching of the muscle fibres and the frequent contacts of their processes with adjacent fibres, probably indicative of myomyal transmission. *co*, massive collagen support characteristic of the aorta. × 16,000. Bar represents 1 μ.

of the muscle fibre and damage to the Schwann cell beneath the nerve fibres is clear evidence of the hypertonic nature of the solution.

In Fig. 6, a ventricular muscle fibre is shown in transverse section. The regular, hexagonal arrangement of thick filaments with their orbits of thin filaments, the columns of mitochondria and the longitudinal sarcotubules suggest again that this is a fairly fast contracting, phasic type of muscle. Throughout the ventricle there is a marked absence of nerve fibres. The latter appear in sections as the ventricle apex is approached (Fig. 7). There is also an increase in the quantity of collagen and mucopolysaccharide, seen more clearly in Fig. 8.

At this level there is a progressive change in the muscle fibres, which develop branching processes as the aorta is approached. Endothelial cells appear in the sections (Fig. 9A) and occasionally a myo-neural junction is seen (Fig. 9B).

At the junction of the ventricle and the aortic bulb (Fig. 10) many nerve fibres are visible. These contain large numbers of electron dense granules, the character of which differs from those seen in auricular nerve fibres. The former appear to be larger and less regular in shape and density, characteristics that require further analysis.

In the aortic bulb there is a further change in structure (Fig. 11). The quantity of supporting collagen is increased and there is now a paucity of nerve fibres. Extensive branching of the muscle cells gives rise to finer processes that make contact with adjacent fibres and may be nexuses. Although this 'unitary' type of muscle is known to characterize the heart of the oyster[32], nexuses have not been observed in the cardiac muscle of Achatinidae.

Features of special interest in this account are the following:

(a) A heavy innervation of the auricle, the nerve fibres lying at the lumenal surface. Each nerve fibre is filled—even distended—with electron dense granules.

(b) A surprising paucity of cholinergic endings.

(c) Marked lack of nerve fibres or myo-neural junctions in the ventricle.

(d) Reappearance of nerve fibres, containing electron dense granules, at the junction of the ventricular apex and the aortic bulb.

The suggestion has been made by G. A. COTTRELL (personal communication during this Symposium) that nerve endings in the auricle—at least, in *Helix*—may be neurosecretory. If this is so, then there are possibly three types of nerve fibres entering the auricle, viz. those containing electron dense granules and ending at myo-neural junctions[4], those containing electron dense granules but releasing their transmitter into the lumen, and those containing cholinergic vesicles and ending at myo-neural junctions[4].

Thus, within the heart, there would be both neurohumoural and neurosecretory transmission, as defined by WELSH[12]. The hypothesis is attractive because of the poor innervation of the ventricle and its powerful response to 5-hydroxytryptamine (p. 63). The small number of cholinergic endings found in the auricle may be due to our failure to discover the correct region. The quantitative differences may be real, however, in which case it is necessary to consider a role for endogenous acetyl choline other than that of cardio-inhibitor (p. 66).

Understanding of the role of the nerve fibres at the ventriculo-aortic junction is not helped by present failure to discover more than a few axon terminals. Although there are suggestions of qualitative differences in the granules from those of the auricle, a neurosecretory function would be unlikely to relate them to the heart.

Experimental Analysis

Methods

Responses of the auricle, ventricle and whole heart to tension, perfusion pressure and to drugs were obtained by perfusion in an isolated organ bath. For the whole heart (Fig. 1) the cannula was inserted into the pallial vein and tied with its tip at the entrance to the auricle. A second short cannula was inserted into the anterior aorta, care being taken to tie this distal to the aortic valve. The pallial vein and aortae were cut and the preparation was mounted vertically in the bath with the pallial entrance ventral and connected to a Marriotte bottle. The aortic cannula was tied to a fine frontal writing stylus at which point the tension exerted by a counterbalancing weight was 90 mg. A hook was attached to the counterbalance for the addition of further weights. The bath was filled to a standard height of 1 cm above the cardiac outflow and this height was maintained by an overflow tube connected to a drop recorder. A manometer was connected in parallel with the perfusion tube. Waste pipes were connected to the bath for rapid changes of the saline. For an auricular preparation the *second* cannula was tied into the auriculo-ventricular junction. Similarly, a ventricular preparation was made by tying the *first* cannula into the same junction. Ventricle preparations were also made by tying the second cannula at the ventricle apex and aortic junction. When drugs were perfused these were passed into the perfusion tube at a standard distance from the tip of the first cannula. The volume of this length of tube was 2 ml. All the experiments were carried out at temperatures between 21 and 23°C.

Results

Perfusion was first tried with the *Cochlitoma* saline of DIVARIS and KRIJGS-MAN[25]. Rhythmic activity tended to decline rather quickly, possibly due to the hypertonicity revealed by the electron micrographs (Fig. 5). Thereafter, the *Lymnaea* physiological solution of CARRIKER[27] was used, with excellent results. In this solution, preparations remained viable and responsive for three days or more. The molarities of the solutions are as follows:

Cochlitoma solution   0.117 M

Lymnaea solution   0.065 M

(Note that systole is indicated by downward movement of the stylus.)

Responses to tension and to perfusion pressure are shown in Fig. 12 A–E. Fig. 12A shows the responses of the unperfused auricle and ventricle to suddenly imposed tensions of 0.5 g and 1 g respectively. The ventricle responded with a single twitch followed by tonic relaxation, returning to its resting length when the weight was removed. The auricle, after a single twitch, established a rhythm of contractions and relaxations at a frequency of 8/min, returning to its resting length on removal of the weight. Thus, the auricle shows a stretch sensitivity that is absent from the ventricle up to the maximum tension (5 g) that the latter would withstand. Because of the statement by DIVARIS and KRIJGSMAN[25] that an accelerating pacemaker exists in the ventricular-aortic junction, the experiment was repeated with the aorta included—with precisely similar results to those shown.

In response to a perfusion pressure of 3 mm Hg (flow rate 40 drops/min), the auricle developed a rhythmic diastole/systole at a frequency of 38/min, increase of the pressure and flow leading to positive inotropy. A similar response was shown by the ventricle (Fig. 12 C), although the pressure and flow required were greater and the frequency of the response was lower. The inclusion of the aorta in this preparation gave closely similar results (Fig. 12 D).

In the whole heart preparation (Fig. 12 E) distension by the perfusate was followed by auricular systoles of increasing amplitude, followed in turn by ventricular systoles also of increasing amplitude but at a lower frequency. This established a typical cardiac cycle of auricular and ventricular contractions asynchronous to each other. Pertinent questions to ask are whether the rhythms are due to the stretch sensitivity of the auricular and ventricular muscle or whether there is neurohumoural-neurosecretory involvement.

The cholinergic responses of the whole heart preparation were then examined (Fig. 12 F–H and Fig. 13 K). Although the results were not entirely consistent from one animal to another the traces shown from an *A. panthera* in an extended (12-h) experiment are of considerable interest. The perfusion of 1 ml 10^{-12} M ACh into a heart showing contractions of the auricle and a quiescent ventricle (Fig. 12 F) gave rise to a positive inotropic and positive chronotropic response of the auricle followed by a similar response of the ventricle and finally a fairly close synchrony of the activity by the two chambers. When the response had subsided to the slow auricular beat, perfusion of 1 ml of 10^{-6} M ACh (Fig. 12 G) was followed by complete inhibition. The alternate doses were then continued and trace (Fig. 12 H) shows the 10th repetition of 1 ml 10^{-12} M ACh. The positive inotropic and positive chronotropic responses of auricle and ventricle are still obtained but their magnitude, compared with that in Fig. 12 F, has diminished. In Fig. 13 K the trace shows a barely perceptible response to the 36th repetition of 1 ml 10^{-12} M ACh. A following dose of 1 ml 10^{-6} M ACh completely inhibited the cardiac contractions but, after washing out the bath, perfusion of 1 ml

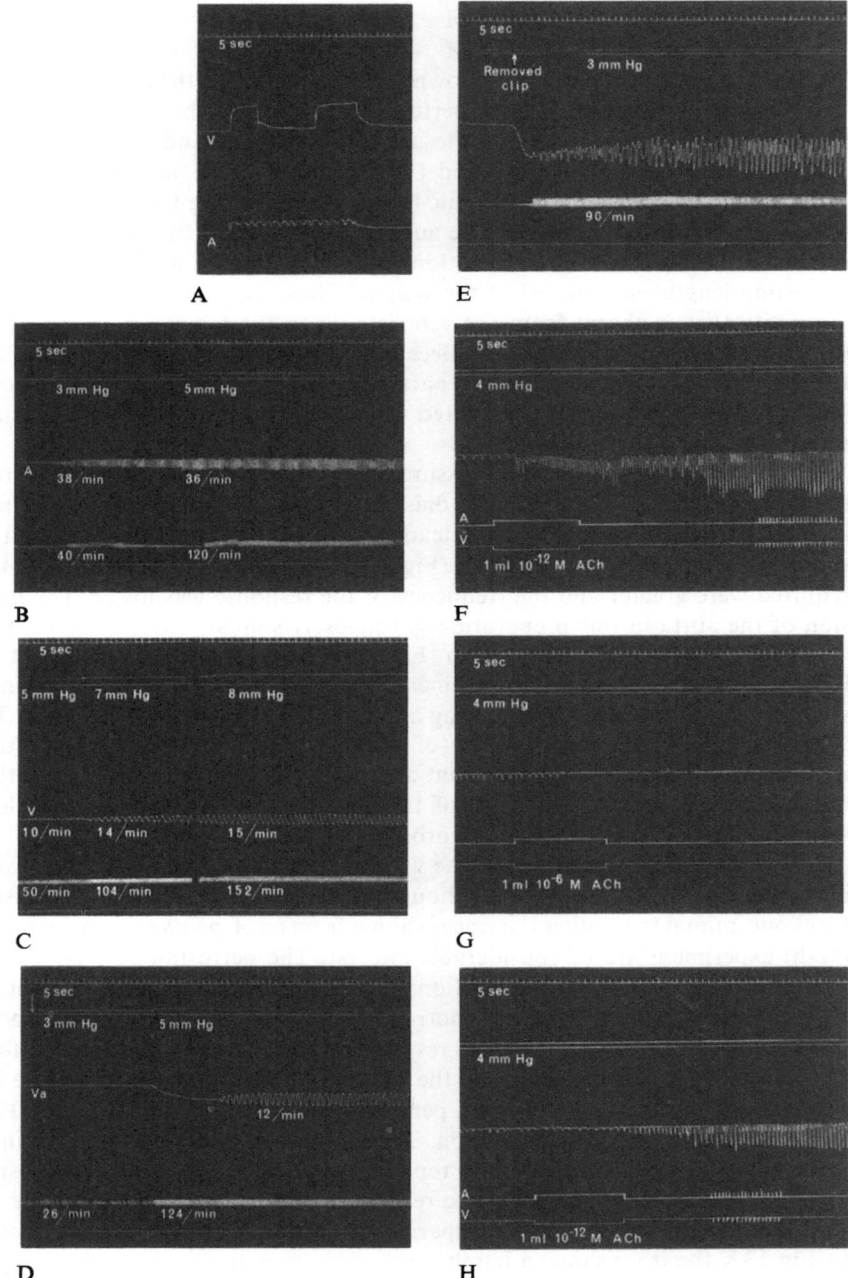

Fig. 12

10^{-6} M 5-HT (Fig. 13 L) gave rise to powerful positive inotropy and chronotropy of both auricle and ventricle. Furthermore, as the response to 5-HT subsided the perfusion of 1 ml 10^{-12} M ACh (Fig. 13 M) restored its earlier magnitude with closer synchronization of auricular-ventricular systoles and diastoles. Perfusion in 1 ml of the equivalents of 1 ml 10^{-6} M 5-HT and 1 ml 10^{-12} M ACh (Fig. 13 N) also gave a powerful positive inotropic and chronotropic response, with auricular and ventricular contractions well synchronized. For comparison, in the same preparation (Fig. 13 P), the responses to 1 ml 10^{-6} M 5-HT followed

◁   Fig. 12.

A. marginata. (A) Responses of passive auricle and ventricle to suddenly imposed tension of 0.5 and 1 g counterweights respectively. No perfusion flow. Upper trace, time marker (5 sec—in all records). Middle trace ventricle (*V*), lower trace auricle (*A*).
The ventricle responds with a single twitch, then lengthens: the auricle responds with a twitch, then establishes a rhythm of contractions and relaxations. Removal of counterweights results in return to resting length of both auricular and ventricular muscle.
(B) Response of auricle to perfusion pressure. Opening of clip on perfusion tube gives pressure of 3 mm Hg, a flow of 40 drops/min, dilatation of auricle followed by rhythmic diastole and systole (38/min). Increasing perfusion pressure to 5 mm Hg (120 drops/min) leads to a positive inotropic response from the auricle.
Lowest line = signal marker.
(C) Response of ventricle (without aorta) to perfusion pressure is slower and more sluggish, requiring a pressure of 5 mm Hg (50 drops/min) to initiate very small contractions (10/min). Raising pressure to 7 mm Hg (104 drops/min) is followed by positive inotropic and slight positive chronotropic response. 8 mm Hg pressure (152 drops/min) gives rise to a further positive inotropic response.
(D) Response of ventricle (including aorta). On raising the perfusion pressure from 3 to 5 mm Hg (flow rate from 26 to 124 drops/min) ventricular contractions commenced and reached a steady amplitude within 1 min. Re-

sponse is essentially similar to that of Fig. 12 C.
(E) Whole heart preparation (perfusion through cannulated pallial vein: exit of perfusate through aortae). Removal of clip initiates flow at 3 mm Hg. Distension of heart is followed by auricular systoles of increasing amplitude, stretching the ventricular muscle. Ventricular systoles follow, also increasing in amplitude but at a lower frequency. This establishes a cardiac cycle in which auricular and ventricular diastole and systole changes from asynchrony, to synchrony, to asynchrony again.
(F) *Achatina panthera.* Whole heart preparation. Early response to 1 ml of 10^{-12} M ACh. Start of trace shows auricular contractions (4 mm Hg perfusion pressure). Positive inotropic and chronotropic response, first by the auricle, then asynchronous entry of ventricle. Finally a fairly close synchrony of auricular and ventricular diastole and systole. Lower signal markers indicate auricle (*A*) and ventricle (*V*). Addition of ACh to perfusate is shown on both: later, auricular and ventricular systoles are signalled.
(G) *A. panthera* (same preparation as F). Heart beat has subsided to auricular diastole/systole, as in first part of trace (F). Addition of 1 ml of 10^{-6} M ACh to perfusate is followed by inhibition of contraction.
(H) *A. panthera* (as in F). Later addition (c. 10th repeat) of 1 ml of 10^{-12} M ACh. Positive ino- and chronotropic response of auricle followed by ventricular systoles, as in 12 F, but magnitude of response has declined.

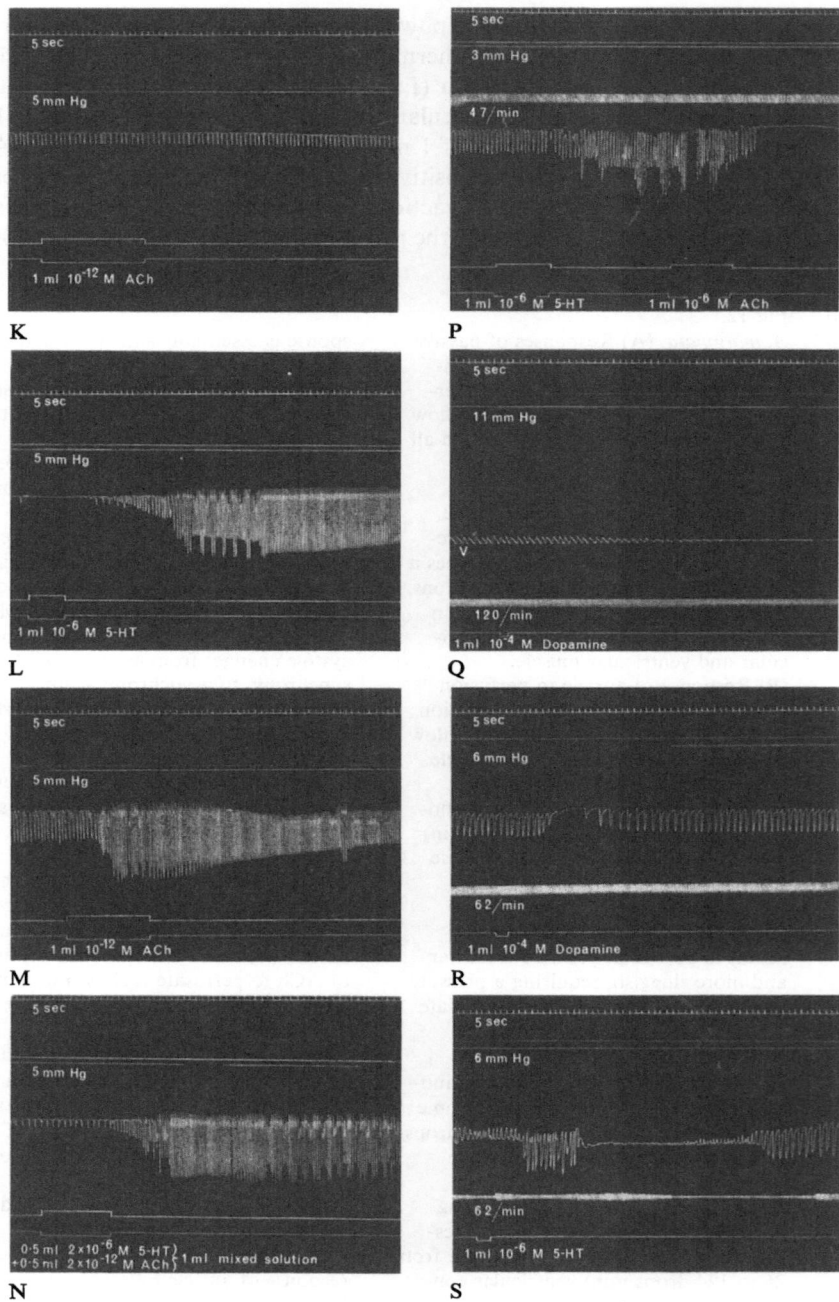

Fig. 13

by 1 ml 10^{-6} M ACh are shown. This concentration of acetyl choline inhibited any level of activity.

The foregoing results give rise to a doubt as to whether or not cholinergic transmission is inhibitory in the Achatinid heart. If it is not inhibitory, a transmitter other than 5-HT may be present in electron dense granules of some of the nerve fibres. The effect of dopamine on the heart has, therefore, been examined. The results are recorded in the last three traces.

Perfusion of 1 ml 10^{-4} M dopamine into the ventricle only (Fig. 13 Q) was followed by a delayed but long-lasting inhibition. Perfusion of the whole heart with a similar quantity of dopamine caused inhibition, but on repetition of the dose the heart showed tachyphylaxis (Fig. 13 R). After thorough washing, a following perfusion with 1 ml 10^{-6} M 5-HT (Fig. 13 S) gave rise to a very powerful response with maximum positive inotropy and chronotropy of both auricle and ventricle. The apparently stationary stylus was balanced between the systoles and diastoles of auricle and ventricle which were in exactly opposite phases until, as the response subsided, the contractions became synchronous.

Discussion

The present work has posed more questions than it has answered. The muscle of the isolated ventricle appears to be insensitive to an externally applied

◁ Fig. 13.

(K) *A. panthera*. The same preparation as in Fig. 12 F–H, c. 10 hours after 12 F and the 36th repetition of 1 ml of 10^{-12} M ACh. Positive response is barely perceptible.

(L) *A. panthera* (same preparation). After check of the inhibitory effect of 10^{-6} M ACh and washing of preparation, 1 ml of 10^{-6} M 5-HT is added to the perfusate. This is followed by a powerful positive inotropy and chronotropy.

(M) *A. panthera*. As the response to 10^{-6} M 5-HT (shown in 13 L above) declined, 1 ml of 10^{-12} M ACh was added to the perfusate. Note the restoration of the powerful response but with closer auricular-ventricular synchrony.

(N) *A. panthera* (as above). Equivalents of 1 ml of 10^{-6} M 5-HT and 1 ml of 10^{-12} M ACh added to the perfusate. This again shows a powerful positive ino- and chronotropic response with auricle and ventricle well synchronized.

(P) *A. marginata*. Whole heart. Response to 1 ml of 10^{-6} M 5-HT, followed by 1 ml of 10^{-6} M ACh. This concentration of ACh inhibits any level of cardiac activity.

(Q) *A. marginata*. Ventricle only. Perfusion with 1 ml of 10^{-4} M dopamine is followed by a delayed cardiac inhibition.

(R) *A. marginata*. Whole heart. Third repeat of a perfusion with 1 ml of 10^{-4} M dopamine. First response was similar to 13 Q but heart now shows tachyphylaxis.

(S) *A. marginata*. Whole heart. Addition of 1 ml of 10^{-6} M 5-HT after perfusion with dopamine in 13R and after washing preparation. Apparently this is a residual effect of dopamine. Auricle and ventricle reach maximum diastolic-systolic amplitude and frequency. Their contractions pass into opposite phases, the opposing balance of tensions maintaining the stylus tip at a nearly constant level until, as the response slowly declines, the contractions become synchronous.

tension. The auricle responds readily to tension but it is possible that tension releases excitatory transmitter from the nerve fibres. On the other hand, both the auricle and the ventricle respond to distension and flow, the latter less readily than the former. This seems to be reasonable evidence for stretch sensitivity in the muscle fibres themselves, in view of the poor innervation of the ventricular muscle. The augmented response seen in Fig. 12 E, however, may result from some degree of local neurosecretory activity.

There appears to be a high degree of ventricular autonomy. This can be observed in the whole animal (p..48 and Fig. 2) and also in the isolated heart. It has not yet been possible to trace the paths of the nerve fibres seen at the aortic-ventricular junction. Whether or not they enter the ventricle is not known. The responses of the isolated ventricle to distension continue in the absence of the aorta.

In the whole heart synchronous rhythm of auricular and ventricular contractions is not maintained. Either chamber may contract several times between successive contractions of the other. Normally, there appears to be a continuous phase change in their contractions, usually due to the slower cycles of the ventricle.

The role of the cholinergic nerve endings is a subject of great interest. GREENBERG[14] quotes CORDA (1955) who recorded temporary increases in the amplitude of the heart beat of *Helix aspersa* to low concentrations (10^{-17} to 10^{-9} g/l) of ACh and who suggested that ACh, in small concentrations, has a role in the control of rhythmicity of *Helix* heart. GREENBERG considers that, with the exception of oyster and cockle hearts, our knowledge of the excitatory effect of low concentrations of ACh on molluscan hearts is based on descriptions of phenomena which, with few exceptions, are infrequent, unpredictable events. He nevertheless remarks that 'not to be neglected is the possibility that 5-HT is involved in the induction and augmentation of beat by ACh... If the neurogenic threshold of ACh is lower than the negative myogenic threshold, then beating might be induced or augmented at low doses.' HILL[17] has shown that ACh and 5-HT together will induce a rhythmicity in the *Busycon* radular protractor comparable to the automatic rhythmicity of the heart. He is tempted to speculate that the radular protractor 'beat' might originate in the presence at the cell surface of the opposing neurohumours in the right proportions for alternate action.

TWAROG[10] has shown that ACh applied to the ABRM of *Mytilus* causes depolarization, a train of spike-like potentials with increments of tension that was sustained after removal of the ACh. On application of 5-HT 10^{-7} M the tension fell abruptly. On the other hand, if 5-HT 10^{-7} M remained in the bath and ACh was added, large spike-like potentials and corresponding tension increments were prominent; the total tension development was much increased but rapid relaxation occurred when the muscle was washed.

GREENBERG[14] has objected to HILL's suggestion, that judicious mixtures of ACh and 5-HT might be the basis of the normal mechanism of rhythmical activity in molluscan muscle, on the grounds that the hypothesis rests on the ability of 5-HT to relax muscle contracted by ACh and this has not been observed in bivalve hearts. Nevertheless, GREENBERG has himself suggested the possibility that endogenous ACh might release 5-HT by stimulating serotonic nerves in the heart. The paucity of cholinergic nerve endings in Achatinid hearts, together with the experimental evidence outlined above, lends support to this suggestion.

Summary

Evidence on the gross and fine morphology of Achatinid hearts, together with experimental analysis and observational studies, suggests that:
1. The whole heart is sensitive to pressure and to flow, the auricle much more so than the ventricle.
2. The auricle is driven by the distension pressure of returning blood and its contractions act as a stimulus to ventricular contractions.
3. The ventricle is similarly driven by distension pressures but is slower both in its response and in its frequency, although the reverse may be seen to occur.
4. The auricle and ventricle are subject to firm extrinsic neural control, both excitatory and inhibitory, the latter being particularly significant in relation to movements of the whole animal.
5. Excitatory control may be neurohumoural in the auricle and neurosecretory in the ventricle.
6. Inhibitory control may be due to an, as yet, undetected transmitter substance.
7. It is possible that synchronous augmentatory activity is determined by a balanced release of ACh and 5-HT.
8. The aorta appears not to influence cardiac responses.
9. The function of the nerve fibres at the junction of the aorta and ventricle apex is not known.

Acknowledgments

Our grateful thanks are due to Mr. J. T. GUNNER, Miss D. G. LONSDALE, Mrs. G. PARK and Mr. D. WOULDS for much help in the completion of this work.

The work was supported in part by the U.S. Department of Health, Education and Welfare Public Health Service Grant No. HD 01476 from the National Institutes of Health, Bethesda, Maryland, U.S.A.

References

1   R. H. NISBET, Proc. R. Soc. [B] *154*, 267, 309 (1961).
2   E. C. AMOROSO, M. I. BAXTER, A. D. J. CHIQUOINE and R. H. NISBET, Proc. R. Soc. [B] *160*, 167 (1964).
3   M. I. BAXTER and R. H. NISBET, Proc. malac. Soc., Lond. *35*, 167 (1963).
4   R. H. NISBET and J. M. PLUMMER, Proc. malac. Soc., Lond. *37*, 199 (1966).
5   D. E. PHILPOTT, M. KAHLBROCK and A. G. SZENT-GYÖRGY, J. ultrastruct. Res. *3*, 254 (1960).
6   J. HANSON and J. LOWY, Proc. R. Soc. [B] *154*, 173 (1961).
7   J. C. RÜEGG, R. W. STRAUB and B. TWAROG, Proc. R. Soc. [B] *158*, 156 (1963).
8   J. LOWY, B. M. MILLMAN and J. HANSON, Proc. R. Soc. [B] *160*, 525 (1964).
9   J. C. RÜEGG, Proc. R. Soc. [B] *160*, 536 (1964).
10   B. M. TWAROG, J. Physiol., Lond. *152*, 236 (1960).
11   R. L. C. PILGRIM, J. Physiol., Lond. *125*, 208 (1954).
12   J. H. WELSH, Am. Zool. *1*, 267 (1961).
13   R. E. LOVELAND, Comp. Biochem. Physiol. *9*, 95 (1963).
14   M. J. GREENBERG, Comp. Biochem. Physiol. *14*, 513 (1965).
15   J. H. WELSH, J. Mar. Biol. Ass. U.K. *35*, 193 (1956).
16   J. H. WELSH and M. MOORHEAD, Science *129*, 1491 (1959).
17   R. B. HILL, Biol. Bull. Woods Hole *115*, 471 (1958).
18   H. IRISAWA, M. KOBAYASHI and T. MATSUBAYASHI, Jap. J. Physiol. *11*, 162 (1961).
19   R. B. HILL and H. IRISAWA, Life Sciences *6*, 1691 (1967).
20   R. B. HILL, Comp. Biochem. Physiol. *23*, 1 (1967).
21   R. B. HILL and P. J. SCHUNKE, Experientia *23*, 570 (1967).
22   R. B. HILL, Experientia *23*, 772 (1967).
23   B. J. KRIJGSMAN and G. A. DIVARIS, Biol. Rev. *30*, 1 (1955).
24   R. B. HILL and J. H. WELSH, *Physiology of Mollusca* (Academic Press, New York and London 1966), vol. 2, 125.
25   G. A. DIVARIS and B. J. KRIJGSMAN, Arch. int. Physiol. *62*, 211 (1954).
26   R. H. NISBET, *Structure and Function of the Buccal Mass in Monodonta lineata* (*da Costa*) (Thesis for Ph. D., University of London 1953), p. 114.
27   M. R. CARRIKER, Biol. Bull. Woods Hole *91*, 88 (1946).
28   M. R. CARRIKER, Trans. Wisc. Acad. Sci. *38*, 9 (1946).
29   B. A. CHADWICK, *Some Aspects of the Physiology of Molluscan Circulation* (Thesis for Ph. D., University of Reading 1962), p. 72.
30   J. RIPPLINGER, Annls. scient. Univ. Besançon, Zool. et Physiol. *8*, 3 (1957).
31   J. S. ALEXANDROWICZ, Q. Jl. microsc. Sci. *75*, 181 (1933).
32   H. IRISAWA, XXIVth Proc. Int. Un. Physiol. Sci. VI, 150 (1968).

Theory of Step-Wise Excitation in Gastropod Hearts

by Katalin S. Rózsa
Biological Research Institute, Hungarian Academy of Sciences,
Tihany, Hungary

As a result of recent investigations on transmitter phenomena, new interpretations of the effect of active substances are being suggested which differ from widely accepted theories based on earlier researches. The increasing number of active substances isolated from living systems makes it necessary to investigate the effect of any substance not as an independent agent but in interaction with other substances so as to build up the whole picture of humoral regulation from the separate details. This approach seems especially justified in the investigations of excitatory effects on molluscan hearts, on which nearly all the biologically active amines[1]*, some nucleotides[2] and even active peptides[3-5] cause excitation. These data support the suggestion that in this phylum the effects of excitatory transmitters must be somewhat unusual[6].

In this paper, based on our studies on gastropod hearts, I shall present a scheme for the step-wise production of excitation.

Methods

Investigations were conducted on the isolated hearts of two Gastropod species, *Lymnaea stagnalis* L. and *Helix pomatia* L. Isolated heart preparations were studied[1] for their mechanical responses, and homogenates were prepared for enzymatic studies. The localization of biologically active monoamines was studied with the fluorescence histochemical method of Falck and Owman[7,8].

Phosphorylase *a* and phosphorylase *b* were assayed at 30 °C according to the method of Cori and Illingworth[9]. The procedure used was as follows: Treated and untreated hearts were homogenized in a solution containing 0.02 M Na-glycerophosphate, 0.002 M EDTA, 0.1 M NaF, pH 6.8. The reaction mixture contained 0.064 M glucose-1-phosphate and 4% glycogen. When adenosine-5-monophosphate (AMP) was used, it was present at 0.004 M. 4–8 mg of heart tissue was used in 0.8 ml reaction mixture to give a concentration of enzyme which ensured first-order reaction kinetics. The reaction was terminated by the addition of 3.2 ml of 5% trichloroacetic acid. After filtration, inorganic phosphate (Pi) was determined[10]. The release of Pi from glucose-1-phosphate in the

---

* Numbers refer to References, p. 77.

absence of added AMP represents 'active phosphorylase', and in the presence of AMP it represents 'total phosphorylase'.

During the experiments the following agents were used: 5-hydroxytryptamine (5-HT) Fluka; dopamine (DA) Fluka; adrenaline (A) Rhone-Paulenc; noradrenaline (NA) Calbiochem; tyramine, Fluka; tryptamine, Fluka; glutamine, Fluka; histamine, Calbiochem; acetylcholine, Calbiochem; adenine, adenosine triphosphate (ATP), adenosine diphosphate (ADP), adenosine monophosphate (AMP) Reanal; Cyclic $3',5'$-AMP Sigma; guanine, guanosine triphosphate (GTP), guanosine diphosphate (GDP), guanosine monophosphate (GMP), uridine, uridine triphosphate (UTP), uridine diphosphate (UDP), uridine monophosphate (UMP), cytidine, cytidine monophosphate (CMP) Koch-Light; cytidine triphosphate (CTP), cytidine diphosphate (CDP) Sigma.

The agents were applied in MENG[11] and RINGER solutions in concentrations of 10^{-10} to 10^{-2}M at constant perfusion rates, and the experiments were carried out at 20–25 °C.

Results

The Effects of Biologically Active Amines and Nucleotides on the Isolated Hearts of Gastropods

Because of the contradictory results reported for the effects of amines on different molluscan species and because, among the nucleotides, only the adenine derivatives have been studied[12,13], these experiments were performed.

The response, threshold concentration and maximal effectiveness of applied amines are summarized in Table 1. These results are unexpected in two respects: (1) with the exception of acetylcholine, all amines examined produced an

Table 1. Effects of biologically active amines on *Lymnaea* and *Helix* hearts.

Agent	Response	Molar concentration for threshold effect	Maximal percentage of change of beat amplitude by agent
Acetylcholine	inhibition	10^{-9}	100
Dopamine	stimulation	10^{-9}	40
Noradrenaline	stimulation	10^{-9}	40
Adrenaline	stimulation	10^{-9}	40
5-hydroxytryptamine	stimulation	10^{-10}	90
5-methoxytryptamine	stimulation	10^{-10}	20
	inhibition	10^{-6}	20
Tryptamine	stimulation	10^{-10}	30
Tyramine	stimulation	10^{-9}	10
	inhibition	10^{-4}	50
Glutamine	stimulation	10^{-9}	10
Histamine	stimulation	10^{-7}	8

excitatory effect, and (2) the threshold concentrations of different amines are surprisingly near each other; for example the threshold concentration of the most effective agent, 5-HT, is only three orders lower than that of histamine, the most ineffective cardio-accelerator.

Thus one cannot distinguish between the amines by their types of action and threshold concentrations. Since all the catecholamines acted at the same threshold and produced about the same maximum excitatory effect, our results do not support the claims for the outstanding role of dopamine among the catecholamines[6].

Similar results showing excitation were obtained with nucleotides (Table 2) and inhibition was seen only for GMP, GDP, GTP and UTP in autumn just before hibernation. The pyrimidine and purine bases (A, G, U and C) also caused stimulation of the heart but only at concentrations greater than 10^{-4} M.

The interaction between 5-HT and various nucleotides was next investigated.

Our experiments showed that none of the adenine nucleotides (ATP, ADP, AMP) nor the monophosphates (AMP, GMP, UMP, CMP) influenced the 5-HT effect. Even after preincubation of up to 1 h with concentrations of 10^{-4} to 10^{-2} M, the threshold to 5-HT was unchanged at 10^{-10} to 10^{-9} M.

On the other hand, treatment of the heart with 10^{-6} to 10^{-5} M UDP or CDP stimulated the heart, but in these cases addition of 5-HT caused no further increase in stimulation. If the heart was repeatedly washed out with physiological saline after the UDP treatment, then 5-HT would again produce

Table 2. Effects of purine and pyrimidine bases and nucleotides on the isolated hearts of *Lymnaea* and *Helix*.

Agent	Type of effect	Threshold in Moles	Note
Adenine	stimulation	10^{-4}	
AMP	stimulation	10^{-6}	
ADP	stimulation	10^{-6}	
ATP	stimulation	10^{-9}	
3',5'-AMP	stimulation	10^{-8}	
Guanine	stimulation	10^{-3}	
GMP	inhibition	10^{-5}	in autumn
GDP	inhibition	10^{-8}	in autumn
GTP	stimulation	10^{-7}	in spring
	inhibition	10^{-7}	in autumn
Uridine	stimulation	10^{-4}	
UMP	no effect		
UDP	stimulation	10^{-8}	
UTP	stimulation	10^{-8}	in spring
	inhibition	10^{-8}	in autumn
Cytidine	stimulation	10^{-4}	
CMP	stimulation	10^{-10}	
CDP	stimulation	10^{-10}	
CTP	stimulation	10^{-5}	

its stimulatory effect. These interactions were most pronounced at about the threshold concentrations of UDP, GDP and 5-HT.

Pretreatment with the triphosphates GTP, UTP and CTP dramatically reversed the effect of 5-HT. These experiments were performed during the season of the year when these nucleotides, as well as 5-HT, produced a stimulatory effect on the heart. The effect of UTP on the 5-HT response is shown in Fig. 1: after 15 min of treatment with 10^{-5} M UTP, the addition of 5-HT at a concentration of 10^{-9} M arrested the heart. The heart later escaped from the inhibitory effect, but the original beat amplitude was not reached (Fig. 1 B). A similar effect was produced with 10^{-5} M CTP (Fig. 1 C). The addition of 5-HT alone always stimulated the control hearts (Fig. 1 A).

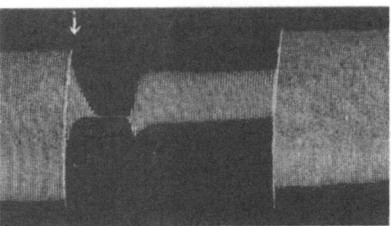

B

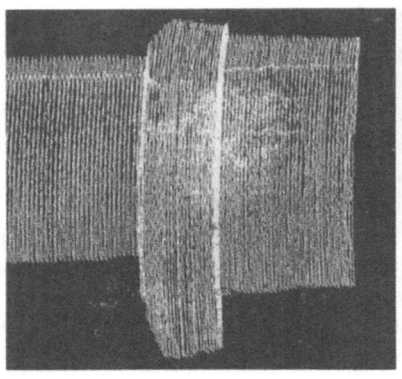

A

Fig. 1.
Effect of triphosphonucleotides modulating the action of 5-HT on the contraction of *Lymnaea* heart.
(A) Effect of 5-HT on a control heart.
(B) Effect of 5-HT on a heart pretreated with UTP.
(C) Effect of 5-HT on a heart preincubated with CTP.
In B and C the 5-HT at 10^{-9} M was added at the arrow.

C

Similar changes had been found in an earlier experiment in relation to cyclic 3′,5′-AMP and to theophylline[14]. High concentrations of 3′,5′-AMP from 10^{-3} to 10^{-2} M made 5-HT ineffective, while the theophylline at 10^{-5} M changed the 5-HT effect from stimulation to inhibition. However, low concentrations of $3',5'$-AMP and theophylline potentiated the stimulatory effect of 5-HT.

These results show that some stimulatory agents may abolish the effects of other stimulatory agents, so their effectiveness will depend on the totality of other factors present.

Localization of Monoamines in the Hearts of Gastropods

Since RÓZSA and NAGY[8] found nerve cells in the *Lymnaea* heart, the distribution of the monoamines was examined in this species. 5-HT was not found in nerve cells or axons of the heart at levels detectable by FALCK's method. The fluorescence in the neurones and their branches was always green, so they presumably contained only catecholamines (Fig. 2 A). After 20 min stimulation of the extracardiac nerve at 5-10 Hg, 3 msec, 5–10 V, this catecholamine content of the nerve cells completely disappeared (Fig. 2 B). At the same time the density of green axons increased, especially in the auricle (Fig. 2 C), indicating that the component of the cytoplasm displaying green fluorescence may have moved down into the axons as a result of stimulation.

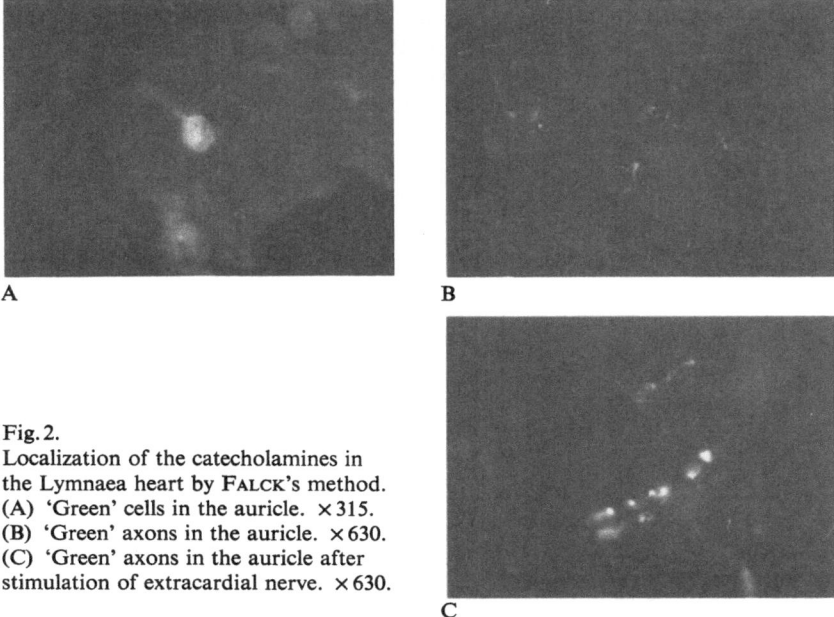

A

B

Fig. 2.
Localization of the catecholamines in the Lymnaea heart by FALCK's method.
(A) 'Green' cells in the auricle. × 315.
(B) 'Green' axons in the auricle. × 630.
(C) 'Green' axons in the auricle after stimulation of extracardial nerve. × 630.

C

Yellow fluorescence was found inside the myocardial cells, indicating the presence there of 5-HT. This emphasizes that the 5-HT content of the heart is located in the muscle elements and not in the neural structures of the heart.

The localizations of 5-HT and catecholamines in the snail heart pose two questions: (1) How can 5-HT play a transmitter role in the classical sense?, and (2) What kind of effect can 5-HT produce on the myocardial cells themselves?

We will try to answer the second question here and will consider the first question in the Discussion.

The Effect of 5-HT on the Myocardial Cells

The fact that 5-HT is lacking in the nerve cells of the heart precludes the possibility that its excitatory effect is produced by its release at the post-synaptic membrane of the myoneural junction. If we preclude the influence of 5-HT on the membrane, then we must search for its effect in more general metabolic processes. We have, therefore, chosen to study its action on the regulation of phosphorylase activity as the limiting factor in glycolysis[15].

We found that both phosphorylase a and phosphorylase b are detectable in the *Lymnaea* heart. In control hearts, the level of phosphorylase a varied between 49 and 51% of the total phosphorylase activity. The phosphorylase activity of the hearts treated with 5-HT was then measured. At a concentration of 10^{-7}M, 5-HT transformed the phosphorylase b to phosphorylase a to an extent between 91 and 94%. Higher concentrations of 5-HT, from 10^{-6} to 10^{-4}M, increased the transformation to 100% within 20 min after the application (Fig. 3). The level of phosphorylase a in the presence of 10^{-6}M 5-HT as a function of time is shown in Fig. 3. Each value is the average from 5 measure-

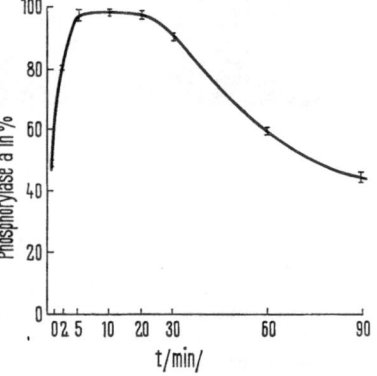

Fig. 3.
The proportion of phosphorylase present in *Lymnaea* heart as phosphorylase a following the addition of 10^{-6}M 5-HT to the heart. Each point is the mean of 5 determinations ± S.E.

ments. It can be seen that under the influence of 5-HT, phosphorylase b was transformed very quickly to the a-form. After only 2 min treatment of the heart with 5-HT, the phosphorylase a activity was increased to 91% and, after 5, 10, and 20 min treatment, practically all the phosphorylase was present in the a-form. However, treatment of the heart for longer than 20 min resulted in a decrease of the level of phosphorylase a, and after 90 min this value fell to the control level (Fig. 3). The time course for the increase in the enzymatic activity coincided with that for the physiological effect of 5-HT.

The effects of catecholamines on the phosphorylase transformation were also studied. The results in Table 3 show that all the agents increased the total

Table 3. Effects of 5-HT and catecholamines on the activation of inactive phosphorylase.

Agent		Phosphorylase Total	a	$\% a$	Increase of $\% a$ as compared with the control (in 10 min)
Control		0.645 ± 0.025	0.332 ± 0.023	51	–
5-HT	$10^{-5} M$	1.059 ± 0.132	1.059 ± 0.132	100	49
DA	$10^{-5} M$	1.124 ± 0.025	0.761 ± 0.049	67	16
NA	$10^{-5} M$	1.052 ± 0.010	0.785 ± 0.025	74	23
A	$10^{-5} M$	0.991 ± 0.038	0.809 ± 0.040	81	30

phosphorylase activity. Of the catecholamines, adrenaline was the most effective agent, and increased the proportion of phosphorylase a by 30%. Dopamine was even less active in this respect than noradrenaline, confirming the results of the physiological experiments (Table 1) which showed no prominent role for dopamine in the regulation of the gastropod heart.

Discussion

According to our investigations there are many excitatory factors in the regulation of the molluscan heart. Considering the actions of the three main groups of substances studied, active amines, nucleotides and peptides[1-5] on isolated hearts, the stimulatory effect predominated.

There may be two reasons for the appearance of excitatory effects caused by different substances: (1) the presence of identical fundamental molecular mechanisms, which may be activated by different substances to different degrees; and/or (2) the presence of different modes of regulation operating with different chemical substances and enzymes.

It is hardly to be expected that the first possibility is widespread, although it may be valid in the case of amines which may influence the same enzyme system. However, it is unlikely that for similar actions by nucleotides and peptides the molecular backgrounds would be identical.

It seems likely that in the excitatory process more than one single transmitter substance is involved[16], and there may well be interaction between the different substances.

The existence of a step-wise mechanism, in accordance with anatomical regulatory levels, is therefore postulated. The localization of different amines supports this hypothesis. On the postsynaptic membranes related to extra- and intracardiac nerves of the heart, catecholamines are present in the nerve fibers and cells, as seen by the green fluorescence, and they may be liberated by stimulation of the extracardiac nerve[8]. These nerve structures do not contain detectable amounts of 5-HT, although the myocardial localization of 5-HT is more certain. For this reason, the liberation of significant amounts of 5-HT following stimulation of the extracardiac nerve in the heart of gastropods[17] suggests that the effect of serotonin is not on the myoneural junction.

In the interpretation of the action of serotonin one must take into consideration the fact that it characteristically increases the activity of phosphorylase b, the enzyme limiting the rate of glycolysis (Table 3). Catecholamines are less active in this respect. On the basis of the present results we cannot decide whether the 5-HT acts through cyclic $3',5'$-AMP and adenyl cyclase in gastropods as it does in other species[15,18].

Taking all the data into account, we believe that regulation of molluscan heart activity can occur at central neural, peripheral neural and myocardial levels, with amines and peptides acting on the central neurones, catecholamines acting at neuromuscular levels, and 5-HT acting at myocardial levels.

At the present time there are no reliable data on the chemical specificity of the central neurones involved in the control of heart activity, but the peripheral neural regulation may be outlined: excitation of the extracardiac nerve releases a transmitter from the presynaptic terminal, which in turn liberates catecholamines stored in nerve cells situated in the heart. These catecholamines act on the postsynaptic membrane of the muscle fibers, liberating 5-HT stored in the muscle fibers. This 5-HT later influences the metabolism of the muscle fibers to give the stimulatory effect.

In addition to the vertical organization of the regulatory process linking CNS to intracardiac neurones to myocardium, horizontal organization must also exist at each level. The effect of nucleotides to inhibit or reverse the excitatory action of 5-HT takes place in such a system, and it is probable that nucleotides play a more significant role in maintaining the normal function of biologically active agents than we realize at the present.

Summary

All biological amines, except acetylcholine, have predominantly stimulatory effects on the heart of *Helix* and *Lymnaea*. The effects of purine and pyrimidine bases are similar to that of the amines.

The gastropod hearts are protected by UDP and CDP against the stimulatory effect of 5-HT, while GTP, UTP and CTP transform the stimulation into an inhibition.

The neural elements of the heart of gastropods contain catecholamines but no 5-HT demonstrable by FALCK's method. The material with yellow fluorescence, indicating 5-HT, is localized in fine granules of the myofibrils.

Under the influence of 5-HT the inactive phosphorylase of the *Lymnaea* heart is completely transformed to the active form ($\%$ a = 100). Catecholamines are less effective in this respect (DA = 67$\%$ a, NA = 74$\%$ a, A = 81$\%$ a).

Dopamine does not play a prominent role among catecholamines in the regulation of the gastropod heart.

It is suggested that the excitatory effect of extracardial nerves and the excitatory processes as a whole is a step-wise phenomenon: stimulation of extracardial nerve liberates transmitter from its endings, which induces the release of catecholamines stored in the nerve cells of the heart, which in turn acts on the postsynaptic membrane of the myocardium to produce excitation both by the activation of 5-HT and by its intracellular action on metabolism.

References

[1]   K. S.-Rózsa and T. Pécsi, Annls Biol. Tihany *34*, 59 (1967).

[2]   K. S.-Rózsa and T. Pécsi, Annls Biol. Tihany *35*, 61 (1968).

[3]   G. A. Kerkut and M. S. Laverack, Comp. Biochem. Physiol. *1*, 62 (1960).

[4]   G. A. Cottrell, Comp. Biochem. Physiol. *17*, 891 (1966).

[5]   J. H. Welsh and N. Frontali, IIIrd Int. Pharmacol. Congr. S. Paulo, Brasil, 24–30 July (1964).

[6]   E. Florey, Ann. Rev. Pharmac. *5*, 357 (1965).

[7]   B. Falck and C. Owman, Acta univ. Lund. *2*, 1 (1965).

[8]   K. S.-Rózsa and I. Zs.-Nagy, Comp. Biochem. Physiol. *23*, 373 (1967).

[9]   F. C. Cori and B. Illingworth, Biochem. biophys. Acta *21*, 105 (1956).

[10]   H. H. Taussky and E. Shorr, J. Biol. Chem. *202*, 675 (1953).

[11]   K. Meng, Naturwissenschaften *19*, 470 (1958).

[12]   D. A. Sakharov and S. N. Nistratova, Physiol. J. USSR *49*, 1475 (1963).

[13]   T. Aikava and S. Ishida, Comp. Biochem. Physiol. *18*, 797 (1966).

[14]   K. S.-Rózsa, Symp. Invert. Neurobiol. Plenum Press. Corpor. 238 (1968).

[15]   T. E. Mansour, Federation Proc. *26*, 1179 (1967).

[16]   J. H. Welsh, Ann. N.Y. Acad. Sci. *66*, 618 (1957).

[17]   K. S.-Rózsa and L. Perényi, Comp. Biochem. Physiol. *19*, 105 (1966).

[18]   E. W. Sutherland, T. Øye and R. W. Butcher, Recent. Prog. Horm. Res. *21*, 623 (1965).

The Energy Output of Normal and Anoxic Cardiac Muscle

by C. L. GIBBS
Physiology Department, Monash University, Clayton, Victoria, Australia

Abstract

A myothermic technique has been used to investigate the energy output of rabbit papillary muscles that were either hypoxic or metabolically inhibited by iodoacetic acid and nitrogen. At room temperature, 18–22°C, hypoxia reduced resting heat production from a mean of 30.1 to 18.4 mcal/g muscle·min. Tension development fell from 3.29 to 2.48 (g) and work output under a 1.0 g load decreased from 0.069 to 0.044 g.cm. The mean isometric heat coefficient rose from 3.6 to 5.3. 30 min exposure to 0.5 mM IAA followed by 10 min of anoxia, oxygen tension < 2 mm Hg, caused the resting heat production to fall from a mean of 29.8 to 16.5 mcal/g muscle·min and developed tension to fall from 3.18 to 1.97 g. The total heat per isometric contraction fell from a mean of 2.5 to 0.83 mcal/g muscle and the activation heat decreased from 0.48 to 0.18 mcal/g muscle. The mean isometric heat coefficient rose from 3.7 to 6.7. Evidence is presented that the fast phase of heat production associated with the contractile event is unchanged by hypoxia or anoxia and this phase may therefore be taken to be the equivalent of skeletal muscle initial heat.

Introduction

Only in recent years has it become possible to investigate the energetics of cardiac muscle at a biochemical level[1]*, and at a myothermic level[2,3]. As the heat studies were made under conditions where both aerobic and anaerobic metabolism could take place whilst the chemical studies were made under conditions where both anaerobic and aerobic metabolism were inhibited, it seemed desirable to examine the heat production of cardiac muscle under similar conditions.

The actual experimental design used was similar to that described by HUKUDA[4], for sartorius muscles of frogs. The energy output of rabbit papillary muscles was examined under normal conditions, under conditions where oxidative energy metabolism was impaired by periods of hypoxia and under con-

* Numbers refer to References, p. 91.

ditions where both aerobic and anaerobic energy production was inhibited by iodoacetic acid (IAA) and nitrogen.

Methods

Papillary muscles were taken from the right ventricles of rabbits. Eighteen preparations were used in the present series of experiments. They had a mean weight of 6.2 mg and a mean length of 6.8 mm under a resting tension of either 0.5 or 1.0 g. The muscles were suspended vertically in a chamber containing 55 ml of Krebs-Henseleit solution. The solution was aerated with 95% O_2 and 5% CO_2 and had a pH of 7.4. The experiments were run at room temperature which ranged from 18.0 to 22.3°C.

The muscles were tied at either end with braided non-capillary silk (Ethicon 50) and the ventricular end of the preparation was positioned at the bottom of the thermopile. The other end of the muscle was connected through a thin stainless-steel wire to an isometric transducer, Sanborn FTA-100, or an isotonic one, a modified Brush Metripak.

Experimental Design

Results were first obtained under normal conditions, the total energy output in a block of 15–20 contractions being measured at a stimulus rate of 15/min. Most experiments were isometric ones but in six preparations the energy output in a block of after-loaded isotonic contractions was also examined. Results in this paper are always expressed in terms of a single contraction, i.e. the total energy output in a block of contractions has been divided by the number of stimuli.

After the aerobic results had been obtained the same experiments were repeated either under conditions of hypoxia or after treatment with iodo-acetate and nitrogen. In the hypoxia experiments muscles were stimulated at the standard rate under solution. At zero time the O_2/CO_2 aeration was stopped and aeration began with 95% N_2 and 5% CO_2.

Experiments with the same solution volume and the same flow rates showed that in 20 min the oxygen tension in the bath was in the range 20–60 mm Hg. These measurements were made with a Beckman 39065 polarographic sensor and Model 777 oxygen analyser. By the end of the equilibration period all muscles showed a depressed tension development or a decreased work output although there was often an initial increase in cardiac contractility. The chamber was then drained and the heat experiments took place in an atmosphere of N_2 and CO_2 which was continually passed into the chamber. Because the solution in the bottom of the chamber still contained oxygen and because there was the possibility of some oxygen entering around the connection to the transducer the atmospheric oxygen tension was not zero but was in the range of 4–8 mm Hg (experiments with a comparative set-up). After the experiments were concluded the muscles were returned to a normally aerated solution. Six preparations which almost recovered their original contractility were later used in subsequent IAA/N_2 experiments.

In experiments where all energy metabolism was inhibited the muscle was left unstimulated in oxygenated solution containing 0.5 mM IAA (iodoacetic acid, Sigma Chemicals). After 30 min this oxygenated solution was replaced by one of the same composition that had previously been equilibrated with 95% N_2 and 5% CO_2. The muscle was now bathed in this solution for 10 min before heat studies were commenced. Only isometric studies were undertaken. Twelve muscles were examined in this way and although six of these had been used previously in the hypoxia studies the results seemed to be essentially the same. By using a previously equilibrated solution in which the oxygen tension was less than 1 mm Hg, and by using a slightly higher N_2 flow rate into the drained chamber to prevent any convective entry of air, the oxygen tension level was kept below 2 mm Hg.

Heat Measurements

Two thermopiles were used in the present experiments. All were made using the method devised by N. V. RICCHIUTI[2]. Both had about 100 active silver-constantan junctions and outputs of 2.71 mV/°C and 3.48 mV/°C. Their signals were taken to an Astrodata 120 nV amplifier. A filter network on the amplifier output limited the frequency response to 20 Hz. The average heat loss from the muscle thermopile system was 10.6%/sec and was corrected for electronically.

The difficulty of making an absolute calibration of heat production has been emphasized by HILL and WOLEDGE[5], and by WILKIE[6]. The problems encountered with the small cardiac preparations are much greater than with skeletal muscle. For reasons outlined previously[3] an individual calibration on each muscle was carried out by liberating a known amount of energy into a muscle. Experiments have shown the accuracy and repeatability of this step to be within a few per cent. Two major problems remain. First, there is probably non-uniform heating of the muscle plus its adhering solution[7]. This effect is more likely to be serious in cardiac muscle because the stimulating electrodes have to be positioned on top of the muscle and not in the plane of the thermopile as occurs in most skeletal muscle heat experiments. If the muscle nearer the thermopile is cooler than the top surface the measured temperature rise will be too small and will subsequently cause an overestimation of the heat produced when a muscle contracts. Placing the electrodes elsewhere presents several technical problems so for the moment the possibility of this error can only be acknowledged. Second, there is the problem of accurately assessing the weight of the preparation plus adhering solution. It is difficult to accurately cut the papillary muscle at the position of the ventricular tie and it is difficult to only blot up the solution in the central thermopile groove that is associated with the muscle. Fortunately solution drainage appears to be good and the error here is much less than one would intuitively suspect. Once again an error introduced here will lead to an overestimate of the actual amount of heat produced by the muscle. All the heat records shown in this paper are the actual records obtained in the experiments and no attempt has been made to correct for heat conduction lags or delays. Comparative studies of the rate of heat production under the same experimental conditions are valid but the mechanical records should not be compared directly with the heat traces—the latter are undoubtedly too slow.

Heat Measurements under Conditions of Anoxia

When a muscle is stimulated in the absence of oxygen metabolic end-products accumulate and raise the osmotic pressure of the muscle. This in turn lowers its vapour pressure and allows water vapour present in the thermopile chamber to condense upon the muscle with the attendant production of heat[8]. Stimulation of the muscle therefore causes an apparent rise in resting heat production. As this heat rate provides the 'baseline' in the heat experiments any increase consequent upon stimulation will produce a false result and this would produce large errors in the block experiments. This increment in base-line can however be measured and a correction made for its effect upon the measurement of active heat production[9].

In the present experiments the osmotic effect was of minor importance in experiments where the muscle was stimulated and equilibrated in the nitrogen bubbled solution for 20 min before making the heat recordings. There are three probable reasons: (a) the 'relatively' large amount of solution that adheres to the muscle in the cardiac experiments tends to keep it in osmotic equilibrium with the chamber solution, (b) the oxygen tension level was not negligible, 4–8 mm Hg, (c) the extended period of hypoxia-anoxia has already reduced the mechanical output of the muscle so that lactic acid production, for example, would be less than it would be in an unstimulated but previously oxygenated preparation. When the muscle is not stimulated beforehand but is equilibrated in oxygen plus IAA the vapour pressure effect is generally obvious in later oxygen-free studies. A positive baseline shift often occurs and unfortunately it probably also interferes with the measurement of the anaerobic resting heat. In skeletal muscle, HILL and KUPALOV[8] showed that, provided they did not stimulate the muscle, the rate of heat production was the true (i.e. metabolic) rate—this may be the case in skeletal muscle where the resting heat forms a small proportion of the total energy output of the muscle but in cardiac muscle the resting metabolism is itself a large fraction of the total[2,10,11]. Thus metabolic end-products probably accumulate rapidly even in the absence of stimulation. It was noticeable that upon draining these muscles the resting heat tended to rise slowly but continuously and sometimes reached a value in excess of that obtained in the hypoxia experiment.

Stimulation

The muscles were stimulated via two platinum electrodes cantilevered from the thermopile frame. The stimulus voltage used was 5–7 V and the stimulus duration (generally 0.3–1.0 msec) was adjusted to produce a maximal response. The stimulus heat artifact was determined as outlined elsewhere[12]. Anoxia did produce increases in threshold and it was essential that a correction be made for any changes in this artefact.

Results

Hypoxia

The effects of 20 min exposure to a solution in which the oxygen tension was gradually lowered to the 20–60 mm Hg level are given in Table 1. It was

Table 1. Effects of 20 min hypoxia.

Measurement	No.	Normal			Treated		
Resting heat mcal/g·min	10	30.1	± 3.0	S.E.	18.4	± 1.4	S.E.
Tension (g)	12	3.29	± 0.43	S.E.	2.48	± 0.36	S.E.
Work (g. cm)	6	0.069	± 0.014	S.E.	0.044	± 0.009	S.E.
Pl/H	9	3.6	± 0.3	S.E.	5.3	± 0.3	S.E.

not uncommon for cardiac contractility to increase during the early stages of aeration with N_2 and CO_2. This increase could be manifested either in the ability of the papillary muscle to develop more tension or to do more work. Similar effects have been reported in the literature most recently by POOL, COVELL, CHIDSEY and BRAUNWALD[13]. Following this, however, there was usually a gradual decrease in contractility and by the time the first heat readings were made, developed tension had dropped from a mean of 3.29 g to 2.48 g. There was a noticeable fall in resting heat production, from 30.1 to 18.4 mcal/g·min. A typical effect of hypoxia upon cardiac heat production is shown in Figs. 1 and 2. In both these muscles, hypoxia did not produce a great tension decline but their heat output was noticeably lower, more so in Fig. 1 than in Fig. 2. It will be noticed that the tension independent heat component (the probable equivalent of skeletal muscle activation heat) has declined. This was a general

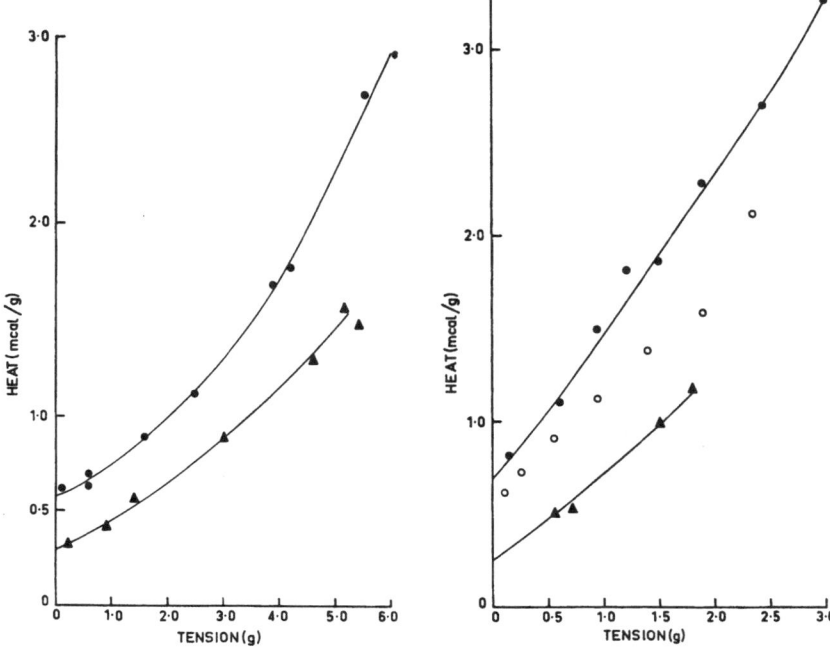

Fig. 1.
Heat versus tension plot obtained in oxygen (•) and after 20 min of hypoxia (▲).
Muscle wt. = 14.1 mg, length = 8.2 mm, bath temp. 18.7 °C.

Fig. 2.
Heat versus tension plot obtained in oxygen (•), after 20 min hypoxia (○) and after I AA/N₂ treatment (▲). Muscle wt. = 7.1 mg, length = 8.5 mm, bath temp. = 20.2 °C.

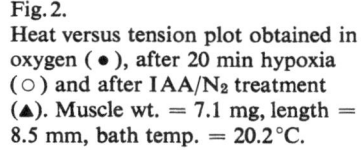

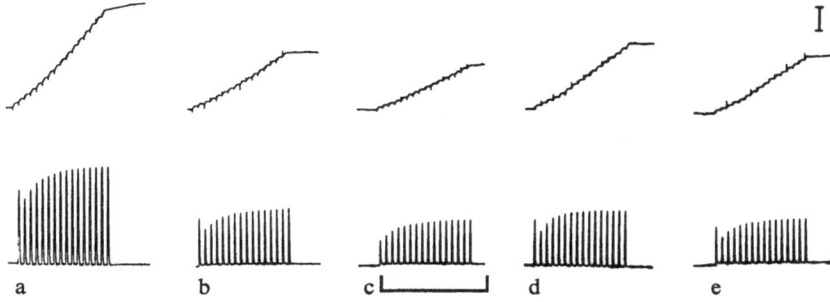

Fig. 3.
Block experiments showing summed heat (top traces) and developed tension (bottom traces). Calibration bar: heat traces = 1.57 g.cm *a*, *b* and *c*, 0.79 g.cm *d* and *e*, tension = 0.5 g. Time markers 1 sec apart.
Muscle wt. = 3.3 mg, bath temp. = 22.3 °C. Muscle length *a* = 6.8 mm, *b* = 6.5 mm, *c* = 6.1 mm, *d* = 6.8 mm and *e* = 6.5 mm.

finding in all muscles although the extent of the fall varied considerably. In Fig. 3 the heat outputs of isometric contractions at different muscle lengths are shown first in oxygen (traces *a*, *b* and *c*) and then after 20 min of hypoxia. Note that even when the tension level is about the same (traces *b* and *d*) the heat output is greatly reduced after exposure to nitrogen (traces *d* and *e* were obtained at a higher heat sensitivity).

In Figs. 4 and 5 the heat outputs associated with single cardiac contractions are shown. In Fig. 4 both readings were taken with the muscle at the same

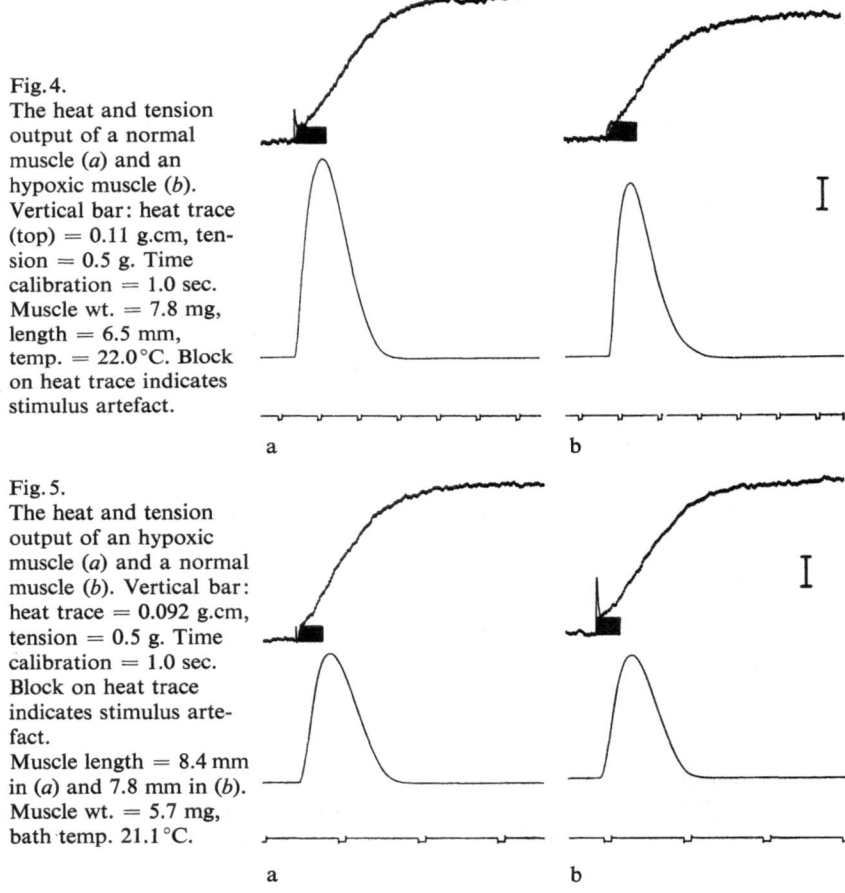

Fig. 4.
The heat and tension output of a normal muscle (*a*) and an hypoxic muscle (*b*). Vertical bar: heat trace (top) = 0.11 g.cm, tension = 0.5 g. Time calibration = 1.0 sec. Muscle wt. = 7.8 mg, length = 6.5 mm, temp. = 22.0 °C. Block on heat trace indicates stimulus artefact.

a   b

Fig. 5.
The heat and tension output of an hypoxic muscle (*a*) and a normal muscle (*b*). Vertical bar: heat trace = 0.092 g.cm, tension = 0.5 g. Time calibration = 1.0 sec. Block on heat trace indicates stimulus artefact.
Muscle length = 8.4 mm in (*a*) and 7.8 mm in (*b*). Muscle wt. = 5.7 mg, bath temp. 21.1 °C.

a   b

length and at the same time interval after a previous block of 15 contractions. It will be noticed that although hypoxia has decreased developed tension it has not significantly altered the rate at which the heat is produced or unduly lowered the magnitude of the fast phase of heat production. In Fig. 5 tension development has been kept the same under both conditions by recording the energy output at a shorter muscle length in oxygen. It is quite apparent once again that any changes in the fast phase of heat production are minimal both in regard to rate and magnitude.

Six preparations were used where the isotonic energy output of muscles was examined under both normal and hypoxic conditions. Similar results were obtained. If the work output under a 1.0 g load is taken as a point of comparison work output averaged a 36% decline and heat output fell 33%. Typical results are shown in Fig. 6, where the upper traces show the external work done and

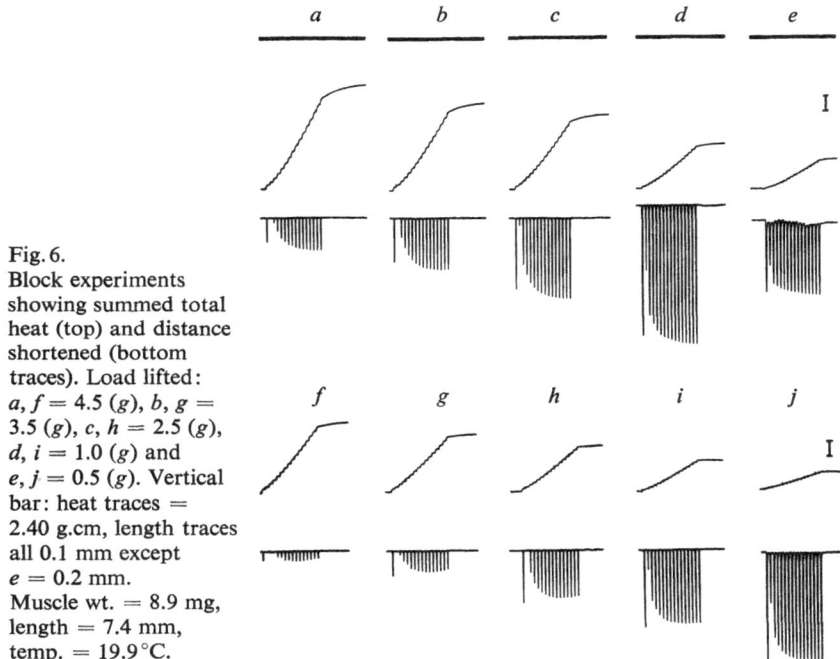

Fig. 6.
Block experiments showing summed total heat (top) and distance shortened (bottom traces). Load lifted: $a, f = 4.5$ (g), $b, g = 3.5$ (g), $c, h = 2.5$ (g), $d, i = 1.0$ (g) and $e, j = 0.5$ (g). Vertical bar: heat traces = 2.40 g.cm, length traces all 0.1 mm except $e = 0.2$ mm. Muscle wt. = 8.9 mg, length = 7.4 mm, temp. = 19.9 °C.

the heat produced under a series of isotonic loads. The lower traces show the work and heat outputs after the standard period of hypoxia. Notice in these experiments that when stimulation ceases the heat records reach a plateau level (more quickly after nitrogen treatment where the recovery heat is less) and compare these records with the heat trace shown in Fig. 7 where there is no plateau because of a baseline shift (see next section).

Iodoacetate and Nitrogen

The results obtained after treatment with iodoacetate and subsequent exposure to oxygen-free nitrogen are given in Table 2. The difficulties associated

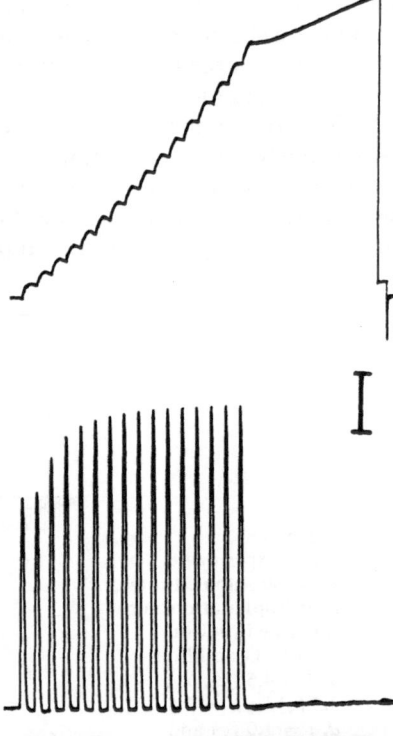

Fig. 7.
Block experiment showing summed
heat (top) and developed tension
(bottom) after IAA/N₂ treatment.
Note that when heat loss correction
device turned off heat baseline rose
(readjusted to correct level to show size
of increment). Vertical bar: heat trace
= 1.07 g.cm, tension trace = 0.5 g.
Muscle wt. = 3.3 mg, length = 6.7 mm,
temp. = 20.6 °C.

Table 2. Effects of IAA + N₂.

Measurement	No.	Normal	Treated
Resting heat mcal/g·min	12	29.8 ± 3.1 S.E.	16.5 ± 1.6 S.E.
Tension (g)	12	3.18 ± 0.40 S.E.	1.97 ± 0.34 S.E.
Activation heat mcal/g·contraction	12	0.48 ± 0.06 S.E.	0.18 ± 0.03 S.E.
Total contraction heat mcal/g·contraction	12	2.50 ± 0.13 S.E.	0.83 ± 0.09 S.E.
Pl/H	12	3.7 ± 0.2 S.E.	6.7 ± 0.4 S.E.

with making many of the heat measurements after this treatment have been
discussed in the Methods section.

Normally some 30–80 contractions could be elicited before the muscle went
into a state of contracture.

Resting heat production fell from 29.8 to 16.5 mcal/g·min and there are
reasons for believing that the latter figure is probably too high (see Methods).
Peak tension development dropped from a mean of 3.18 to 1.97 g and the iso-

metric heat coefficient rose from a mean of 3.7 in oxygen to a mean of 6.7 in IAA/N_2. More will be said about this latter measurement in the Discussion. The tension independent heat production fell from a mean of 0.48 mcal/g·contraction to a mean of 0.18. The heat production, in an isometric contraction at the normal muscle length used, fell from a mean value of 2.5 mcal/g·contraction to 0.83. Part of this fall is consequent upon the decrease in developed tension and part is caused by a decrease in recovery heat.

Fig. 7 shows an example of the osmotic baseline shift first described by HILL and KUPALOV[8]. After stimulation had ceased heat production continued to rise rapidly and when the heat loss correction device was switched off a baseline shift was apparent. This shift did not decline with time, as it would if it were normal heat loss decline or a recovery heat effect, instead it usually got worse. An allowance was made for this effect and it was minimized by re-zeroing the amplifier between each block of contractions and by reducing the number of stimuli in each block. This latter procedure must be carried out anyway as one wishes to obtain several readings over a range of tension outputs because only a small number of contractions may be generated before contracture occurs.

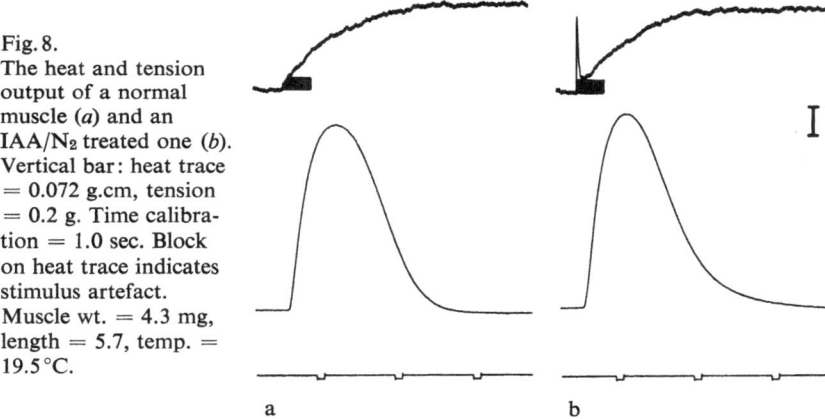

Fig. 8.
The heat and tension output of a normal muscle (*a*) and an IAA/N_2 treated one (*b*). Vertical bar: heat trace = 0.072 g.cm, tension = 0.2 g. Time calibration = 1.0 sec. Block on heat trace indicates stimulus artefact. Muscle wt. = 4.3 mg, length = 5.7, temp. = 19.5 °C.

a   b

In Fig. 8 the heat and tension output of a normal and iodoacetate/nitrogen-poisoned muscle are compared. This muscle was one where there was little tension decline after treatment and it can be seen that there is little change in the rate or magnitude of the fast-phase heat production. Heat versus tension plots obtained under normal and IAA/N_2 conditions are shown in Figs. 9 and 2.

Discussion

An encouraging feature of the present results is that there seems to be reasonable agreement between results obtained myothermically and those obtained biochemically. In Table 3 a comparison is made of some of the findings.

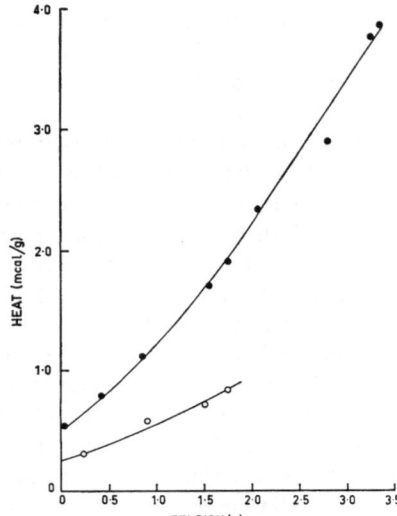

Fig. 9.
Heat versus tension plot obtained in
oxygen (●) and after IAA/N₂ treatment (o). Muscle wt. = 5.7 mg,
length = 8.4 mm, bath temp. = 21.1 °C.

Table 3. Comparison of myothermic with biochemical results.

Measurement	Present Study	From POOL and SONNENBLICK 1967*
Resting energy mcal/g·min	16.5	6.8–10.7
Activation energy mcal/g·contraction	0.18	0.21
Total contraction energy mcal/g·contraction	0.83	1.41
Pl/M g·cm²	270	360
Pl/H	6.7	7.9

* Assumed 1 mol CP = 10 kcal. P = tension, l = muscle length, H = heat.

A valid criticism of the present results would be that no biochemical tests
have been carried out to show that energy production has been inhibited under
the experimental conditions used. Numerous studies have however now been
made using this combination of inhibitors[1,14,15] and the reported mechanical
results are very similar to those in the literature.

The normal mechanical performance of the muscles was less than we have
reported previously (380 g/cm² compared with 470 g/cm²)[2]. This is no doubt
mainly caused by the reduction in external calcium from 5.0 mM used previously to 2.5 mM used in the present study. This figure is less than the 500
g/cm² reported by POOL and SONNENBLICK[1]. The difference is probably primarily caused by additional system compliance that is inevitable in the myothermic experiments because the muscle cannot be clamped at its ventricular
end. Other contributing factors may be the different animals used or the fact
that POOL and SONNENBLICK searched for the optimal length to carry out their
experiments whereas it has been our practice to keep the resting tension constant at 0.5 or 1.0 g. The results of IAA/N₂ treatment seem to be the same in

both preparations as the fall in tension was about the same in both studies, being 71 and 62 % respectively. Although some of this fall may be related to the 30-min rest period this cannot be the complete explanation as the hypoxia experiments produced a similar fall although the preparations were stimulated continuously. Several groups of workers have now shown that there is poor correlation between existing levels of ATP, CP and the force of contraction. Alternative suggestions have been made[13] that anoxia might interfere with a separate store of energy needed for the maintenance of cell membrane function or with excitation-contraction coupling. These suggestions are rendered more likely by the effect of nitrogen upon the electrical threshold of the papillary muscle and the tension-independent heat. Recent work has shown that the latter component is calcium dependent[12], and is liberated during the rising phase of tension development. POOL and SONNENBLICK reported an 'activation energy' of 0.83 mcal/g·contraction but this included the resting heat contribution which if subtracted out would have left a figure of 0.21 mcal/g·contraction which would be quite comparable to the 0.18 mcal found in this study. There must be some doubt, however, as to whether these two values obtained by quite different methods, refer to the same physiological process. These authors also reported that the total energy (exclusive of the resting energy contribution) in an isometric twitch was 1.41 mcal/g and this should be compared with 0.83 mcal/g reported in Table 2. As the tension per cross sectional area is less in the muscles studied here and the decline in mechanical performance somewhat greater after the IAA-N_2 treatment the values are not that widely different. In their paper POOL and SONNENBLICK have in certain instances somewhat misleadingly compared their biochemical results obtained with metabolically inhibited preparations with myothermic data obtained on normal muscles where the heat associated with ATP regeneration is also included.

The largest discrepancy between both sets of results occurs in the estimations of the isometric heat coefficient. This coefficient, which is simply the tension developed (g) x muscle length (cm) divided by the heat produced (g.cm), is often taken to express a fundamental connection between the thermal and mechanical response of muscle. The measured value reported here is lower than expected. HILL's[16] data obtained on skeletal muscle would suggest that the coefficient should be about 2.1 times higher when measured anaerobically as when measured aerobically. This would suggest a value of about 7.6 rather than the 6.7 obtained. The former figure would be more in keeping with 7.9 obtained by POOL and SONNENBLICK. Initially one might suspect that the heat calibration is incorrect (see Methods) but the aerobic value is a reasonable one and the discrepancy really arises because the 2.1:1.0 ratio (total heat/initial heat) found in skeletal muscle has not been observed. Since the skeletal muscle measurements can be made with more accuracy and since considerable technical difficulties arise when making cardiac myothermic measurement during anoxia it would seem reasonable at present to place more reliance on the biochemical estimate of this coefficient.

The fast-phase heat traces obtained in nitrogen confirm the skeletal muscle results[16-19] that the magnitude and even the time course of heat production associated with the mechanical event is little altered by the absence of oxygen. This suggests therefore that fast-phase heat production is the equivalent of skeletal muscle initial heat.

There is one relevant and important point that this paper has not examined. It relates to the time sequence of the recovery and initial heat phases in cardiac muscle. Originally we decided to divide active cardiac heat production into a fast phase, associated with the contractile event, and a slow phase occurring after mechanical activity had ceased. This division was not made to burden the literature with new terminology but because we were not certain about the tightness of any coupling of the recovery mechanisms to the initial ones in a preparation which is normally beating constantly. It was thought that it might not be possible to observe the classical time separation of the initial and recovery heats in cardiac muscle. Now it is possible to record the fast phase of cardiac heat production with reasonable accuracy and repeatability but the measurement of the slower phase of heat production is a frustrating experience because of baseline instability. It was soon discovered, however, that if one measured the magnitude of the fast-phase heat production in a single contraction and compared it with the total heat (initial plus recovery) produced in a contraction of a repetitive series, where the same tension was being produced, less heat was liberated in the single contraction. Upon analysis the difficulty arises that there is no firm proportionality between the two heats. In some preparations the fast-phase heat is as low as 50% of the total heat measurement and in others it is as high as 80%. Thus although in some preparations it might appear that the fast phase of heat production is the exact equivalent of skeletal-muscle initial heat in other preparations it appears to be too large a proportion of the total heat. Why do we see varying ratios of steady-state total heat to single-contraction fast-phase heat? At the moment I can only suggest one possibility. Many experimenters[20,21] have shown that cardiac oxygen consumption can be related to the tension time integral and I have recently found that cardiac heat production can be linearly related to the tension integral (unpublished observations). Now although the tension developed in a single contraction may be exactly the same as the tension developed in a steady-state series of contractions the tension integral of the single contraction varies from being slightly greater in some preparations to being more than 50% larger in others. In the latter muscles it may not be fair to make a comparison between the heat produced in an isolated contraction and in a contraction that is one of steady-state series.

The results presented here have added to our knowledge of cardiac energetics and more importantly they seem to be in reasonable agreement with biochemical studies made on cat papillary muscles, when allowance is made for differences in the tension generated per cross-sectional area. In general the results are similar to those found in skeletal muscle by A. V. HILL and his colleagues. A major area of uncertainty, the temporal coupling of initial and recovery heats and their exact magnitudes in cardiac muscle, still remains unsolved.

Acknowledgment

This work was supported by a Grant-in-aid from the National Heart Foundation of Australia.

References

[1]   P. E. POOL and E. H. SONNENBLICK, *The Mechanochemistry of Cardiac Muscle. I. The Isometric Contraction*, J. gen. Physiol. *50*, 951 (1967).

[2]   N. V. RICCHIUTI and C. L. GIBBS, *Heat Production in a Cardiac Contraction*, Nature *208*, 897 (1965).

[3]   C. L. GIBBS, W. F. H. M. MOMMAERTS and N. V. RICCHIUTI, *Energetics of Cardiac Contractions*, J. Physiol. *191*, 25 (1967).

[4]   K. HUKUDA, *The Energy Liberated in Total Exhaustion of Frog's Muscle*, J. Physiol. *72*, 438 (1931).

[5]   A. V. HILL and R. C. WOLEDGE, *An Examination of Absolute Values in Myothermic Measurements*, J. Physiol. *162*, 311 (1962).

[6]   D. R. WILKIE, *Heat Work and Phosphorylcreatine Break-Down in Muscle*, J. Physiol. *195*, 157 (1968).

[7]   A. V. HILL, *The Heat of Shortening and the Dynamic Constants of Muscle*, Proc. R. Soc. [B] *126*, 136 (1938).

[8]   A. V. HILL and P. S. KUPALOV, *The Vapour Pressure of Muscle*, Proc. R. Soc. [B] *106*, 445 (1930).

[9]   D. K. HILL, *The Anaerobic Recovery Heat Production of Frog's Muscle at 0°C*, J. Physiol. *98*, 460 (1940).

[10]   R. J. BING and G. MICHAL, *Myocardial Efficiency*, Ann. N.Y. Acad. Sci. *72*, 555 (1959).

[11]   K. S. LEE, *The Relationship of the Oxygen Consumption to the Contraction of the Cat Papillary Muscle*, J. Physiol. *151*, 186 (1960).

[12]   C. L. GIBBS and P. V. VAUGHAN, *The Effect of Calcium Depletion upon the Tension-Independent Component of Cardiac Heat Production*, J. gen. Physiol. *52*, 532 (1968).

[13]   P. E. POOL, J. W. COVELL, C. A. CHIDSEY and E. BRAUNWALD, *Myocardial High Energy Phosphate Stores in Acutely Induced Hypoxic Heart Failure*, Circ. Res. *19*, 221 (1966).

[14]   F. D. CARLSON and A. SIGER, *The Mechanochemistry of Muscular Contraction. I. The Isometric Twitch*, J. gen. Physiol. *44*, 33 (1960).

[15]   P. PADIEU and W. F. H. M. MOMMAERTS, *Creatine Phosphoryl-Transferase and Phosphoglyceraldelyde Dehydrogenase in Iodo-Acetate Poisoned Muscle*, Biochim. biophys. Acta *37*, 72 (1960).

[16]   A. V. HILL, *Tails and Trials in Physiology* (Arnold, London 1965), p. 161.

[17]   V. WEIZSACKER, *Myothermic Experiments in Salt-Solutions in Relation to the Various Stages of a Muscular Contraction*, J. Physiol. *48*, 396 (1914).

[18]   A. V. HILL and W. HARTREE, *The Four Phases of Heat Production of Muscle*, J. Physiol *54*, 84 (1920).

[19]   A. V. HILL, *The Recovery Heat Production in Oxygen after a Series of Muscle Twitches*, Proc. R. Soc. [B] *103*, 183 (1928).

[20]   S. J. SARNOFF, E. BRAUNWALD, G. H. WELCH JR., R. B. CASE, W. N. STAINSBY and R. MACRUZ, *Hemodynamic Determinants of Oxygen Consumption of the Heart with Special Reference to the Tension Time Index*, Am. J. Physiol. *192*, 148 (1958).

[21]   R. H. MCDONALD, *Developed Tension: A Major Determinant of Myocardial Oxygen Consumption*, Am. J. Physiol. *210*, 351 (1966).

II
Excitation and Transmission in Cardiac Tissues

Introduction

by NICK SPERELAKIS

Department of Physiology, University of Virginia School of Medicine, Charlottesville, Va.

There is controversy as to the mechanism of transmission of excitation from cell to cell in vertebrate myocardium. Some of the electrophysiological data support the hypothesis of a simple electrical syncytium (one-, two-, or three-dimensional) and some support the hypothesis that there is no DC electrical coupling between contiguous myocardial cells.

The properties of the intercalated disc are very important both from theoretical and practical points of view, as illustrated by the following examples. The interpretation of many potential and resistance measurements in cardiac muscle depends on the three-dimensional electrical architecture assumed, e.g. there is doubt concerning the voltage dependency of g_K[1,2]*. The interpretation of some voltage-clamp experiments presupposes that the membrane is uniformly clamped. It is also conceivable that one or more components of the intracellularly recorded cardiac action potential may be influenced by the intercalated discs, i.e. the junctional membranes. The normal functional interactions of one cell with another across the disc may be altered during fibrillation, and the conduction disturbances produced under various conditions could be due to effects on the junctional membranes. The process of so-called 'healing-over' which occurs during focal mechanical or metabolic injuries to cardiac muscle probably involves the intercalated discs[3], i.e. contiguous cells may become functionally disconnected.

I wish to discuss here briefly some of the non-electrophysiological observations dealing with the possible functions of the intercalated discs of vertebrate hearts as an introduction to this session on excitation transmission.

Possible Mechanisms for Transmission of Excitation from Cell to Cell in Myocardium

Some of the possible mechanisms for transmission of excitation from cell to cell in vertebrate myocardium are summarized in Fig. 1. The old hypothesis

* Numbers refer to References, p. 100.

of a branching morphological syncytium, i.e. a fusion of cells like that occurring in the embryogenesis of skeletal muscle fibers, has been ruled out by electron microscopy which shows that the intercalated discs mark cell boundaries and that there is no cytoplasmic or membranous continuity from one cell to the next[4]. A more recent hypothesis is that the heart is a branching three-dimensional electrical syncytium with very low resistance discs. Proponents of an electrical syncytium presume that the so-called 'tight-junction' regions of the intercalated discs, in which the contiguous membranes are in close apposition without much intervening interstitial fluid, are very low in resistance and serve to connect the

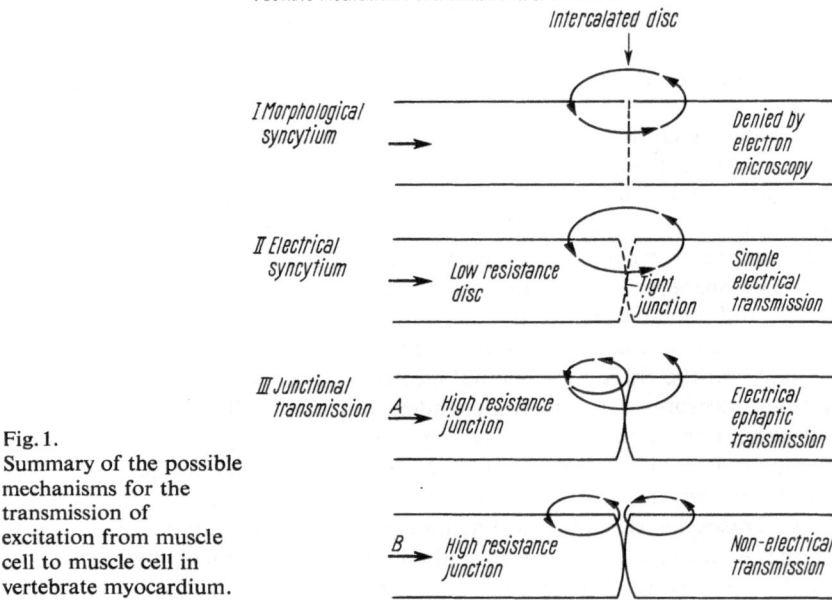

Fig. 1.
Summary of the possible mechanisms for the transmission of excitation from muscle cell to muscle cell in vertebrate myocardium.

cells electrically. On the other hand, some electrophysiological experiments suggest that the heart is not a two- or three-dimensional electrical syncytium and that some form of junctional transmission between cells may occur. A junctional potential, representing the excitatory interaction of an activated cell upon its contiguous neighbor, may be present. However, a junctional potential could result from electrical ephaptic transmission across high-resistance discs rather than from chemical synaptic transmission. In electrical transmission, the current source for depolarization of the postjunctional cell comes from the prejunctional cell, whereas in chemical transmission, the source of the depolarizing current is the postjunctional cell itself when activated by a chemical mediator from the prejunctional cell. Measurements of propagation velocity alone do not allow distinction between an electrical or a chemical junction.

The possibility of a mechanical mechanism, whereby one cell pulls on its contiguous neighbor to depolarize it to threshold, is unlikely because cardiac muscle propagates action potentials in the complete absence of mechanical activity[5]. However, in some smooth muscles, there is some evidence that a mechanical distortion of the membrane of a cell leads to changes in membrane ion conductances, depolarization, and excitation[6,7].

The possibility of a nervous mechanism for vertebrate myocardium, i.e. a neuromuscular junction on each myocardial cell, is unlikely for several reasons including the fact that denervated cultured heart cells propagate impulses[5]. However, there is good evidence that some invertebrate cardiac muscles and vertebrate smooth muscles do not transmit excitation from cell to cell, and activation of any cell is solely controlled by its innervation. In smooth muscle electrophysiology, more and more muscles are being found to be directly activated by motor nerves at discrete neuromuscular junctions, and EPSPs and IPSPs can be recorded[8,9]. In the case of *Limulus* heart, it is clear that there is no electrical coupling between cells and the cells are activated by neuromuscular junctions; the plateau of the action potential appears to be fusion of EPSPs by temporal summation[10]. It would be interesting to learn something about the morphology of adjoining regions of contiguous muscle cells in such cases where there are clearly no functional myo-myo junctions.

Summary of Some Morphological and Histochemical Data Dealing with the Intercalated Discs of Vertebrate Hearts

Some of the morphological and histochemical evidence consistent with the hypothesis that the intercalated discs are special myo-myo junctions are summarized in the table and in Fig. 2. Electron microscopy has shown that the discs are cell boundaries with a standard junctional gap of 200–300 Å along most of

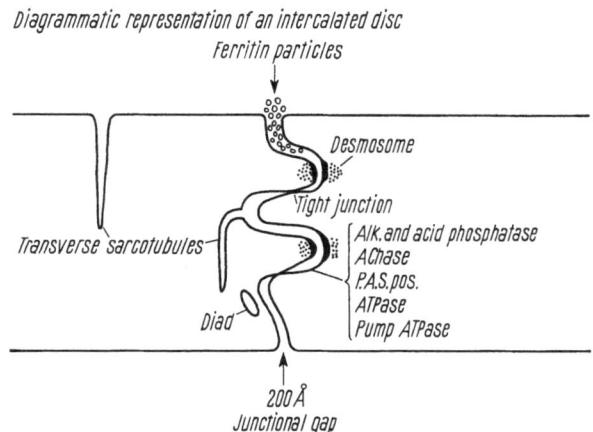

Fig. 2.
Diagrammatic representation of an intercalated disc which summarizes some of the morphological and histochemical evidence consistent with the hypothesis that the intercalated discs of vertebrate myocardium are special myo-myo junctions.

its path[4]. Although the basement membrane, which dips into the transverse sarcotubules, supposedly does not enter the junctional gap of the discs, ferritin particles readily penetrate into the junctional gap at the discs as well as into the transverse sarcotubules[11]; therefore fluid in the junctional gap is continuous with the interstitial fluid. The transverse tubules also open into the junctional gap at the disc as well as onto the cell surface. In addition, flattened cisternae of the sarcoplasmic reticulum analogous to the lateral elements of triads, are present close to the intercalated discs, forming 'diad' structures[12,13]. Enzymes which split ATP and ADP are also present in these cisternae at the discs. Therefore, the intercalated disc may function in excitation-contraction coupling as does the remaining cell surface.

Histochemical studies show that many enzymes are localized, sometimes specifically, at or near the intercalated discs. Included are the following: alkaline and acid phosphatases[14,15], 5-nucleotidase[15], acetylcholinesterase[16], ATPase[17,18], and specific transport ATPase[19]. The discs are also positive for periodic acid Schiff reagent which stains the mucopolysaccharides present at cell surfaces[20]. Granules, perhaps containing catecholamines, have been found at the intercalated discs[21]. Although synaptic-like vesicles have been described at the discs[4], it may be difficult to resolve them because of the myofilaments.

These ultrastructural and cytochemical studies emphasize that the intercalated discs must be considered as cell surfaces and as cell-to-cell junctions. Further, a light microscopic study has shown, by fixation during fibrillation, that the impulse to contraction is blocked at the intercalated disc; often a cell on one side of a disc is fixed in a contracted state whereas its adjoining neighbor is in a relaxed state[22]. This means that a presumed cable has to be capable of being interrupted at the disc.

Ultrastructural Changes of the Intercalated Discs Coincident with Depressed Transmission

The morphology of the intercalated disc is altered by some changes in the physiological state of the heart, including (a) hypoxia, (b) hypothermia, and (c) hypertonicity. Hypoxia causes the junctional gap of the intercalated disc to increase by 10- to 20-fold in dog heart[23]. A graded junctional potential becomes very prominent in intracellular recording from guinea-pig ventricle during hypoxia; some of the cells fail to fire action potentials, leaving only the junctional potential (R. RUBIO, personal communication).

Hypothermia (15 °C) slows conduction velocity in rat heart and produces ectopic foci associated with widening of the cell membranes at the fascia adherens and macula adherens regions of the intercalated discs; no changes occur in the interspaces, including the regions of fascia adherens and macula occludens[24]. In contrast, in the ground squirrel (a hibernator) coordinated heart function continues at body temperatures of 5 °C, and an increase in the intracytoplasmic density occurs at the desmosomal junctions of the intercalated discs, which the authors interpreted as suggesting increased chemical activity[24].

Hypertonicity (perhaps because of cell shrinkage) increases the junctional gap of the intercalated disc (SCHEYER, quoted in ref. 25). It was also reported[26] that tight junctions are pulled apart reversibly in hypertonic solution concomitant with block of transmission. However, ROUILLER and FORSSMANN[27], and GIRARDIER[28] observed that the tight junctions at the intercalated discs do not become separated in hypertonic solutions (sucrose or NaCl), and TOMITA[29] found that twofold hypertonic (sucrose) Ringer does not affect propagation velocity in guinea-pig *Taenia coli*. GIRARDIER suggests that one of the effects of hypertonicity on conduction velocity is due to increased shunting of action current by dilated transverse tubules.

These observations indicate that when the junctional gap increases with hypoxia and hypertonicity, the transmission of excitation from one cell to the next is profoundly affected. Hence, the structure of the junctions is critically important for the transmission process. If the 'tight junction' can be pulled apart reversibly, then it must not represent true fusion of the contiguous cell membranes.

In the production of cultured heart cells, the myocardium is dissociated into individual cells by trypsin action in a Ca^{++}-free, Mg^{++}-free solution, each cell being capable of firing spontaneous action potentials and contracting immediately after such dissociation[5]. The cells separate at the intercalated discs. New intercalated discs are reformed upon re-association of the cells in monolayer cultures[5,30-32].

Absence of Tight Junctions at Intercalated Discs of Some Hearts

Proponents of the hypothesis of an electrical syncytium presume that the tight junction regions of the intercalated disc are very low in electrical resistance and hence serve to connect contiguous cells electrically, thereby giving a long length constant. A case has been made for a correlation between the presence of tight junctions in various muscles with the presence of functional transmission from muscle cell to muscle cell[33]. There is good evidence that true tight junctions in various epithelia, endothelia and gland cells prevent the extracellular passage of ions in the *longitudinal* direction across sheets of cells. In the case of some gland cells, there also is evidence that even quite large molecules can pass easily from one cell to the next across special regions[34]; however, even these leaky membranes exhibit high resistance when they are directly exposed, either intracellularly or extracellularly, to the Ca^{++} concentration of plasma[34].

However, there is no direct evidence that the tight junctions in cardiac and smooth muscles are low in resistance, allowing easy passage of ions from one cell to the next. It has been argued that since there may be no extracellular fluid at the outer surface of the cell membranes at tight junctions in muscle, the double-layered membrane has high $[K^+]_i$ on both sides, hence no trans-membrane resting potential in this region. It is further argued that the absence of a 'resting potential' leads to a very low resistance because of delayed rectifi-cation. However, the R_m of frog sartorius fibers bathed in 83 mM $[K^+]_0$ is 1.5 $K\Omega$-cm^2 compared to the control value of 3.8 $K\Omega$-cm^2 [35]; similarly, cultured heart cells bathed in 65 mM $[K^+]_0$ have an R_m about 20% of that of the con-trol[36]. These decreases in R_m at high $[K^+]_0$ can be accounted for by the increase in g_K due to the concentration effect. Thus, the decrease in R_m produced by high K^+ is much too small to be significant for the very low cell-to-cell resistance necessary to account for successful electrical transmission from one cell to the next. Compared to the usual R_m values of about 500–5000 Ω-cm^2 for muscle membrane, the effective resistance of the disc membranes would need to be lower than 2 Ω-cm^2.

Another proposal is that the resistance of the tight junction region is low because this double membrane is exposed only to low Ca^{++} concentration ($\sim 10^{-7}$ M in myoplasm) on both of its sides[3]. However, REVEL and KARNOV-SKY[37] have evidence that there are no true tight junctions with zero extracel-lular gap in vertebrate cardiac and smooth muscles. Instead, they find regions of 'close gaps' in which the gap separation of about 20–40 Å is readily pene-trated by La^{+++}. Hence the assumption of their low electrical resistance based solely on the supposed absence of extracellular fluid is untenable.

Furthermore, tight junctions, similar to those of the intercalated discs of mammalian heart, were not seen in frog ventricular muscle by STALEY and BENSON[38]. Therefore, if transmission in frog cardiac muscle is purely electrical, it must not be related to tight junctions.

These observations indicate that the suggested role of the tight junctions in transmission is equivocal. Furthermore, even if tight junctions were low in resistance, the important parameter which determines whether successful trans-mission of excitation is possible by purely electrical means is the total fraction of the action current generated by the prejunctional cell that passes into the postjunctional cell. For this fraction to be large, there would have to be a large number of zero-resistance small 'holes' between the cells, i.e. the important parameter is the overall resistance of the total junctional membranes. (See ref. 37 for a discussion of hexagonal arrays of subunits found at some junctions.) The portion of the total junctional area covered by tight junctions is relatively small[33].

Summary of Some Electrophysiological Experiments

Although many investigators intuitively feel that the resistance of the inter-calated discs should be very low, they often disagree with one another with regard to some of the pertinent parameters, such as length constant and input resistance. The data obtained with cardiac Purkinje fibers are not necessarily applicable to myocardial cells. It should not be too surprising if it turns out that the myo-myo junctions are not electrical junctions since the vast majority of neuro-neural synapses and neuro-muscular synapses in vertebrates are chemical junctions. Even the good example of a vertebrate electrical synapse, the chick ciliary ganglion, also has a parallel chemical transmission[39]. In this case, the morphology particularly favors electrical coupling, the termination of the prejunctional cell (calyx) nearly completely surrounding the soma of the postjunctional cell. The morphology certainly is not nearly as favorable in the case of cardiac and smooth muscles.

Our electrophysiological experiments have demonstrated that myocardial cells do not form a three-dimensional or a two-dimensional electrical syncytium, i.e. there is no electrical coupling between cardiac cells lying side by side. The degree of electrical interaction between cells lying end to end in a fiber tract is not unequivocally resolved because the interpretation of the data rests on probability arguments as to the location of the electrodes. However, although this question as to whether a series of cells stacked end to end forms a simple one-dimensional cable is not fully resolved, the following additional data suggest

Table. Summary of some morphological and histochemical evidence consistent with the intercalated discs being special cardiac myo-myo junctions.

I.   Electron microscopy
   (a) Discs represent cell boundaries; junctional gap of ~ 250 Å[4]
   (b) Ferritin particles readily penetrate into the junctional gap[11]
   (c) Transverse sarcotubules open into the junctional gap[11]
   (d) The 'tight junctions' may be pulled apart in hypertonic solutions[25,26]; no morphological evidence that they are low in resistance
   (e) Increase in the junctional gap with hypoxia[23] and hypertonicity[25,26]
   (f) Absence of true tight junctions in at least some cardiac muscles[37,38]
   (g) Intercalated discs form in cultured heart cells[15,30-32]

II. Light microscopy
   Fixation during fibrillation shows impulse to contraction blocked at disc, i.e. cell on one side of disc contracted whereas its adjoining neighbor did not[22]

III. Histochemistry
   (a) Alkaline phosphatase[14]
   (b) Acid phosphatase, 5-nucleotidase[15]
   (c) Acetylcholinesterase[16]
   (d) Adenosine triphosphatase (ATPase)[17,18]
   (e) Transport ATPase (cultured cells)[19]
   (f) Periodic acid Schiff positive (mucopolysaccharides)[20]
   (g) Catecholamine granules[21]

Tight junctions were not found in chicken myocardium also [J. R. SOMMER and R. L. STEERE, Fed. Proc. 28, 328 (1969)] (see Table, If).

that the cable may be terminated at each intercalated disc because of their high resistance: (a) recording of junctional potentials representing excitatory inter-actions between contiguous cells[5,40-42], (b) observations of a labile partial block of transmission from cell to cell[5,40], (c) measurement of a high input resist-ance[5,40,43-46], (d) measurements of a short apparent length constant[43-46], and (e) measurements of a high capacitance of transversely-oriented membranes[47]. Therefore, it is possible that transmission of excitation may not be accounted for strictly on the basis of local-circuit action current, and that some sort of a 'transmitter' ion or molecule may be involved at such bidirectional myo-myo junctions. Because of these findings, the syncytial hypothesis for the trans-mission of excitation from cell to cell must be re-evaluated.

References

1   E. A. JOHNSON and J. TILLE, J. gen. Physiol. *44*, 443 (1961).
2   D. NOBLE, Biophys. J. *2*, 381 (1962).
3   J. DÉLÈZE, *Electrophysiology of the Heart* (Ed. B. TACCARDI and G. MARCHETTI; Perga-mon Press, London 1965), p. 147.
4   D. W. FAWCETT and C. C. SELBY, J. biophys. biochem. Cytol. *4*, 63 (1958).
5   N. SPERELAKIS, *Electrophysiology and Ultrastructure of the Heart* (Ed. T. SANO, V. MIZU-HIRA and K. MATSUDA; Bunkodo Co., Tokyo 1967), p. 81.
6   G. BURNSTOCK and C. L. PROSSER, Am. J. Physiol. *198*, 921 (1960).
7   E. BÜLBRING, J. Physiol. *125*, 302 (1954).
8   E. BÜLBRING and T. TOMITA, J. Physiol. *189*, 299 (1967).
9   N. C. R. MERRILLEES, G. BURNSTOCK and M. E. HOLMAN, J. Cell Biol. *19*, 529 (1963).
10   B. C. ABBOTT, F. LANG, I. PARNAS, W. PARMLEY and E. SONNENBLICK (this volume).
11   W. G. FORSSMANN and J. GIRARDIER, Z. Zellforsch. mikrosk. Anat. *72*, 249 (1966).
12   E. PAGE, J. Ultrastruct. Res. *17*, 72 (1966).
13   J. ROSTGAARD and O. BEHNKE, J. Ultrastruct. Res. *12*, 579 (1965).
14   G. H. BOURNE, Nature, Lond. *172*, 588 (1953).
15   E. B. BECKETT and G. H. BOURNE, Acta anat. *33*, 289 (1958).
16   F. JOÓ and B. CSILLIK, Nature *193*, 1192 (1962).
17   J. M. DE BEYER, J. C. H. DE MAN and J. P. PERSIJN, J. Cell Biol *13*, 451 (1962).
18   E. ESSNER, A. B. NOVIKOFF and N. QUINTANA, J. Cell Biol. *25*, 201 (1965).
19   A. WOLLENBERGER, personal communications.
20   D. J. GOLDSTEIN, Anat. Record *134*, 217 (1959).
21   E. PAGE, J. Ultrastruct. Res. *17*, 63 (1967).
22   M. IMCHANITZKY, Arch. Int. Physiol. *4*, 1 (1906).
23   F. BÜCHNER and S. ONISHI, Naturwissenschaften *54*, 1 (1967).
24   M. L. ZIMNY, M. SHERMAN and C. C. ROMANO, Cryobiology *4*, 317 (1968).
25   J. W. WOODBURY and W. E. CRILL, *Nervous Inhibition* (Ed. E. FLOREY; Pergamon Press, London 1961), p. 124.
26   L. BARR, M. M. DEWEY and W. BERGER, J. gen. Physiol. *48*, 797 (1965).
27   CH. ROUILLER and W. G. FORSSMANN, Verh. schweiz. naturf. Ges. *145*, 171 (1965).
28   J. GIRARDIER, *Electrophysiology of the Heart* (Ed. B. TACCARDI and G. MARCHETTI; Pergamon Press, New York 1965), p. 53.
29   T. TOMITA, J. Physiol. *183*, 450 (1966).
30   N. JELLINEK, N. SPERELAKIS, L. M. NAPOLITANO and T. COOPER, J. Neurochem. *15*, 959 (1968).

31   I. HARARY and B. FARLEY, Expl. Cell Res. *29*, 466 (1963).
32   M. J. HOGUE, Anat. Rec. *99*, 157 (1947).
33   B. P. LANE and J. A. G. RHODIN, J. Ultrastruct. Res. *10*, 470 (1964).
34   W. R. LOEWENSTEIN and Y. KANNO, J. Cell Biol. *22*, 565 (1964).
35   N. SPERELAKIS, M. F. SCHNEIDER and E. J. HARRIS, J. gen. Physiol. *50*, 1565 (1967).
36   A. J. PAPPANO and N. SPERELAKIS, Expl. Cell Res. *54*, 58 (1969).
37   J. P. REVEL and M. J. KARNOVSKY, J. Cell Biol. *33*, C7 (1967).
38   N. A. STALEY and E. S. BENSON, J. Cell. Biol. *38*, 99 (1968).
39   A. R. MARTIN and G. PILAR, J. Physiol. *168*, 443 (1963).
40   N. SPERELAKIS and D. LEHMKUHL, J. gen. Physiol. *47*, 895 (1964).
41   N. SPERELAKIS and H. K. SHUMAKER, J. Electrocardiol. *1*, 31 (1968).
42   T. HOSHIKO and N. SPERELAKIS, Am. J. Physiol. *201*, 873 (1961).
43   N. SPERELAKIS, T. HOSHIKO and R. M. BERNE, Am. J. Physiol. *198*, 531 (1960).
44   M. TARR and N. SPERELAKIS, Am. J. Physiol. *207*, 691 (1964).
45   M. TARR and N. SPERELAKIS, Am. J. Physiol. *212*, 1503 (1967).
46   D. LEHMKUHL and N. SPERELAKIS, J. cell. comp. Physiol. *66*, 119 (1965).
47   N. SPERELAKIS and T. HOSHIKO, Circulation Res. *9*, 1280 (1961).

Electrical Transmission between the Cells of Vertebrate Cardiac Muscle

by L. BARR
Department of Physiology and Biophysics, Woman's Medical College of Pennsylvania, Philadelphia, Pennsylvania 19129

The heart is a favorite organ for demonstrating the all-or-none law because, among other reasons, an electrical response spreads throughout it with little change in form. Early electrophysiological studies of the spread of activity in the heart led to the belief that it was a syncytium, and classic electrocardiography used this interpretation extensively[1, 2]*. Light-microscopists agreed with this hypothesis because the intercalated disc appeared to them to be quite different from the sarcolemma.

With the advent of high resolution electronmicroscopy, the non-syncytial nature of cardiac muscle was demonstrated. The intercalated discs were shown to be regions where neighboring cell membranes are in close apposition. While a detailed analysis of this intercellular area has not as yet been possible, such information is essential for understanding the mechanism of propagation of action potentials. For example, the presence of low resistance saline gaps between cell membranes at the intercalated discs would provide shunt pathways for current; thus current spread along heart fibers would be attenuated. This would make electrically mediated propagation more difficult.

As a result of the earliest electronmicroscopic studies a controversy has arisen regarding the mechanism of propagation and a chemical as well as electrical transmission process has been suggested between cardiac muscle cells. However, as described in Dr. DEWEY's chapter in this volume, vertebrate cardiac muscle cells apparently are not completely separated along the intercalated discs, and in certain regions the apposing membranes fuse to form nexuses.

There are three characteristics of the nexus which would lead one to consider that, theoretically, the resistance of the nexal membrane is lower than its equivalent plasma cell membrane. They are: (1) nexal membranes separate solutions of similar composition so that there can be no diffusion potential across the nexus; (2) nexal membranes are bathed on both sides by high potassium solutions; and (3) they are exposed on both sides to low calcium solutions. These conditions in other circumstances lead to low membrane resistance and there is considerable evidence for significant electrotonic spread of current along cardiac muscle bundles, in frog atrial fibers[3, 4], rat atrium[5], and sheep Purkinje fibers[6].

---

* Numbers refer to References, p. 109.

Two questions which Drs. BERGER, DEWEY and I considered answerable were: (1) Are the electrotonic currents generated by frog and guinea-pig atrial fibers sufficient to effect transmission between cells? and (2) What are the intercellular connections along the internal pathway for current?

Electrical Transmission

A means of obtaining evidence of electrical transmission between cells was provided by a modified sucrose gap technique[7,8]. This involved the electrical isolation of a region in the middle of a bundle of cardiac fibers and the control of the extracellular current past that region. The experimental arrangement is shown schematically in Fig. 1. A small muscle bundle with pacemaker tissue

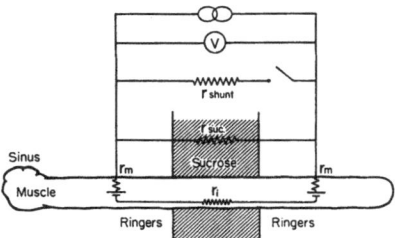

Fig. 1.
A schematic representation of the experimental arrangement.

intact was excised from frog or guinea-pig atria and threaded through holes in thin rubber membranes. Isotonic sucrose solution flowing between the membranes in the center compartment replaced the ions in the extracellular space of the bundle. This raises the extracellular resistance and blocks local circuit currents and propagation.

The experimental question posed was whether or not a resistor, connected between the end compartments to provide an 'artificial extracellular space', would allow enough current flow for excitation to be transmitted past the inactive region in the gap. Demonstration of tissue-generated current sufficient for propagation would be strong evidence for the hypothesis of electrical transmission between cells.

Fig. 2 shows the effect of the shunting resistance on the amplitude of the action potentials measured across the gap. When the shunting resistance was infinite, no propagation occurred across the gap and a large monophasic action potential (curve 1) was recorded. As the shunting resistance was decreased, the recorded potential also decreased, but remained monophasic down to a critical value (curves 2 and 3). Beyond that value the waveforms became biphasic in an all-or-none way indicating that propagation across the shunted gap had occurred. These results were reversible in the sense that, if the resistance were increased again, monophasic action potentials would be recorded indicating that propagation past the gap was not occurring.

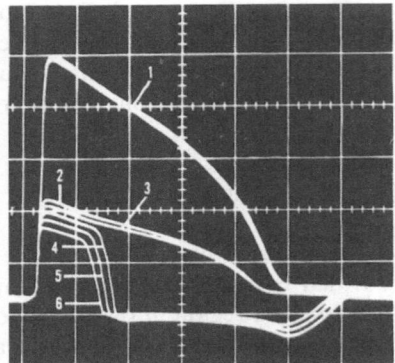

Fig. 2.
Records of frog atrial action potentials
recorded with the arrangement shown in
Fig. 1. Amplitude and waveform are a
function of shunt resistance. Propagation of action potential across the sucrose gap is indicated by the change from
monophasic to diphasic action potential. Curves 1–6 are with r_{sh} infinity,
10^5, 9×10^4, 8×10^4, 7×10^4, and
6×10^4 ohms, respectively. Divisions
mark 20 mV and 100 msec. Gap width
400 μ.

Gaps longer than a millimeter or so apparently will not allow action potentials to pass even when they are shunted with straight wire connections between
electrodes (total resistance about 2,000 ohms). This is probably because the
excitatory current is attenuated strongly by the internal resistance of the cells in
the sucrose gap.

These experiments have been repeated on pectinate and trabecular bundles
from guinea-pig atria, with similar results. This is significant in terms of structure,
since guinea-pig atrial muscle has true intercalated discs while frog atrial muscle
does not.

Hypertonic Solutions

As may be seen from the analog circuit in Fig. 1, the observed amplitude of
an action potential across the sucrose gap approaches true amplitude as the
resistance of the extracellular space in the gap becomes much larger than the
internal resistance of the bundle running through the gap. Therefore, if the
internal resistance of the bundle running through the gap were increased, the
observed action potential would become smaller. In addition, if current pulses
were passed from one physiological saline pool to the other through the gap,
current would flow in parallel through the bundle and the sucrose of the extracellular fluid. Therefore, if the internal resistance of the tissue in the gap increases, the voltage observed due to current flow will increase with the internal
resistance of the bundle. Furthermore, if a technique were found to break the
nexuses with a resultant increase in the internal resistance of the bundle, then
this would be evidence that the nexuses are relatively low resistance connections
between cell interiors. Although it may be possible to rupture the nexuses in a
number of ways, e.g. by removal of calcium, many of these may affect the plasma
membrane deleteriously and may not be reversible. It was found that sucrose
solutions of three times the normal osmolarity caused an increase in the internal
resistance and prevented action potentials from jumping across the gap when

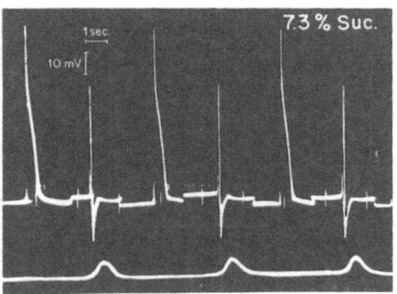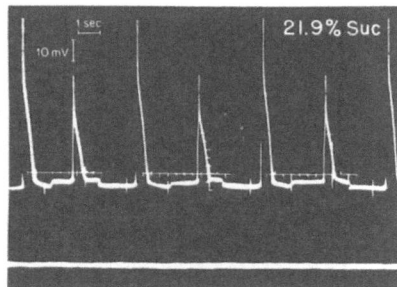

Fig. 3.
Isosmotic sucrose solution was in the gap when left hand record was obtained. In this record of frog atrial activity the upper trace shows alternating monophasic and diphasic action potentials according to whether the shunt resistance was absent or present. Lower trace shows contraction of the postgap muscle when the action potential is propagated past the gap.

In the right hand record, the upper trace shows only monophasic action potentials when hypertonic sucrose was in the gap. The lower trace shows that when the propagation of the action potential was blocked, no mechanical activity was seen beyond the gap. Gap width 400 μ.

the straight wire shunt was connected. Fig. 3 shows the block of propagation in an experiment in which a solution with three times isotonic sucrose concentration replaced the isotonic sucrose flowing through the center region. In the control experiment (left hand record), the shunting resistor was alternately switched in for every second action potential, yielding alternate monophasic and biphasic action potentials. The lower trace records tension developed by the muscle in the post-gap region. The tension developed indicates that the action potential spread across the entire bundle. When a hyperosmotic sucrose solution was substituted for the isosmotic one, propagation no longer occurred when the shunt resistor was switched in. This propagation block was completely reversed upon reintroduction of the isotonic sucrose. Since the hypertonicity was restricted to the gap, only the connections between cells were altered. Membrane areas which generate the action potentials were unaffected.

The increase of internal resistance was calculated from the monophasic action potentials in Fig. 4, recorded while repetitive current pulses were applied across the sucrose gap. The pulses give the action potential records a doubled appearance. On the basis of the difference in the amplitude of the action potentials and of the size of the induced voltage pulses, the internal resistance in this experiment was calculated to change by a factor of about 2.7. The increase in internal resistance was reversible and it returned to almost normal when isosmotic sucrose again flowed through the center region. The resistance change was calculated according to the equation

$$R_i(\text{hyper})/R_i(\text{iso}) = AP_{obs}(\text{iso}) \times V_P(\text{hyper})/AP_{obs}(\text{hyper}) \times V_P(\text{iso})$$

where R_i is internal resistance, AP_{obs} is the amplitude of the observed action potential and V_P is the voltage deviation due to a constant current pulse. It

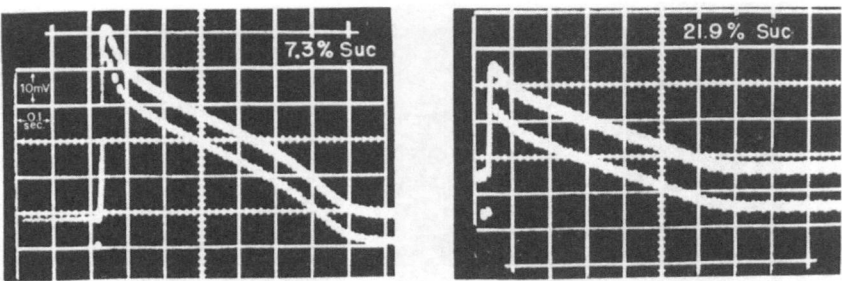

Fig. 4.
Monophasic action potentials from frog atrium recorded while 8 msec rectangular
pulses of 5 mA were passed through the preparation. Left hand record was obtained
with an isosmotic sucrose solution in the gap while the right hand record was obtained
with a hypertonic sucrose solution in the gap. The decreased amplitude of the action
potential and the increased pulse height show the increase of longitudinal resistance.
Gap width 600 μ.

would appear that in these experiments the effective resistance of the membrane
across which the potentials were recorded is small relative to the internal
resistance. This conclusion derives from two observations: the size of the
voltage pulse did not vary throughout the course of the action potential nor
did the calculated resistance of the bundle change when KCl was applied to one
of the Ringer pools.

If a hyposmotic sucrose solution replaced the normal isosmotic sucrose in
the gap, propagation past the gap was facilitated. This may be seen in Fig. 5
where, in the left hand traces, the shunt resistance was set a little too high to

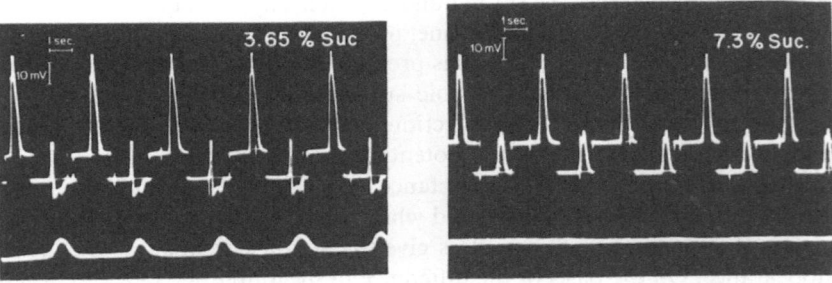

Fig. 5.
Isosmotic sucrose solution was in the gap when the left hand record of frog atrial
activity was obtained. The shunt resistance was set a little larger than that allowing
propagation. The jump of the upper trace when the shunt was switched in was due
to imbalance across the sucrose gap. The upper trace shows only monophasic action
potentials and the lower trace shows no postgap contraction.
In the right hand record, about 10 sec later, diphasic action potentials on the upper
trace indicate that propagation occurred when an isosmotic sucrose solution was
replaced by a hyposmotic one. Lower trace shows mechanical activity paralleled the
action potentials. Gap width 400 μ.

allow propagation across an isosmotic sucrose gap, but in the right hand traces the same resistance became adequate for propagation when hyposmotic sucrose solution had flowed for less than 15 sec. This experiment, like the earlier one, was readily reversible.

When action potentials propagate across the sucrose gap in the presence of a shunt resistor, they do so under quite adverse conditions. Therefore, the blockade due to hypertonic sucrose in this experiment does not allow one to conclude that hypertonic solutions would also block propagation along a bundle in physiological saline. An experiment to examine this directly was performed (Fig. 6), and activity was blocked very quickly in a solution made hypertonic by adding sucrose to a normal Ringer's solution. When hypotonic Ringer's solutions were used (Fig. 7), propagation velocity in frog atrial bundles was slightly increased, indicating that the ability to transmit activity from one cell to another may be enhanced by hypotonicity.

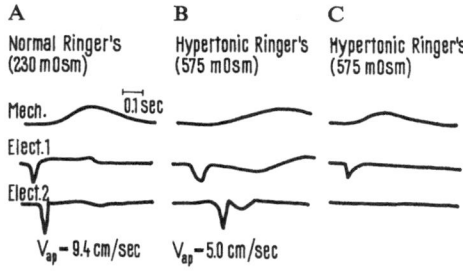

Fig. 6.
Electrical and mechanical record of activity in a small bundle of frog atrial muscle under oil. Upper trace. Tension developed by bundle. Two lower traces. Voltage between pairs of small platinum wires about 0.5 mm apart. A. Control. B. Propagation slowed 2 min after substituting hypertonic (575 mOsm) Ringer's solution for normal. C. Propagation blocked 5 min after exchange. Note: In this experiment pacemaker activity is still recordable. The propagation block is reversed by returning the bundle to normal Ringer's solution.

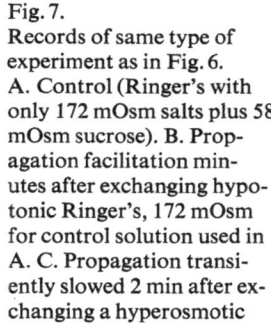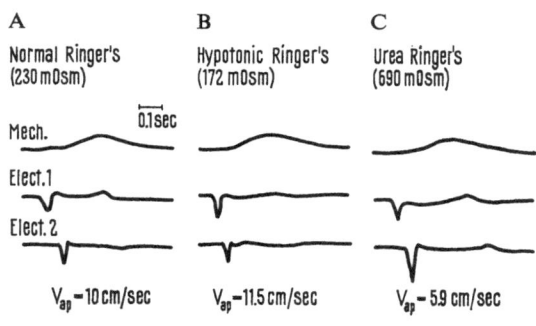

Fig. 7.
Records of same type of experiment as in Fig. 6. A. Control (Ringer's with only 172 mOsm salts plus 58 mOsm sucrose). B. Propagation facilitation minutes after exchanging hypotonic Ringer's, 172 mOsm for control solution used in A. C. Propagation transiently slowed 2 min after exchanging a hyperosmotic but isotonic solution (460 mOsm urea in normal 230 mOsm Ringer's solution). After 5 min in this urea-Ringer's solution the propagation velocity returned to normal.

The question as to whether the activity of the water (the osmolarity) or the volume of the cells (the tonicity) was responsible for increasing the longitudinal resistance and blocking propagation was answered by making solutions very hyperosmotic by adding relatively permeant urea to the Ringer's solution. Fig. 7C shows, after 2 min, a striking decrease in the propagation velocity. However, the propagation velocity returned to normal in less than 5 min. Therefore, it is the tonicity which causes blockade. Perhaps as the cells shrink the nexuses are mechanically ruptured, and this in turn causes a break in the conducting pathway between the cell interiors.

From these experiments we have concluded that cardiac muscle is an electrical syncytium, and therefore proceed to describe the electrotonic spread of current in a syncytium in three dimensions from a point or nearly point source[9].

Consider a coaxial double microelectrode penetrating a cell in a tissue such that the inner microelectrode resides in the cell interior while the outer coaxial electrode is in the extracellular fluid immediately outside the membrane. In the equations below, V_o and V_i are the potential differences between a distant reference point in the tissue bath and points in the extracellular fluid and the intracellular fluid respectively, due to current flow; V_m is the transmembrane potential variation due to current flow; R_o and R_i are the specific resistances of the extracellular and intracellular pathways respectively; r_m is the transmembrane resistance in an element of volume; a_o and a_i are fractions of a spherical surface falling extracellularly or intracellularly respectively; I_o and I_i are the currents in the extracellular and intracellular pathways, respectively; I_m is the transmembrane current and the prime values of all the variables are derivatives with respect to radial distance R away from the site of current injection. Then

$$V_m = V_i - V_o .$$

From Ohm's Law it may be deduced for the intracellular pathway that

$$I_i = (4 \pi a_i / R_i) r^2 V_i' .$$

Similarly, for the extracellular pathway

$$I_o = (4 \pi a_o / R_o) r^2 V_o'$$

and also for the transmembrane path, any spherical element

$$I_m = (4 \pi / r_m) r^2 V_m .$$

From conservation of current we obtain

$$I_m = I_i' = (4 \pi a_i / R_i) d(r^2 V_i')/dr$$
$$I_m = - I_o' = - (4 \pi a_o / R_o) d(r^2 V_o')/dr.$$

Letting

$$\lambda^2 = r_m/(R_i/a_i + R_o/a_o)$$

and combining and rearranging we have

$$V_m'' + 2 V_m'/r - V_m/\lambda^2 = 0 .$$

The solution is

$$V_m = A e^{-r/\lambda}/r .$$

This treatment is highly simplified. Nonetheless, some insights may be gained from this case which would be obscured by a more accurate but more geometrically complicated treatment. The most obvious is that at short distances electrotonic current may fall off more rapidly than for an exponential relationship such that close-up measurements showing very large decrements in electrotonic current need not indicate electrical isolation of cells.

Summary

The fact that action potentials can jump across portions of frog atrial bundles in sucrose demonstrates that transmission between atrial cells is electrical, i.e. the interiors of these cells are connected by pathways of resistance low enough to allow electrical transmission of action potentials with a large safety factor. This is demonstrated by experiments wherein action potentials jump over as much as 600 μ. In these cases, the excitatory current density is much smaller than in vivo. The excitatory current density is low not only because of the segment in the sucrose gap, but also because of an increased internal resistance of segments exposed to sucrose due to appreciable loss of intracellular ions to the sucrose solution. The segment in the sucrose gap shows no electrical or mechanical response. In fact, if soaked long enough it loses its ability to recover when returned to Ringer's solution. The all-or-none nature of the propagation past the gap as the shunt resistance is decreased makes it difficult to imagine an intermediate chemical transmission step. On the contrary, the segment of muscle in the sucrose gap behaves as a core conductor connecting segments in the Ringer's solution pools.

That the longitudinal resistance increases markedly during the same treatment that causes the nexuses to be broken, argues strongly that the nexuses are the effective low-resistance connections between the cells. This conclusion is strengthened by the observation that both structural and functional alteration in hypertonic solutions is reversed upon return to normal solution. Moreover, when tissues are exposed to Ringer's solution to which enough urea has been added to increase the osmolarity more than three times, there is no permanent effect on the structure of the nexuses and propagation of action potentials continues indefinitely.

In short, the hypothesis that the nexus is a specialized structure that allows current flow between cell interiors is corroborated by the structural changes in the nexus and changes of electrical coupling between cells following soaking in solutions of abnormal tonicity.

References

[1] F. N. WILSON, A. G. MACLEOD and P. S. BARKER, *The Distribution of the Currents of Action and Injury Displayed by Heart Muscle and Other Excitable Tissues* (Univ. Michigan Press, Ann Arbor 1933).

[2] H. SCHAEFER and H. G. HAAS, *Handbook of Physiology*, Section 2, on Circulation (Ed. W. F. HAMILTON; American Physiological Society, Washington, D.C. 1962), p. 323.

[3] W. TRAUTWEIN, S. W. KUFFLER and C. EDWARDS, J. gen. Physiol. *40*, 135 (1956).

[4] J. W. WOODBURY and A. M. GORDON, J. cell. comp. Physiol. *66*, 35 (1965).

⁵   J. W. WOODBURY and W. E. CRILL, *Nervous Inhibition* (Ed. E. FLOREY; Pergamon Press, London 1961), p. 124.
⁶   S. WEIDMANN, J. Physiol. *118*, 348 (1952).
⁷   R. STÄMPFLI, Experientia *10*, 508 (1954).
⁸   L. BARR, M. M. DEWEY and W. BERGER, J. gen. Physiol. *48*, 797 (1965).
⁹   L. BARR and M. M. DEWEY, *Handbook of Physiology*, Section 6, on Alimentary Canal (Ed. C. F. CODE; American Physiological Society, Washington, D.C. 1968), p. 1733.

This work was supported by Public Health Research Grant HE-10084.

Spread of Excitation and Injected Current in the Tunicate Myocardium

by Mahlon E. Kriebel

Department of Anatomy, Albert Einstein College of Medicine, Bronx, New York 10461

Present address: Department of Physiology, State University of New York, Upstate Medical Center, Syracuse, New York.

Introduction

The structure of tunicate hearts is apparently simple when compared to higher chordate hearts because these simple tubes have no structural valves and are only one cell layer thick. Yet action potentials recorded from single cells in the tunicate heart are very similar to those of mammalian heart cells and there appears to be only one electrophysiological cell type[1-3]*. Pacemaker regions at the ends of the tunicate heart have not yet been investigated with microelectrodes but since most cells can develop pacemaker properties when isolated from the primary pacemakers there may be only a quantitative difference between all cells[4].

Since there is no specialized conduction system, impulse spread is from cell to cell[2,3]. Transmission is not chemical since conduction time per cell during impulse spread in the heart axis is only 0.3 msec and since the maximal rate of rise of the action potential is only 10 V/sec, there would not be enough time for transmitter release, diffusion and postsynaptic action[2]. However, the cells are in electrical continuity through tight junctions which have a specific membrane resistivity of about 1,000 times less (0.2 Ω cm^2) than that of the cell membrane[5]. There are no desmosomes or intercalated discs. These observations strongly suggest that impulse spread is by local current flow through tight junctions as in vertebrate hearts[6-11].

A useful property of this heart is that excitation does not cross the raphe[2] so that the heart can be opened to form a relatively large sheet of tissue (6 mm $\times$ 44 mm) in respect to its thickness (10 μ). Thus an impulse front can be delineated by a simple curve. The gross shape of the heart tube does not affect impulse spread or alter its electrical characteristics[2] so observations on opened hearts would closely represent those in situ.

Perhaps the most unique property of the tunicate cardiovascular system in comparison to vertebrates is that of the periodic reversal of heart beat[3,12-15]. Many tubular invertebrate hearts reverse the direction of contraction[16,17], and retrograde conduction or different pathways of conduction can occur in the mammalian heart[18-21]. Thus, the fact that hearts may reverse in the direction

---

* Numbers refer to References, p. 133.

of conduction does not reflect any unusual cellular properties. The uniqueness of the tunicate heart lies in its long tubular structure which by its geometry permits bidirectional contraction and in the valveless circulatory system which does not appear to show a directional preference for blood flow.

In the tunicate heart conduction velocity in both directions of conduction is the same, which demonstrates that the heart is not electrically polarized in its long axis; however, the front moves perpendicular to the cell axis in both directions, which strongly suggests a greater conduction velocity in the cell axis[2]. In mammalian hearts conduction velocity is greater in the cell axis than against the cell axis[22,23]. However, the cells of vertebrate hearts are usually associated in a complex array of branching cords, and cells in adjacent cords may be electrically isolated[24-29]. Thus, differences in conduction velocity in vertebrate hearts may reflect the gross morphology of the tissues and not a difference at the cellular level. Moreover, it is usually difficult to determine the direction of propagation in thick walled hearts with specialized conductile tissue so that only the apparent conduction velocities can usually be determined[21,30,31]. Another difficulty with thick walled hearts in determining the effect of cell shape on impulse spread is that the area of tight junctions has not been described in detail[32-34]. Since the tunicate heart is one cell layer thick and since all cells are oriented the same way, electron micrographs of cross sections were sufficient to demonstrate that tight junctions account for 15 % of the border between cells[5].

The cell axis and the relative positions of the undifferentiated line of cells and raphe were determined for the entire length of tunicate hearts. These studies as reported here have permitted a complete description of the temporal position of an impulse moving along the heart and moving from a point stimulus.

Analogous to the study of impulse spread from a point stimulus is a study of the passive spread of injected current from a point source. A technique has been developed for holding the tissue in order to visually observe cell impalement with intracellular microelectrodes. The voltage attenuation perpendicular to the cell axis has been determined in cells up to 250 μ from a current electrode[35]. The tunicate heart is an ideal preparation for studies concerned with the spread of injected current in a thin sheet of tissue, since a simple cable equation was found to describe voltage decrement at distances somewhat removed from a point current source. As in chick-heart tissue cells[36] and rat atrium[37] and in other thin walled mammalian hearts[38], the tunicate heart functions electrically like a single thin cell.[39]

Methods

A. Determination of the Cell Axis

The test of large *Ciona* (6–10 cm long, from California) was first removed and then the pericardium was exposed by cutting through the body wall. Hearts ranged in size from 24 to 44 mm in length and from 1.8 to 3.0 mm in diameter. Most experiments were performed on 30–36 mm long hearts.

Large, adult *Chelyosoma productum* (from Washington) weighing from 18 to 40 g were also used. The bottom part of the test was removed leaving the siphon surface attached to the body wall (Fig. 1 A). Since the pericardium lies close to the body wall, a single incision through the body wall into the pericardial cavity exposed the heart. Hearts ranged in size from 30 to 45 mm in length and from 2.0 to 3.5 mm in diameter.

Hearts of both species were removed from the pericardium by cutting the heart wall along the attachment (raphe) which connects the heart to the pericardium. The cut edges were separated with a jet of sea water delivered from an eye dropper.

Hearts were pulled onto a microscope slide by partially submerging it in the sea water bath. The cell axis was determined from photomicrographs obtained with a $40 \times$ phase contrast reflecting objective (N.A. 0.57).

B.   Determination of the Temporal Position of Impulse Fronts

Isolated hearts were transferred with an eye dropper to a 5 ml bath which contained a plexiglass plate with suction electrode holes in it. A jet of sea water was used to wash the tissue free of adhering blood cells and to flatten the heart. This procedure stretched the tissue. Therefore, in order to establish a reference condition of the tension, hearts were allowed to beat 1 or 2 min before their edges were secured with hooks to the top surface of the plexiglass electrode plate. In this condition, the heart wall was slack and there was little or no movement during contraction. It was necessary to secure the edges of the heart to the top surface of the plexiglass plate to prevent too much tissue from being 'sucked' into the electrode openings. The plexiglass plate contained several holes of 30–40 μ diameter either spaced linearly at 1.00 mm intervals or placed around a central hole on a 2 mm radius. Each hole was connected to a hydraulic system made from a micrometer and a tubing adapter. A wire bolted to the micrometer spindle was sealed into a piece of polyethylene tubing with the tubing adapter so that the wire served as a piston. The hydraulic system and much of the tubing connecting it to the electrode were filled with oil to prevent grounding when operating the micrometer and to reduce antennal pick-up. A stainless steel needle was inserted into the tubing to serve as the electrode terminal. It was connected to one input terminal of a Tektronix 122 preamplifier. The second terminal of the preamplifier and the bath were grounded. Four signals could be displayed on a Tektronix 565 oscilloscope with 3A72 dualtrace amplifiers and recorded on film with a Grass kymograph camera.

Tektronix pulse and wave form generators were used to stimulate the heart tissue with 8 msec pulses.

All wave front analyses were carried out at 10 °C. Experiments concerned with the effect of acetylcholine or adrenaline were carried out at room temperature (21 °C) and at 10 °C. Solutions of the drugs in sea water at a temperature equal to that of the bath temperature were added with a pipette. Mixing was accomplished within a few seconds with the aid of an eye dropper. At least 100 ml of sea water, at bath temperature, were perfused through the bath to wash out a drug. The drugs used were: L-epinephrine bitartrate (Nutritional Biochemicals Corporation) and acetylcholine chloride (Merck).

C.   Determination of Electrical Coupling between Cells

Hearts were dissected as described in section A and then transferred to a 5 ml bath which was fixed to the stage of a Vicker Patholux microscope. Hearts were positioned with the lumen surface down and were usually allowed to adjust to normal dimensions before their edges were secured with small hooks. A few hearts were stretched in all dimensions by 20% to completely immobilize them. Muscle cells were directly observed with a phase contrast $40 \times$ reflecting objective during penetration with microelectrodes.

Two microelectrodes were held at an angle of about 25° to the surface of the heart by sliding plate micromanipulators. Penetration was accomplished by advancing the tip of the microelectrode to the apical surface of a cell close to the cell border and then gently tapping the manipulator. 30–60 MΩ glass microelectrodes were used (3 M KCl).

Resting potentials were measured against ground with D.C. negative capacitance amplifiers (designed by Dr. J. W. Woodbury; input impedance 10^{10} Ω, grid current less than 2.5×10^{-13} A, response time of 40 μsec with 44 MΩ input resistance).

One intracellular electrode was connected to a stimulator with a switch so that current could be passed into the cell following penetration. After passing current the resting potential was routinely measured. An American Electronic Laboratories Stimulator (Model 104A) and an isolation unit were used to deliver 100 msec long square pulses of current. The amount of polarizing current was determined from the voltage developed across a 1 MΩ resistor to ground. To approximate a constant current source, 100 MΩ were placed in series with the electrode.

The second microelectrode was used to measure potential changes in the same cell and in cells at various distances from the current electrode. Interelectrode distances were estimated to within 2 μ with an eyepiece micrometer.

Results and Discussion

A. Orientation of Cell Axis

Ciona and Chelyosoma hearts are one cell layer thick and the effective ionic barrier across the heart wall occurs at the apical membrane where individual cells are electrically coupled[5]. The apical membrane of each cell bulges around the nucleus and these bulges are located in rows oriented 0–30° to the long axis of the heart. These rows of apical bulges give the heart a corrugated appearance under a dissecting microscope and can be easily mistaken for single muscle cells. The lumen surface is smooth. The cell axis in relation to the long axis of the heart gradually changes from 90° at the ends to 65° ($\pm$ 8°) in the middle region of each arm. Near the bend of the U-shaped hearts the long axis again changes to 90° (Fig. 1 A and 1 B).

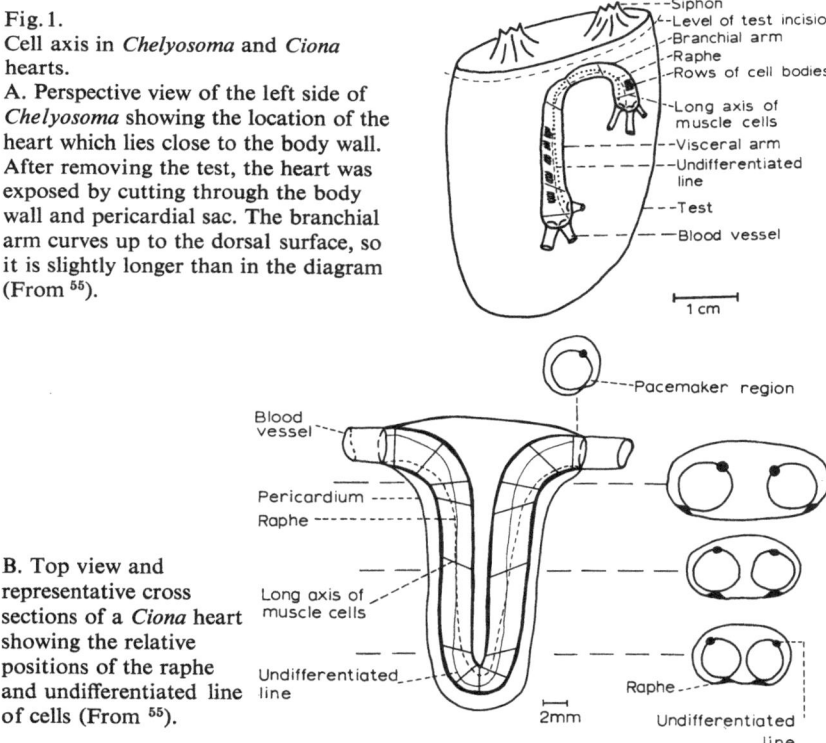

Fig. 1.
Cell axis in *Chelyosoma* and *Ciona* hearts.
A. Perspective view of the left side of *Chelyosoma* showing the location of the heart which lies close to the body wall. After removing the test, the heart was exposed by cutting through the body wall and pericardial sac. The branchial arm curves up to the dorsal surface, so it is slightly longer than in the diagram (From [55]).

B. Top view and representative cross sections of a *Ciona* heart showing the relative positions of the raphe and undifferentiated line of cells (From [55]).

The muscle fibers are parallel on each side of the undifferentiated line which was composed of cuboidal cells from 1 to 10 cells wide[40,41]. In many hearts, large groups of cells, which sometimes interrupted the undifferentiated line, were observed in various stages of differentiation into muscle cells.

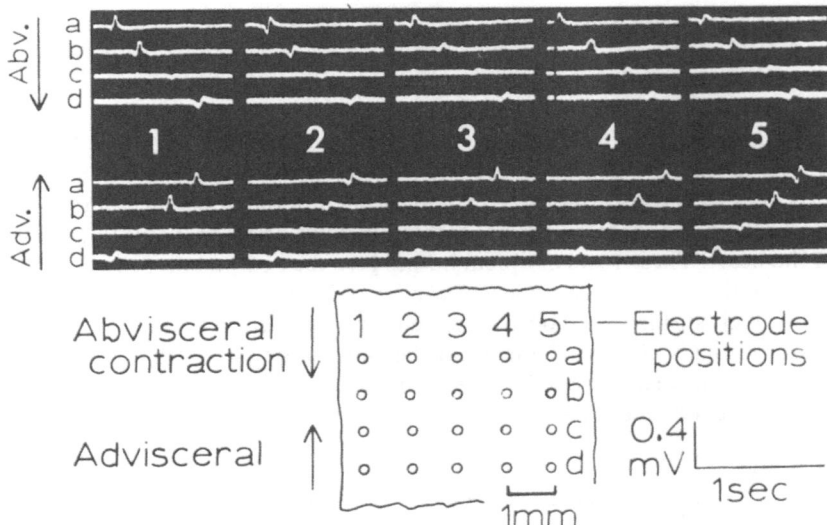

Fig. 2.
Extracellular action potentials of impulses traveling in opposite directions of conduction in the middle region of a *Ciona* heart arm. Each record is composed of 4 traces. The top row of records shows the signals from abvisceral impulses (*Abv.*) and the lower row shows the signals from advisceral impulses (*Adv.*). The 5 recording positions are numbered between the rows of records. The diagram below the records represents the middle region of an opened heart arm. The numbered rows of circles indicate the positions of the line of electrodes. The electrode sequence is lettered and corresponds to that of the traces. The arrows indicate the direction of advisceral and abvisceral contractions (From [55]).

Fig. 3.   ▷
Extracellular action potentials showing impulse spread. The electrode positions (1 mm apart) on the opened hearts are indicated by circles in the diagram below each record. S = stimulating electrode. P = pacemaker region. Arrows = direction of conduction. The curves in the diagrams indicate possible wave fronts.
A. Action potentials recorded during a stimulated impulse have been superimposed on those of a natural contraction. *Chelyosoma*. Time scale 1 sec.
B. Action potentials recorded during a stimulated impulse have been positioned below those of a natural impulse. *Chelyosoma*.
C. Two records are superimposed so that the action potentials occurring on the fourth trace coincide. End of *Ciona* heart. Time scale 1 sec.
D. One record of action potentials from a stimulated wave of contraction. The undifferentiated line is indicated by the chain in the diagram. *Ciona*. Time scale 1 sec.
E. A record of a wave of excitation has been superimposed on a second record so that its traces are slightly lower and so that in both records the signals in trace 1 occur at the same time. Time scale 0.5 sec.
F. Superposition of two records resulting from two successive impulses in opposite directions of conduction. In order to facilitate the interpretation, one of the records has been reversed and displaced downward (read right to left). The signals of the fourth traces were made to coincide. Note that the discrepancy in the time intervals is only between electrodes 1 and 2. *Ciona*. Time scale 1 sec.

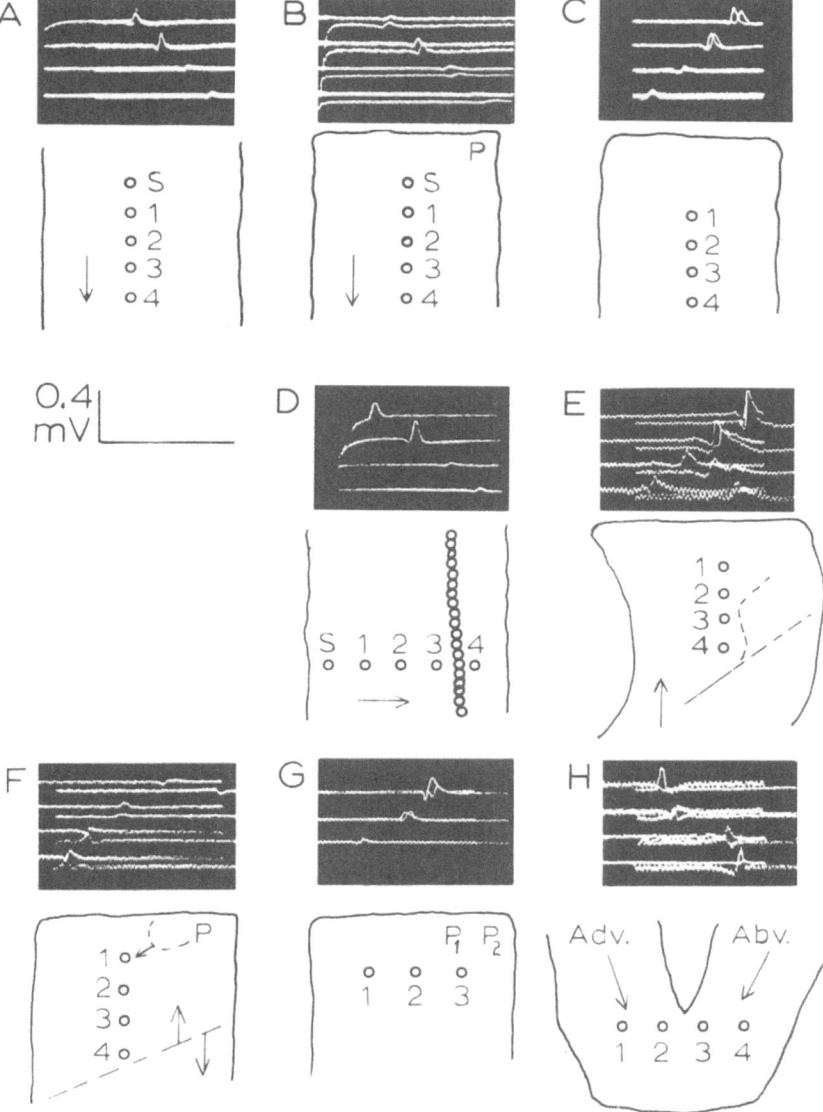

G. Two records of waves of excitation in the same direction have been superimposed so that the signals recorded in the third trace appear to occur at the same time. *Ciona*. Time scale 0.5 sec.

H. A record of an abvisceral contraction has been reversed, superimposed and displaced downward (read right to left) on a record of an advisceral contraction. The conduction times are equal in both directions and are greatest between electrodes 2 and 3. The amplification was slightly different in the two directions of conduction. At bend of a U-shaped *Ciona* heart. Time scale 1 sec (From [55]).

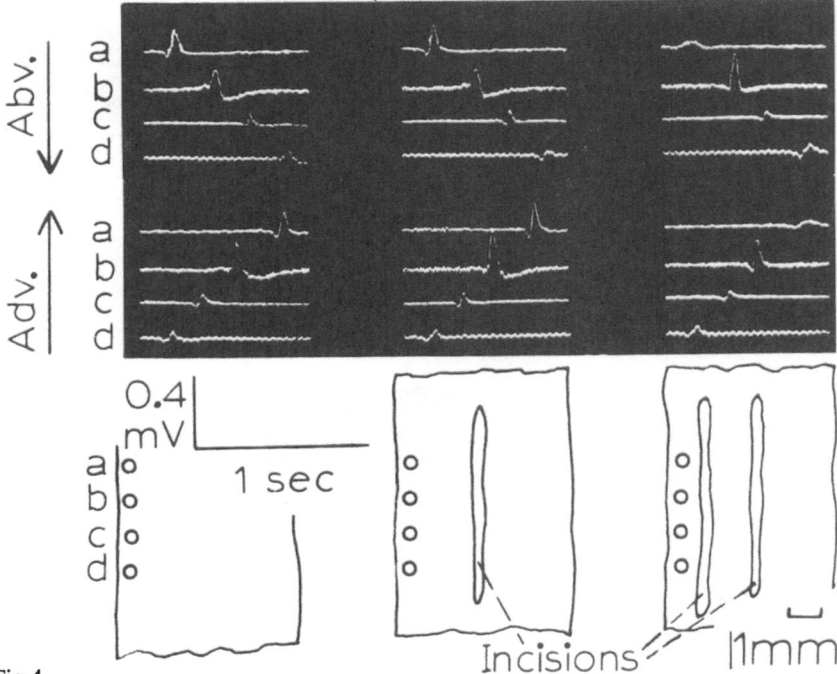

Fig. 4.
Effect of decreasing the heart width on the conduction velocity. The diagram represents the middle region of a *Ciona* heart arm. Incisions were made with a razor blade
and crushed areas were left attached to keep the strips at constant tension. There was
no effect of the first incision (middle diagram). Conduction times were longer following the second incision (right diagram) which left a strip of tissue about 250 μ wide.
However, conduction velocities returned to the control values after a few minutes.
Since the cells are from 50 to 75 μ long and since the cell axis was about 65° to the
heart axis, the remaining strip of undamaged tissue after the second incision was
only about 150 μ (From [55]).

Fig. 5.   ▷
Successive positions of a wave front moving along one arm of a *Ciona* heart.
The conduction velocities for successive millimeter lengths of heart were determined
in the midline of the diagram. The signals recorded at the 8 successive heart place-
ments are shown in the sequence of records (each record has 4 traces) to the left of
the diagram. The conduction velocity at the end of the heart was about 1 mm/sec
and it gradually increased to 6.5 mm/sec at an electrode position 7 mm from the end
and remained at 6.5 mm/sec to the bend region.
The arrival times of the front at various positions perpendicular to the heart axis
(electrode positions *a, b, c, d*) are shown in the sequence of records to the right of
the diagram for 10 heart positions.
The diagram shows the position of a wave front moving from the end pacemaker at
successive 0.2 sec intervals. The conduction velocities determined from the records at
the left were used to determine the central position of each successive wave front line.
The inclinations of the front lines were interpolated from the conduction velocity
and arrival times for the electrode positions to the right of the diagram (dots in wave
front lines were calculated) (From [55]).

B. Determination of the Temporal Position of a Wave Front

Conduction velocity was the same before and after cutting along the raphe to make the opened heart preparation. The experiments reported below were performed on opened hearts which were allowed to adjust to normal dimensions (i.e. they were not stretched).

When hearts were repeatedly lifted from the plexiglass plate containing the electrodes and replaced in the same position the conduction times remained unaltered. Thus, the slack condition in the heart wall is reproducible.

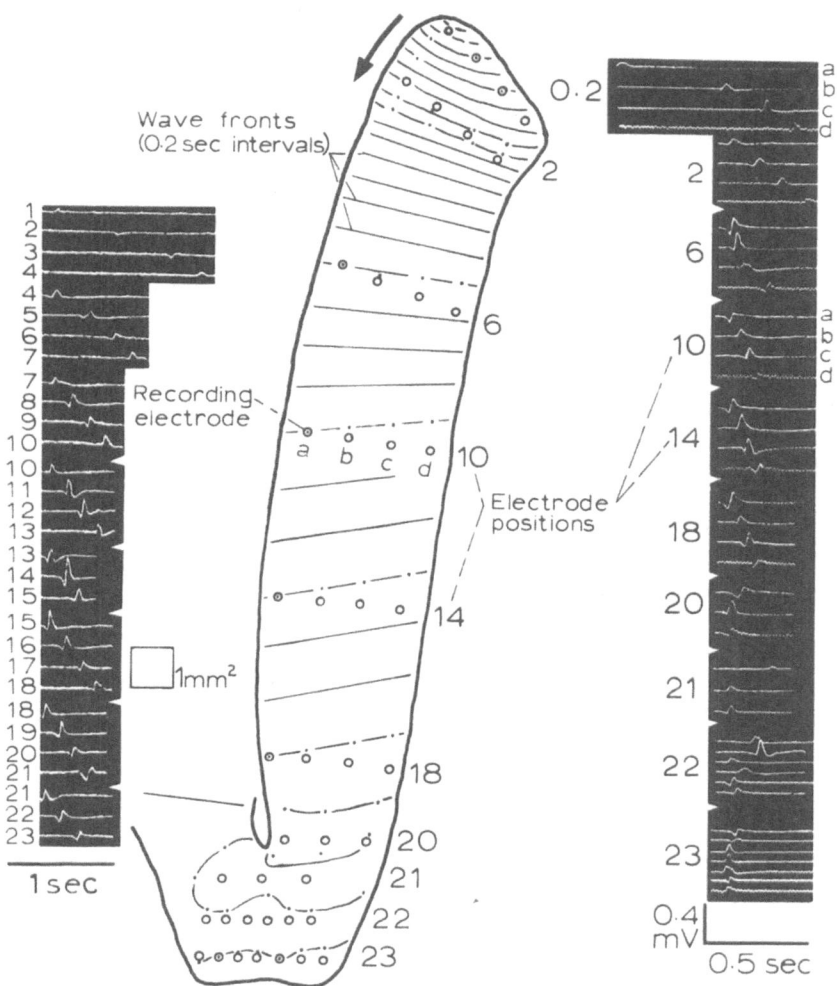

In the middle of the arm of an opened heart, conduction velocity in the long axis of the heart was constant and equal in both directions throughout the width of the tissue (Fig. 2). This was true for both natural and stimulated waves of excitation (Fig. 3A). Since conduction velocity was the same at the edges of the heart (Fig. 2 and 4), and since excitation did not pass the raphe[2], it is concluded that opened hearts are electrically intact. Impulse transmission was sometimes interrupted for several minutes after making incisions as shown in Fig. 4. Since conduction velocity is low there is a low safety margin for conduction and any slight depolarization associated with an incision (which would occur since the cells are in electrical continuity) could inactivate the cells. Recovery could be accounted for by a sealing at the tight junctions (cf. mammal[42,] Salpa[43] (see intracellular investigations below).

In order to determine the conduction velocity along the length of the heart, the heart was positioned over the line of suction electrodes and conduction times determined for each millimeter length of heart (left set of records in Fig. 5). Conduction velocity was the least at the ends of the heart and least in small hearts[2]. Hearts 15 mm long had an average conduction speed of about 6 mm/sec while that of 30 mm long hearts was about 12 mm/sec (15°C). The larger hearts have larger cells which may explain these differences in conduction speed. It was noted that the visceral arm of the heart was sometimes as much as 2 mm longer than the branchial arm. This may account for the apparently greater speed in the branchial arm reported by many earlier authors from visual observations[12–14]. Conduction velocity increased linearly from about 1 mm/sec at the ends of Ciona hearts to about 6.5 mm/sec (5.5 mm/sec in Chelyosoma) at a position 8 mm from the end of the heart and then remained constant to the bend region. EBARA[44,45] has found that (at 20°C) abvisceral waves travel at a velocity of only 0.46 mm/sec and advisceral waves travel at 0.61 mm/sec in the small 1 mm long heart of Perophora orientalis.

The conduction speed around the bend of a Ciona heart depended upon the electrode placement. When the axis of the line of electrodes was placed across the bend as close to the arms of the heart as possible, so that the narrow axis of the cells was between the electrodes, the lowest apparent conduction velocity was recorded[2] (Fig. 3H). When the electrodes were placed across the outer edge of the bend region, the greatest apparent conduction velocity was recorded because the wave front reached the electrodes about simultaneously (Fig. 5, position 23).

As will be shown, conduction velocity is greatest parallel to the long axis of the muscle cells, thus the change in conduction velocity along the heart (except near the ends) can be attributed to the change in the cell axis with respect to the long axis of the heart (see fiber orientations in Fig. 1A and 1B).

Decremental Conduction in the Ends of the Heart

At the ends of the heart a change in fiber orientation is not enough to account for the low conduction velocity. Conduction velocity in the middle region of a heart arm in the cell axis was at least 14 mm/sec in *Ciona* and 12 mm/sec in *Chelyosoma* (Table 1). Yet, at the ends of the heart, the maximal conduction velocity in the cell axis was only 1–2 mm/sec (Fig. 5, position 0.2). Furthermore, conduction velocity in the heart axis at the ends of the heart was only 1–2 mm/sec although conduction velocity in the middle of the heart arm (also perpendicular to the cell axis) was at least 6 mm/sec in *Ciona* and 3 mm/sec in *Chelyosoma*. These comparisons indicate that decremental conduction occurs as an impulse approaches an end of the heart (cf. vertebrates[1,21,46,47]).

Decremental conduction in vertebrates is generally characterized by a progressive decrease in amplitude and an increase in the rise time of the action potential as well as a lower resting potential or diastolic depolarization[1,21,46,48]. In the tunicate heart, it is not known whether these parameters are different in cells at the ends of the heart.

The lower conduction velocity in the ends of the tunicate heart would result in a lower safety margin for conduction. This could explain why contractions were sometimes observed to die out (also observed in *Perophora*[44]) and why local non-propagated responses could be produced by stimulating with suction electrodes which had tips of a small diameter (20 μ) at the ends of the heart (cf. tunicates[49] and vertebrates[50]).

Wave Front Analysis of an Impulse Originating at an End Pacemaker, i.e. a Normal Heart Beat

To reconstruct the position of an impulse moving along the heart at successive time intervals, data of two parameters are needed. First, the conduction velocities in the heart axis for each millimeter length of heart are needed (series of records to the left of diagram in Fig. 5). The second data are the arrival times of the front at points on a line normal to the long axis of the heart for each millimeter length of heart. The latter were measured by orienting the long axis of the heart perpendicular to the line of electrodes (series of records to the right of diagram in Fig. 5).

In the major portion of each heart arm straight lines parallel to the cell axis (65°, ± 8°, to the heart axis) represent positions of the front at successive intervals (as shown in diagram in Fig. 5).

The wave front analysis in Fig. 5 shows that in large hearts, excitation spreads into the second arm of the heart before spreading into all of the bend region.

The undifferentiated line of cells does not distort the wave front (Fig. 2 and 5) or impede conduction (Fig. 3 D). Tight junctions exist between these cells

Table 1. Conduction velocities obtained in vector analyses of an impulse spreading from a point stimulus.

Species	No. of hearts	Length of hearts (mm)	Signals used to calculate velocity	Inter-electrode distance (mm)	Greatest conduction velocity (mm/sec) and degree of rotation	Least conduction velocity (mm/sec) and degree of rotation	Conduction velocity in heart axis (mm/sec)
Ciona	4	36–46	S and 1	1	14 (± 6) at 65° (± 11)	4.2 (± 2.0) at 155° (± 11)	
			S and 4	4	24 (± 10)	6.6 (± 2.5)	
			1 and 4	3	56 (± 28)	8.6 (± 2.6)	
			3 and 4	1	77 (± 27)	8.7 (± 1.8)	6.5 (± 2.0)
Chelyosoma	4	30–45	S and 2	2	12 (± 2) at 70° (± 6.0)	2.9 (± 0.4) at 160° (± 6.0)	5.5 (± 2.0)

In rows S and 1 of the signal column, the average conduction velocities were calculated from the time intervals between the stimulus artefacts (S) and the signals at the first recording electrode (1). In row S and 4, the average conduction velocities were calculated from the time intervals between the stimulus artefacts and the signals at the fourth (4) recording electrode. In row 1 and 4, the average conduction velocities were calculated from the time intervals between the signals recorded by electrodes 1 and 4. Since the impulse reached its maximal velocity before it reached electrode 3, the maximal velocities were calculated from the time intervals between the signals recorded at electrodes 3 and 4. Parentheses give the ranges, 10 °C.

and muscle cells[51], so the muscle cells on each side of the undifferentiated line are in electrical continuity.

The wave of excitation usually started in the end of the heart in the tissue next to the inside edge of the raphe (cf. intact heart of Fig. 1 B to opened heart in Fig. 5). In some hearts a secondary pacemaker center in the end of the heart sometimes became active so that different arrival times were measured for the first few millimeters of heart[4,39] (Fig. 3 G).

The curves delineating the wave fronts at successive intervals were parallel (isochronous lines of guinea-pig papillary muscles are also parallel[30]) in both directions of conduction in the middle 2/3 of each heart arm (for records[52]). At the ends of the heart, the curves delineating a front at 0.2 sec intervals which started from the ipsilateral pacemaker were usually closer together for the first millimeter or two. This is readily explained since the conduction velocity of a wave starting at the pacemaker increased until the wave of excitation reached across the width of the heart[52] (Fig. 3 D). In the example given in Fig. 3 F the wave which started close to the electrodes had not achieved maximal velocity by the time it reached the recording electrodes (the same argument can be applied for Fig. 3 B and 3 G in that the lower velocities were recorded when the wave of excitation started nearer to the recording electrodes). These variations in conduction velocities in opposite directions of conduction occurred only at the ends of the heart where the conduction velocity was lower and decremental in nature (Fig. 3 F). Variations in apparent conduction velocities sometimes occurred at the bend region but these were due to the change in fiber orientation which permitted the impulse to take different pathways (Fig. 1 B).

In the end region of large hearts (36–40 mm long), conduction times of successive waves occasionally differed by as much as a factor of two. Conduction times either gradually changed between the extreme values or only two conduction times were recorded (Fig. 3 C and 3 E). These variations can be attributed to different pathways in the spread of excitation (cf. mammals[22,53]).

The above results demonstrate that care must be employed in placing an electrode at each end of the heart if one wishes to compare the conduction velocities in both arms of the heart. If one electrode is closer to the pacemaker region in one arm of the heart, different conduction times in both directions of conduction would be recorded[2,3].

Vector Analysis of Impulse Spread

Conduction velocities of an impulse starting from a point stimulus in different directions (stimulated wave) were determined from the middle regions of heart arms because the cells are uniformly oriented in respect to the heart axis (Fig. 1 A and 1 B) and the conduction velocity is the same in the heart axis near the cut edge of the tissue as in the midline (Fig. 2 and 4). Conduction velocities were determined by rotating the heart around the stimulating electrode (Fig. 6 B).

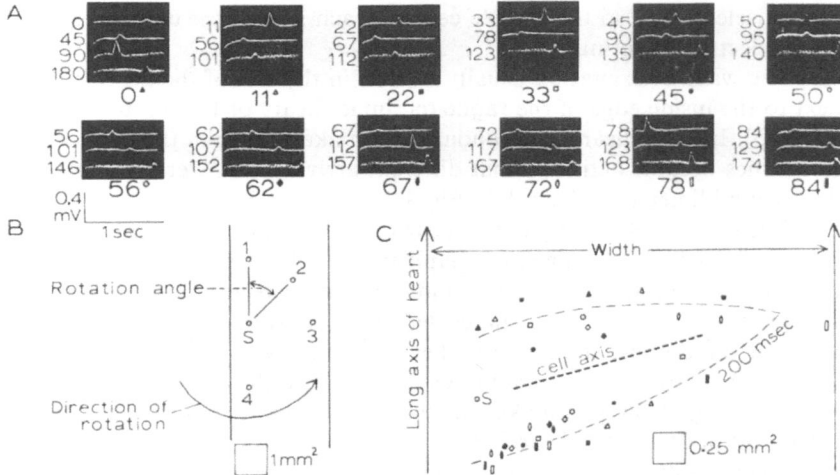

Fig. 6.
Analysis of impulse spread from a point stimulus in a *Chelyosoma* heart. The middle region of the visceral arm was used where there is no change in fiber orientation.
A. Action potentials recorded from different electrode positions. The large numbers below each record identified by a symbol indicate the number of degrees of rotation of the heart. The angle between the long axis of the heart and an imaginary line from each recording electrode to the stimulating electrode is shown to the left of each trace (see B for diagram showing the angle).
B. Diagram of the electrode positions in relation to the heart. The direction in which the heart was rotated is shown by an arrow. The term rotation angle refers to the angle between the heart axis and an imaginary line from each recording electrode to the stimulating electrode (S). In this diagram, the electrode referred to is number 2 and the angle is 45°.
C. Time contour plot showing the impulse front 200 msec after stimulation. The symbols represent point positions which were calculated from each rotational position. The greatest velocity was 12 mm/sec in the cell axis (78° to heart axis) and the least velocity was 2.8 mm/sec (From [55]).

The greatest conduction velocities of stimulated waves were recorded in the cell axis and the lowest velocities were recorded perpendicular to the cell axis (Table 1).

The advantage of the radially arranged electrodes (Fig. 6 B) is that three to four vectors could be determined for each position of the heart. The disadvantage is that only the average conduction velocity was determined for each direction. The advantage of the single row of electrodes is that the velocities between each two electrodes could be used to compare velocities at different distances from the stimulating electrode (Table 1). A natural wave front reached the row of 4 recording electrodes simultaneously at the rotation position that gave the greatest conduction velocities of a stimulated wave. And when the hearts were rotated an additional 90°, the lowest conduction velocities were measured for both stimulated and natural waves of excitation[52].

The conduction velocity increased until the front had spread to about 2 mm in the cell axis. In the direction of the greatest conduction velocity in *Ciona* the maximal velocity (85 mm/sec) was over five times greater than the lowest velocity (14 mm/sec). At the angles that gave the lowest conduction velocities, the largest velocity (9 mm/sec) was about two times greater than the lowest velocity (4 mm/sec). The maximal conduction velocities were about 37% greater than the average conduction velocities (Table 1).

Assuming that impulse spread is by local current flow, the greater conduction velocity in the cell axis indicates that the tissue resistance is less in this axis (cf. mammalian hearts[37]). The internal resistance of the tunicate heart is due to the series resistances of the tight junctions and cytoplasm. Assuming that the cytoplasmic resistivity is uniform, differences in conduction velocities in different directions indicate that the cell shape (and thus the position of the tight junctions) affects the distribution of current flow (see section on injected current).

Effect of Acetylcholine and Adrenaline on the Conduction Velocity

Acetylcholine concentrations of 10^{-7} g/ml stopped the heart beat[54] but ACh concentrations up to 10^{-4} g/ml had no effect on the conduction velocity. Since conduction is by local current flow, this observation is compatible with the result that ACh has no effect on the membrane resistance[55]. ACh decreases the membrane resistance of rat atrial cells[37,56] and decreases the conduction velocity[57].

Adrenaline at concentrations of less than 10^{-4} g/ml had no effect on the conduction velocity although adrenaline at 10^{-5} g/ml stopped the heart beat[54]. At 5×10^{-4} g/ml, adrenaline decreased the conduction velocity to half and greatly diminished the amplitude of the extracellularly recorded action potentials[58]. It is not known if adrenaline at 10^{-4} g/ml has an effect on the membrane resistance[55].

Comparison with Vertebrate Hearts

It is interesting that AGNOLI[31] found that the conduction speed was the same in both endo-epicardial and epi-endocardial directions in the rabbit ventricle. Nevertheless, it is generally true for mammals that retrograde conduction is slower than forward atrioventricular[19-21] and sinus-atrium[59] conduction. The tunicate heart is of interest in this respect since it normally reverses its direction of conduction, i.e. both directions are 'functionally forward' so that the phenomenon of retrograde conduction cannot occur.

A conduction velocity of 6 mm/sec in the tunicate heart axis appears at first to be slow in comparison to those of vertebrate hearts. Conduction velocity in the frog ventricle[60] is 100 mm/sec and in the mammalian ventricle (goat, dog, cat and rabbit)[22] it is from 400 to 800 mm/sec. However, the spread of excitation through the rabbit sinus is 20–60 mm/sec[59] and through the A–V node it is 20–50 mm/sec[61]. The apparent conduction velocity through the chick A–V ring is only 3–5 mm/sec[62]. IRISAWA et al.[20] measured a conduction velocity of only 10 mm/sec through the frog bulbo-ventricular junction which they attributed mainly to the perpendicular orientation of the fibers in relation to impulse spread. In the dog ventricle, SANO et al.[23] found that the conduction velocity parallel to the ventricular fibers was 73–709 mm/sec and 80–100 mm/sec perpendicular to the cells.

Similarly, conduction velocity in the tunicate heart is about 8 mm/sec perpendicular to the fibers and 90 mm/sec in the fiber axis (Table 1). However, conduction velocities in the tunicate heart in different directions are much more uniform in different parts of the heart (except ends and bend) than in mammalian hearts, reflecting the more uniform cellular structure of this endothelial tissue. Only the cell shape affects current flow in the tunicate heart since the cells are not grouped into a branching array of cords.

Rat, guinea-pig and mouse fibers range from 6 to 13 μ in diameter and fibers of the greatest diameter have the greatest conduction velocity[22]. The average diameter of tunicate heart cells is about 7 μ. Since the Q_{10} of the conduction velocity is about two[2] in the tunicate heart (1.6 for frog[63], 1.7 for mammal[64]), there is not much difference between velocities of tunicate hearts and those of small mammal hearts if compared at 30°C. At this temperature, the conduction velocity in the cell axis of tunicates (*Ciona*) would be about 300 mm/sec and it would be about 36 mm/sec across the fiber axis. These velocities are even comparable to those of larger mammalian hearts with cells around 25 μ in diameter because the conduction velocity is proportional to the square of the diameter of the cells (see p. 131). Thus a factor of two times the tunicate velocities may be employed in comparisons to hearts with large cells. These rough comparisons of conduction velocities imply that there is little difference in membrane properties in phylogenetically widely separated hearts (see p. 130).

C.   Intracellular Action Potentials in Chelyosoma

Chelyosoma heart cells exhibited action potentials with 1 to 3 phases of recovery[35] and are similar to those of *Ciona*[1–3]. After a microelectrode penetrated a cell membrane, several seconds were required until a steady resting potential was reached (55 mV ± 8). The initial depolarized condition can be attributed to initial injury of the cell membrane during penetration and the small size of the cells (cf. cultured chick heart cells[65,66]). Recovery must have resulted from a sealing around the microelectrode. After penetration, it was possible to raise the electrode, which was positioned into the apical part of the cell at an angle of 25° to the surface, so that the apical surface was destroyed.

Sometimes a few adjacent cells also deteriorated. However, a clear color change was observed in deteriorated cells and the adjacent normal appearing cells had normal resting potentials suggesting that the tight junctions between normal and deteriorated cells had sealed off (cf. mammal[41]).

Occasionally, small potentials occurred in cells that were isolated by a region of conduction block (resulting from over-stretch) from the rest of the myocardium[3,35]. These potentials resulted from electrotonic currents spreading from depolarizing cells through the low-resistance tight junctions into the penetrated cell[67]. Steps and notches were also seen on the rising phase of action potentials[2] which can be attributed to converging pathways of impulse spread into the cell (cf. mammal[67,68]).

It is useful to note here that impulses have been observed to propagate into a 1 mm² piece of tissue isolated in the middle of an opened heart by a 0.25 mm wide ring of petroleum jelly on each side of the tissue (see KRIEBEL[5] for apparatus, details not published). This observation is similar to that made by BARR and BERGER[7] where action potentials of frog atrial fibers propagated past a sucrose gap shunted with an external shunt resistor.

These results clearly show that the magnitude of electrotonic currents generated by an impulse is more than adequate to account for impulse spread by local current flow.

D. Spread of Intracellularly Injected Current in Chelyosoma Hearts

The electrical coupling ratio between cells (for a given current because the recording electrode was moved from cell to cell) is the ratio of the voltage change in any adjacent cell to the voltage change in the polarized cell ($V_{adjacent\ cell}$/$V_{polarized\ cell}$). The effective input resistance of the tissue between the polarized cell and any distant cell is the transfer resistance ($V_{distant\ cell}$ / $I_{applied\ current}$). The input resistance as measured in the polarized cell is actually a transfer resistance, since the polarizing and recording electrodes were separated by about 5 μ. Therefore, the coupling ratios as presented here are actually the ratios of the transfer resistances in the distant cells to that in the polarized cell.

The input resistance (actually the transfer resistance) of the tunicate heart is about 4 MΩ which is somewhat high compared to that of 200–600 kΩ in the mouse[38] and about 500 kΩ in the rat atrium[37] and much more than that of 29 kΩ of the rabbit ventricle[69] which has larger cells. TARR and SPERELAKIS[25,26] found input resistances of 2–7 MΩ for frog and 7–200 MΩ for the cat. SPERELAKIS[70] also found an input resistance of 13 MΩ for chick heart cells in tissue culture (a monolayer). It should be pointed out here that if a single electrode and a bridge circuit are used to measure an input resistance of a cell, this value would be more than the transfer resistance as measured with a second microelectrode. Note how fast the resistance decreases with distance from the polarizing electrode in Fig. 7. With such parameters being equal as the specific mem-

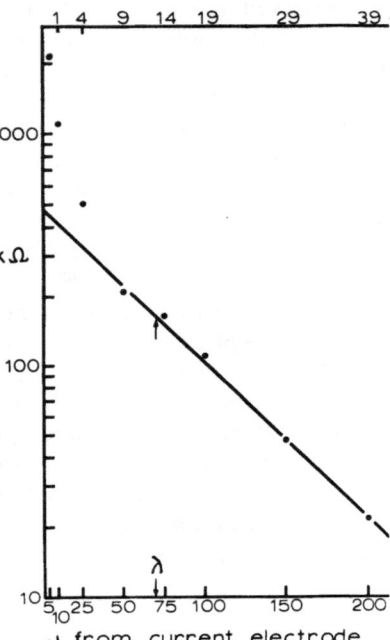

Fig. 7.
Attenuation of membrane voltage
normal to the cell axis, *Chelyosoma.*
The abscissa shows the distance in
microns and the number of cell borders
between the polarized cell and the recorded cell. The ordinate has the dimensions of kohms and is the transfer
resistance. All cells including the stimulated cell had a resting potential of
50–60 mV and showed normal action
potentials. The coupling ratios are given
in Table 3 (From [55]).

brane resistance and internal resistivity of the tissue, a multilayered tissue would
have a lower input resistance than a monolayered tissue since current could flow
from a point source in three dimensions through low-resistance tight junctions.

In an adjacent cell, 5 μ from the current electrode, the coupling ratio was
about 41 % but fell rapidly after the first 50 μ (Table 2). The spatial decrement
of voltage perpendicular to the cell axis was found to be uniform in the middle
region of each arm of the heart, since records of voltage changes from different

Table 2. Transfer resistances and electrical coupling ratios perpendicular to the cell axis
(*Chelyosoma*).

Distance from current source (μ)	No. of cell borders between electrodes	No. of hearts	No. of recorded sites	Transfer resistance (kΩ)	Coupling ratio (%)
5	same cell	4	7	3.9 (± 2.1)	100
5	1	2	2	1.6 (± 0.4)	41
20	4	1	6	0.29 –	7.4
40	8	1	6	0.15 –	3.8
250	50	2	2	0.10 –	1.3

The cells were about 5 μ wide.

Table 3. Voltage decrement perpendicular to a polarized cell (*Chelyosoma*).

Distance from current source (μ)	No. of cell borders between electrodes	Transfer resistance (kΩ)	Coupling ratio (%)
5	same cell	2,200	100
10	1	1,130	50
25	4	500	23
50	9	210	10
75	14	166	7
100	19	110	5
150	29	50	2
200	39	22	1

The data are graphed in Fig. 7.

sites 20 μ or 40 μ from a polarized cell were superposable[35]. The spatial decrement of voltage in several adjoining cells perpendicular to a polarized cell is shown in Table 3. In this preparation, the tissue was stretched (20%) in all directions so that it could contract very little. The action potentials in the polarized cell remained uniform (60 mV) throughout the duration of the experiment and action potentials measured in distant cells were also large. It is felt that this single experiment which is graphed in Fig. 7 is very representative since many points determined from other hearts also fell on or near the curve. However, the coupling ratios were slightly larger in these stretched preparations than in slack hearts as shown in Table 2. A possible explanation may be that in the stretched condition either the internal resistance (cytoplasm and tight junctions) decreased or the surface resistance increased or that both of these parameters changed. With stretch, there could be fewer rows of tight junctions which could decrease the internal resistance although the thickness would decrease so that the cytoplasmic resistance portion of the internal resistance would increase. In addition, there would be less membrane surface, so the membrane resistance per unit area would increase. It is difficult to decide which parameter(s) would have the greater effect.

Fig. 7 shows that current density flowing in the tissue from a point source diminished very rapidly in the first few cells (cf. chick[37]). The entire curve is not exponential as in a nerve or muscle cell or a cord of Purkinje fibers because current was flowing from a point source in two dimensions. Nevertheless, at distances greater than 50 μ from the polarized cell, voltage decrement closely approached an exponential. TANAKA and SASAKI[38] also found that the space constant was short in the mouse (70 μ) when determined close to a point source but increased to 400–700 μ at distances further than 100 μ from the polarized cell. In the tunicate heart, current flow against the cell axis at a distance greater than about 50 μ (Fig. 7) was found to approximate the behavior for one-dimensional current flow so that one-dimensional core conductor formulas may be applicable (see BARR, this volume). The exponential part of the graph in

Fig. 7 gives a space constant of 70 μ. If the current in the tunicate heart approaches one-dimensional flow at relatively great distances from the polarized cell, the measured space constant of 70 μ should approach that of the theoretical space constant for one-dimensional current flow in a core conductor. The theoretical space constant (λ) is given by the following formula where r_e equals the external resistance per unit length (Ω/cm), r_i equals the internal resistance (Ω/cm) and r_m equals the resistance of the membrane (Ω/cm):

$$\lambda = \sqrt{\frac{r_m}{r_e + r_i}}. \tag{1}$$

Since r_e is very small compared to r_i, it may be neglected. The relationship of the specific resistances to the space constant is given by the following equation where R_m is the specific membrane resistance of 230 Ω cm^2, and R_i is the specific internal resistivity (2.8×10^3 Ω cm) which equals the internal resistance (r_i) times the heart thickness (δ, which is 10 μ)[5]:

$$\lambda = \sqrt{\frac{R_m}{R_i}}. \tag{2}$$

These data[5] give a theoretical space constant of 90 μ. However, the internal resistance (r_i) was determined in hearts with the cell axis 70° to the direction of current flow, so the resistivity used in the above calculation would be somewhat less than that perpendicular to the cell axis. It should also be pointed out that the current was still flowing in two dimensions in the graph in Fig. 7 so that the flow only approached one-dimensional flow in the exponential part of the graph. Thus, the measured space constant of 70 μ is somewhat low.

WOODBURY and CRILL[37], and WOODBURY[6] were able to fit similar data from rat atria to a Bessel function to give a space constant of about 90 μ perpendicular to the cell axis.

Since impulse spread in *Chelyosoma* is at least 4 times (10 times in *Ciona*) greater along the cell axis than against the fibers (Fig. 6C and Table 1), the internal resistivity must be a function of the angle of the cell axis to the direction of current flow. Therefore, the spatial decrement of voltage would probably be much less when measured along the cell axis. Assuming that an impulse front traveling at a constant velocity in any given direction in the tunicate heart is similar to an impulse traveling along a nerve, the density of membrane current per unit surface area is given by the formula[71]:

$$I_m = \frac{D}{4 \, R_i} \cdot \frac{d^2v}{dx^2}. \tag{3}$$

If the impulse travels at a constant velocity (θ) then the following relation exists:

$$\frac{d^2v}{dx^2} = \frac{1}{\theta^2} \cdot \frac{d^2v}{dt^2}. \tag{4}$$

Upon substitution of the right term of (4) into equation (3):

$$I_m = \frac{D}{4 \, R_i \, \theta^2} \cdot \frac{d^2v}{dt^2}. \tag{5}$$

For coherence (i.e. for the wave of excitation to stay together), the term $(D/4\,R_i\,\theta^2)$ must be constant. Thus, $\theta \propto \sqrt{1/R_i}$. This relationship permits the determination of R_i along the cell axis since R_i against the cell axis is known and the conduction velocities in both directions of conduction are known. The relationship can be expressed with the following formula:

$$\frac{\theta \text{ along cell axis}}{\theta \text{ against cell axis}} = \sqrt{\frac{R_{i \text{ against cell axis}}}{R_{i \text{ along cell axis}}}}. \tag{6}$$

This expression gives an internal resistivity along the cell axis of $1.4 \times 10^2\ \Omega$ cm. A space constant of about 400 μ along the cell axis can be calculated with the cable equation (2) using this value. This space constant is similar to that of 170 μ in the rat atrium[37], 100–300 μ in the rabbit ventricle[28], 400–700 μ in the mouse[38] and 50–400 μ in frog atria[72].

The tunicate heart space constant is low when compared to that of 1.9 mm in the sheep Purkinje fiber[73] and 1.3 mm of the dog ventricle fibers[29]. Assuming equal fiber lengths and similar membrane properties, the space constants can be compared, since the space constant is proportional to the $\sqrt{\text{dia. tunicate cell/Purkinje cell}}$. This proportionality can be expressed as:

$$\frac{\lambda \text{ tunicate heart (cell axis)}}{\lambda \text{ Purkinje fiber}} = \sqrt{\frac{\text{diameter of tunicate cell}}{\text{diameter of Purkinje fiber}}}. \tag{7}$$

The tunicate space constant in the cell axis is given by $1.9 \times \sqrt{5/70}$ which equals about 480 μ. This compares favorably to the 400 μ calculated with formula (6). The specific internal resistivity as calculated with equation (6) of the tunicate heart is similar to that which WEIDMANN[73] found for Purkinje fibers. However, the specific membrane resistivity of the tunicate heart[5] is about 5 times less than the Purkinje fiber. Nevertheless, this theoretical comparison is in good agreement with that space constant calculated from the cable equation (2) using the theoretical internal resistance along the cell axis (equation 6).

As in the comparisons of conduction velocities (p. 126), membrane characteristics of the tunicate heart are surprisingly similar to those of higher chordate hearts.

The above results support the conclusion that local current flow is responsible for impulse spread in the tunicate heart and that the cell axis imposes electrical anisotropy on the otherwise uniform myocardium. The thin-walled heart is functionally a single planar cell whose resistivity is less along the cell axis.

Summary

Ciona and *Chelyosoma* hearts were severed from the pericardium by cutting along the raphe and then flattened under a coverslip for phase contrast microscopy. The wall of these U-shaped, tubular hearts is composed of a single layer of myoendothelial cells which are oriented perpendicular to the heart axis at the ends and gradually change to 65° ($\pm$ 8°) for most of the middle portion of each arm of the heart. Towards the bend region of the U-shaped heart, the cell axis again gradually changes to 90°.

Conduction velocities were determined by placing the myocardium over a row of fixed suction electrode openings. Conduction velocity linearly increased from 1 mm/sec to 6 mm/sec towards the middle region of a heart arm and remained constant (except at the bend where front orientation changed) until the opposite end was approached where it decreased to 1 mm/sec (10 °C). The conduction velocity was usually the same in both directions of impulse propagation.

Wave fronts propagating along the heart were found to move perpendicular to the cell axis (except at the bend region) in both directions of conduction.

Vector analyses of an impulse spreading from a point stimulus demonstrated that conduction velocities (10 °C) were greatest in the cell axis [*Ciona*, 24 ($\pm$ 10) mm/sec at 65° ($\pm$ 11); *Chelyosoma*, 12 ($\pm$ 2) mm/sec at 70° ($\pm$ 6)] and least perpendicular to the cell axis [*Ciona*, 6.6 ($\pm$ 2.5); *Chelyosoma*, 2.9 ($\pm$ 0.4) mm/sec]. The impulse velocity was found to increase until the front had spread to about 1 mm in the cell axis. These results also show that true decremental conduction occurs at the ends of the heart.

Acetylcholine from 10^{-7} to 10^{-4} g/ml had no effect on the velocity of conduction. Adrenaline at 5×10^{-4} g/ml decreased the conduction velocity to half.

The electrical coupling ratios between cells perpendicular to the cell axis were determined in *Chelyosoma* hearts by polarizing one cell with an intracellular microelectrode and measuring the voltage change in the same and distant cells. In adjacent cells the coupling ratio was 41 %; across 3 cells (20 μ from the current source), it was 7.4 %; and across 7 cells (40 μ), it was 3.8 %. As little as 50 nA of applied current were detected across 50 cells (250 μ, coupling ratio 1.3 %).

Subthreshold potentials occurring during contractions were observed in cells functionally isolated by an area of conduction block. The high coupling ratios suggest that these potentials (also steps and notches on the rising phase of some action potentials) result from electrotonic currents generated by impulses which do not propagate through the blocked region. Impulses were also observed in tissue within a ring of petroleum jelly on each surface demonstrating that electronic currents were able to spread through the tissue covered with petroleum jelly.

Voltage decrement from a point current source perpendicular to the cell axis was exponential at distances greater than 50 μ from the polarizing electrode. This suggests that current flow at relatively great distances from the source approached that of current flowing in one dimension. The region of exponential decrement gives a space constant of 70 μ. The theoretical space constant for one-dimensional current flow calculated with the cable equation,

$$\lambda = \sqrt{[r_m/(r_e + r_i)]}, \quad \text{is } 90 \ \mu.$$

These results demonstrate that the cells are in ionic continuity and that impulse spread is by local current flow. That there is a greater conduction velocity parallel to the cell axis than that perpendicular to the cell axis demonstrates that the tissue is electrically anisotropic.

References

[1] H. H. Hecht, Ann. N.Y. Acad. Sci. *127*, 49 (1965).

[2] M. E. Kriebel, J. gen. Physiol. *50*, 2097 (1967).

[3] M. Anderson, J. expl. Biol., *49*, 363 (1968).

[4] M. E. Kriebel, Biol. Bull. *135*, 166 (1968).

[5] M. E. Kriebel, J. gen. Physiol. *52*, 46 (1968).

[6] J. W. Woodbury, *Handbook of Physiology*, Section 2: *Circulation*, vol. 1, 237 (1962).

[7] L. Barr and W. Berger, Pflügers Arch. ges. Physiol. *279*, 192 (1964).

[8] L. Barr, M. M. Dewey and W. Berger, J. gen. Physiol. *48*, 797 (1965).

[9] J. J. Dreifuss, L. Girardier et W. G. Forssmann, Pflügers Arch. ges. Physiol. *292*, 13 (1966).

[10] K. Kawamura and T. Konishi, Jap. Cir. J. *31*, 1533 (1967).

[11] L. Barr, this volume.

[12] E. v. Skramlik, Ergebn. Biol. *15*, 166 (1938).

[13] B. Krijgsman, Biol. Rev. *31*, 288 (1956).

[14] M. E. Kriebel, Biol. Bull. *134*, 434 (1968).

[15] H. Mislin, Helv. physiol. Acta *24*, c41 (1966).

[16] J. H. Gerould, J. Morphol. Physiol. *48*, 385 (1929).

[17] F. V. McCann, J. gen. Physiol. *46*, 803 (1963).

[18] T. Kanno, Jap. J. Physiol. *13*, 97 (1963).

[19] A. M. Scher, M. I. Rodriquez, J. Liikane and A. C. Young, Circulation Res. *7*, 54 (1959).

[20] H. Irisawa, K. Hama and A. Irisawa, Circulation Res. *17*, 1 (1965).

[21] A. Paes De Carvalho, *The Specialized Tissues of the Heart* (Elsevier Pub. Co., New York 1961).

[22] M. H. Draper and M. Mya-Tu, Quart. J. exp. Physiol. *44*, 91 (1959).

[23] T. Sano, N. Takayama and T. Shimamoto, Circulation Res. *7*, 262 (1959).

[24] N. Sperelakis, T. Hoshiko and R. M. Berne, Am. J. Physiol. *198*, 531 (1960).

[25] M. Tarr and N. Sperelakis, Am. J. Physiol. *207*, 691 (1964).

[26] M. Tarr and N. Sperelakis, Am. J. Physiol. *212*, 1503 (1967).

[27] J. Tille, J. gen. Physiol. *50*, 189 (1967).

[28] M. M. Dewey, this volume.

[29] A. Kamiyama and K. Matsuda, Jap. J. Physiol. *16*, 407 (1966).

[30] C. B. Casella and B. Taccardi, *Electrophysiology of the Heart* (Ed. B. Taccardi and G. Marchetti; Pergamon Press, London, 1964), p. 153.

[31] G. C. Agnoli, *Electrophysiology of the Heart* (Ed. B. Taccardi and G. Marchetti; Pergamon Press, London 1964), p. 163.

[32] M. M. Dewey and L. Barr, J. Cell Biol. *23*, 553 (1964).

[33] E. A. Johnson and J. R. Sommer, J. Cell Biol. *33*, 103 (1967).

[34] J. R. Sommer and E. A. Johnson, J. Cell. Biol. *36*, 497 (1968).

[35] M. E. Kriebel, Life Sciences *7*, 181 (1968).

[36] W. E. Crill, R. E. Rumery and J. W. Woodbury, Am. J. Physiol. *197*, 733 (1959).

[37]   J. W. WOODBURY and W. E. CRILL, *Nervous Inhibition* (Ed. E. FLOREY; Pergamon Press, London 1961), p. 124.

[38]   I. TANAKA and Y. SASAKI, J. gen. Physiol. *49*, 1089 (1966).

[39]   M. E. KRIEBEL, M. S. Thesis: *Studies on the Cardiovascular Physiology of the Tunicate Ciona intestinalis* (Univ. of Washington, Seattle, Washington 1964).

[40]   P. HEINE, Z. wiss. Zool. *73*, 429 (1903).

[41]   R. H. MILLAR, L.M.B.C. Mem. *35* (1953).

[42]   J. DELEZE, *Electrophysiology of the Heart* (Ed. B. TACCARDI and G. MARCHETTI; Pergamon Press, London 1964), p. 147.

[43]   A. EBARA, Tokyo Bunrika Daigaku *7*, 199 (1954).

[44]   A. EBARA, Tokyo Kyoiku Daigaku *8*, 75 (1957).

[45]   A. EBARA, Tokyo Kyoiku Daigaku *8*, 86 (1957).

[46]   B. F. HOFFMAN, A. PAES DE CARVALHO, W. C. MELLO and P. F. CRANEFIELD, Circulation Res. *7*, 11 (1959).

[47]   B. F. HOFFMAN, Ann. N.Y. Acad. Sci. *127*, 105 (1965).

[48]   B. F. HOFFMAN, Circulation *24*, 506 (1961).

[49]   H. SUGI, R. OCHI and M. UDO, Zool. Mag. Tokyo *74*, 45 (1965).

[50]   A. PAES DE CARVALHO and D. F. DE ALMEIDA, Circulation Res. *8*, 801 (1960).

[51]   A. ICHIKAWA, *Sixth International Congress for Electron Microscopy, Kyoto* (Ed. RYOZI VYEDA; Maruzen Co. Ltd. 1966), p. 695.

[52]   M. E. KRIEBEL, in prep.

[53]   M. ITO, M. ARITA, K. SAEKI, M. TANOUE, I. FUKUSHIMA, T. YANAGA and H. MASHIBA, Jap. J. Physiol. *17*, 174 (1967).

[54]   M. E. KRIEBEL, Biol. Bull. *135*, 174 (1968).

[55]   M. E. KRIEBEL, Ph. D. Thesis: *Physiological Studies on the Tunicate Heart* (Univ. of Washington, Seattle, Washington 1967).

[56]   W. TRAUTWEIN and J. DUDEL, Pflügers Arch. ges. Physiol. *266*, 324 (1958).

[57]   W. TRAUTWEIN, Pharmac. Rev. *15*, 277 (1963).

[58]   C. L. SCUDDER, T. K. AKERS and A. G. KARCZMAR, Comp. Biochem. Physiol. *9*, 307 (1963).

[59]   T. SANO and S. YAMAGISHI, Circulation Res. *16*, 423 (1965).

[60]   P. F. CRANEFIELD and B. F. HOFFMAN, Physiol. Rev. *38*, 41 (1958).

[61]   B. F. HOFFMAN, A. PAES DE CARVALHO, W. C. MELLO and P. F. CRANEFIELD, Circulation Res. *7*, 11 (1959).

[62]   M. LIEBERMAN and A. PAES DE CARVALHO, J. gen. Physiol. *49*, 365 (1965).

[63]   A. J. BRADY, *Physiology of the Amphibia* (Ed. J. A. MOORE; Academic Press, New York 1964), p. 211.

[64]   É. CORABŒUF and S. WEIDMANN, Helv. physiol. pharmac. Acta *12*, 32 (1954).

[65]   R. DEHAAN and S. H. GOTTLIEB, J. gen. Physiol. *52*, 643 (1968).

[66]   R. DE HAAN, this volume.

[67]   T. SANO, F. SUZUKI and S. TAKIGAWA, Jap. J. Physiol. *14*, 659 (1964).

[68]   B. F. HOFFMANN, *Biophysics of Physiological and Pharmacological Actions* AAAS, 485 (1961).

[69]   E. A. JOHNSON and J. TILLE, J. gen. Physiol. *44*, 443 (1961).

[70]   N. SPERELAKIS, *Electrophysiology and Ultrastructure of the Heart* (Ed. T. SANO, V. MIZUHIRA and K. MATSUDA; Bunkodo Co. Ltd., Tokyo (1967), p. 81.

[71]   B. KATZ, *Nerve, Muscle, and Synapse* (McGraw-Hill, New York 1966).

[72]   J. W. WOODBURY and A. M. GORDON, J. cell. comp. Physiol. *66*, Suppl. 2, *35* (1965).

[73]   S. WEIDMANN, J. Physiol. *118*, 348 (1952).

Lack of Electrical Coupling between Contiguous Myocardial Cells in Vertebrate Hearts

by Nick Sperelakis
Department of Physiology, University of Virginia School of Medicine, Charlottesville, Va.

Acknowledgments

The author's work reviewed in this report was supported by grants from the U. S. Public Health Service (HE-11 155) and from the American Heart Association.

The measurements on the electrical analog circuit model for diffusion in the extracellular fluid were made by Mr. H. K. Shumaker, and the corresponding theoretical calculations were done by Mr. G. T. Mayer. The probability calculations for the placement of two intracellular electrodes were made by Mr. Mayer.

Studies on Intact Cardiac Muscle

In this paper, I wish to focus on some of the electrophysiological evidence which suggests that there is no electrical coupling between contiguous myocardial cells in vertebrate hearts. Table 1 summarizes most of our published

Table 1. Summary of the electrophysiological evidence consistent with the intercalated discs being special cardiac myo-myo junctions.

(a) Presence of driving junctional prepotentials[5].
(b) Partial block of transmission from cell to cell[4,21].
(c) Not a two- or three-dimensional syncytium; low probability of significant electrotonic interaction between closely spaced electrodes[10,15], and no low-resistance pathways between neighboring cells[7].
(d) Short apparent space constant of about one cell length (150 μ); probability is that there is sharp discontinuity across high-resistance discs (R_{md})[10,15].
(e) High input resistance (r_{in}) of about 10 MΩ[4,10,15,20].
(f) Great sensitivity of tissue resistivity to interspace ion depletion[2].
(g) High C_{md} (frequency-dependent longitudinal tissue impedance)[2].
(h) E_{md} (resting potential across each junctional membrane)[4].
(i) No electrotonic interaction across artificial ephaptic junction between two cultured cells[4].
(j) Lack of spread of injury depolarization to contiguous cells[4,10,20,21].
(k) Synchrony of two cultured cells connected only by a long, narrow process[4,21].
(l) Non-pacemaker cells not stimulated by intracellular depolarizing current[4,20,21].
(m) Quiescent cells adjacent to active ones in hypertonic solution[1].
(n) Induction of unidirectional propagation[5].
(o) Direct stimulation of cells by longitudinal current[6].

electrophysiological evidence consistent with the hypothesis that the intercalated discs are special myo-myo junctions, and some of these data are briefly
discussed below. Emphasis is given to new calculations from various analog
models.

Hypertonic Perfusion

Some cells of the frog ventricle are electrically isolated from their neighbors during hypertonic perfusion[1]*. Quiescent cells having normal resting potentials are often found adjacent to active ones, and some cells fire at a rate some
integral fraction of the surface ECG. Therefore, these cells must be sealed
off by high-resistance membranes under these conditions, otherwise a large
extracellular voltage field or passive electrotonic spread should have been
observed.

High Capacitance of Intercalated Discs

The longitudinal tissue impedance of parallel-fibered cat papillary muscles
and ventricular trabeculae was measured at different sinusoidal frequencies
before and after ion depletion of the interstitial fluid by bathing in isosmotic
sucrose solution. The relative impedance of cardiac muscle at 10,000 cps compared to that at 10 cps is 75% in Tyrode's solution and 48% in sucrose solution
(Table 2)[2]. In contrast, the longitudinal tissue impedance of frog sartorius is
independent of frequency (also see ref. 3). These results indicate the presence
of transverse membranes (intercalated discs) in cardiac muscle of high capacitance (C_{md}) and high resistance (R_{md}).

Change in Specific Resistance of Tissue with Changes in Resistance of Interstitial Fluid

The specific resistances of cardiac muscle and intestinal smooth muscle, in
contrast to skeletal muscle (frog sartorius), undergo relatively great changes
following interspace ion depletion by soaking in 10-fold diluted Tyrode made
isosmotic with sucrose, as summarized in Table 2[2]. This indicates that the
current flow path through the myocardial cells is high resistance in comparison
to that through the extracellular fluid (ECF), suggesting that myocardial cells
are separated from each other by high-resistance membranes. Taking the resistivity of the ECF to be 8-fold greater in 10-fold diluted Tyrode (from Table 3),
the ratio of the resistance of the cellular pathway to that of the extracellular

* Numbers refer to References, p. 164.

Table 2. Summary of measurements of longitudinal tissue resistivity and impedance of cardiac muscle in comparison to smooth muscle and skeletal muscle.

	Cat papillary muscles and ventricular trabeculae	Cat intestinal muscle (circular layer)	Frog sartorius
I. Relative impedance[1] (Z_{10Kc}/Z_{10cps})			
(a) in Tyrode solution	0.75	0.72[4]	1.0
(b) in isosmotic sucrose solution	0.48	–	1.0
II. Specific resistance ($\Omega \cdot$ cm)			
(a) in Tyrode solution (R_t)	268	118	133
(b) in 10-fold diluted Tyrode (R'_t)	1,876	778	367
(c) Ratio R'_t/R_t	7.0	6.7	2.9
III. Calculations[2]			
(a) r_{ecf} ($\Omega \cdot$ cm)	273	122	183
(b) $8 \times r_{ecf}$ ($\Omega \cdot$ cm)	2184	976	1464
(c) r_{cell} ($\Omega \cdot$ cm)	14,640	3,600	488
(d) r_{cell}/r_{ecf}	53.7	29.5	2.7
(e) r_{junc} ($\Omega \cdot$ junc $\cdot$ cm)^3	14,390	3,350	238
(f) R_{md} ($\Omega \cdot$ cm^2)	215	50	–
(g) C_{md} (μF/cm^2)	9.1	–	–
(h) Vol_{ecf} (%) (extracellular space)	16	36	24

[1] Impedance measurements made using an AC bridge and sinusoidal frequencies of 10, 100, 1000, and 10,000 cps (see ref.[2]).

[2] Calculations made assuming the lumped resistances of the extracellular pathway (r_{ecf}) and intracellular pathway (r_{cell}) are in parallel. The following equations were used for these calculations:

(1) $1/R_t = 1/r_{ecf} + 1/r_{cell}$

From Table 3 for ϱ of 0.1 X Tyrode:

(2) $1/R'_t = 1/8 r_{ecf} + 1/r_{cell}$

By subtraction of these simultaneous equations:

(3) $r_{ecf} = 0.875 \, R'_t \cdot R_t/R'_t - R_t$

(4) $r_{cell} = r_{ecf} \cdot R_t/r_{ecf} - R_t$

(5) $Vol_{ecf} = \varrho_t/r_{ecf} = 44 \, \Omega \cdot$ cm$/r_{ecf}$

(6) $C_{md} = 67 \cdot 1/2 \, \pi f X_c$
at $f = 10^4$, $Z = 0.48 \, R_{md}$,
$\therefore \, X_c = 0.543 \, R_{md}$.

[3] Myoplasmic resistivity (R_i) of 250 $\Omega \cdot$cm assumed for room temperature, and this number subtracted from r_{cell} to give r_{junc}. If there are 67 junctions/cm (discs 150 μ apart), then the resistivity of the junctions would have a minimum value of 215 $\Omega \cdot$ cm^2 (even assuming no convolutions and no shunting at the gaps): 14,390 $\Omega \cdot$ junc $\cdot$ cm divided by 67 junc $\cdot$ cm = 215 $\Omega \cdot$cm^2.

[4] From ref.[3].

pathway is calculated to be 53.7 and 29.5 for cardiac and smooth muscles, respectively, and only 2.7 for skeletal muscle. Calculations from the data, explained in the legend of Table 2, give for the intercalated disc an R_{md} of at least 215 $\Omega \cdot$ cm^2 and a C_{md} of 9.1 μF/cm^2. Both the specific resistance and specific capacitance values are close to the values expected for the surface membrane of cardiac cells[4].

Table 3. Summary of resistivity measurements on a glass rod-Tyrode solution analogue of a cardiac muscle fiber bundle for radial compared to longitudinal diffusion in the extracellular fluid.

Concentration of Tyrode solution	ϱ_1 ($\Omega \cdot$cm)	ϱ_1 (relative to normal Tyrode)	$\dfrac{\varrho_1 \text{ without rods}}{\varrho_1 \text{ with rods}}$	ϱ_t/ϱ_1 without rods	ϱ_t/ϱ_1 with rods
0.1 X normal	512	8.0	0.26	1.05	6.0
1 X normal	64	1	0.26	1.12	7.0
10 X normal	8	1/8.0	0.24	–	7.5

Unidirectional Propagation

The unidirectional failure of propagation which can be induced reversibly in frog ventricular strips[5] can be more readily explained by the junctional transmission hypothesis, in which junctions can become polarized in one direction due to fatigue, than by the syncytial hypothesis. If the junctional transmission process is electrical, then the junctions must become rectifying.

Direct Stimulation by Longitudinal Current

Direct and simultaneous excitation of each cardiac cell within parallel-fibered muscle bundles of cat papillary in sucrose solution can occur in longitudinal electric fields over the entire length of the bundle without the necessity of propagation[6]; this does not occur in frog sartorius in which many individual cells extend the entire length of the muscle. Hence, the functional cardiac cell length must be short and radial currents flow in each cell due to the potential differences created across the membrane near the ends of each cell by intercalated discs of high resistance.

Resistance between Two Cell Interiors

By applying constant-current pulses directly between two independent microelectrodes varied from less than 0.5 mm to 10 mm, the interelectrode resistance was measured in frog ventricular muscle when only one or when both microelectrodes were intracellular. The resistance of a single cell is 12 MΩ, and that between two cell interiors is about double and independent of interelectrode distance[7]. Thus, either there are no low-resistance pathways between cells or the length constant is much less than 500 μ. Similar findings have been made in intestinal smooth muscle, except that r_{in} was higher[8,9].

Electrotonic Spread of Current

The most direct method of determining whether current can spread easily from one cell to another is to measure the electrotonic spread of current. This was done in frog and cat ventricular muscles using double microelectrodes cemented at various distances and a bridge circuit to record the change in membrane potential (ΔE_m) also at the site of current injection[10]. At inter-electrode distances of 60 μ or greater (electrodes longitudinally oriented), current flow of up to 16×10^{-9} A never produces a change in E_m at the second electrode, even though large changes in E_m are produced in the cell injected with current; that is, the degree of electrotonic interaction is nearly zero (Table 4). At distances of 45 μ or less, often either a large or no change in potential occurs at the second electrode depending on the impalement. For example, in Fig. 1, the degree of interaction at 7 μ is nearly 100 % in one impalement (Fig. 1 A), whereas in the next impalement the interaction is nearly 0 % (Fig. 1 B). Thus the degree

Fig. 1.
Typical experiments measuring spread of electrotonic current between two closely-spaced intracellular microelec-trodes in intact myocardium at rest. The experiments illustrated were on frog ventricular trabeculum at an inter-electrode distance of 7 μ. In some im-

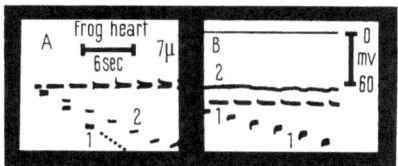

palements (A) the degree of electrotonic interaction during the application of hyperpolarizing pulses (ratio of ΔE_m recorded by each electrode) was high, being nearly 100 % in some cases. In other impalements (B) the degree of interaction was low, approaching 0 %. Seven step increments of current were applied in A and six steps in B; bridge became unbalanced during last four steps in A causing the loss of the potential change at electrode 1, the site of current injection.

of interaction does not depend simply on interelectrode distance. The shorter the distance, the greater the probability of obtaining large interactions. Of course, the probability of impalement of the same cell by both electrodes is greater at shorter distances (see below).

Several criteria, including simultaneous and congruous subthreshold oscil-lations and decline of resting potential due to injury, help to indicate when both electrodes impale the same cell. Another criterion is illustrated in Fig. 2 in which impalements were made in frog ventricle at an interelectrode distance of 11 μ[10]. Two sweeps are superimposed in each photograph to show the degree of inter-action during the resting potential and during the plateau of the action poten-tial. As shown in A, the interaction at rest is zero and does not increase during the plateau. In another penetration shown in B, the electrodes had impaled the same cell which had a low resting potential probably due to injury from the two electrodes; the degree of interaction is large, but it does not increase during the plateau. C and D are sequential photographs from one penetration showing

Fig. 2.
Three typical experiments (A, B, and
C-D) measuring the spread of electrotonic current between two closely-
spaced intracellular microelectrodes in
intact frog ventricular trabeculae at rest
and during the plateau of the action
potential. The interelectrode distance
was 11 µ. Two successive sweeps of the
oscilloscope superimposed in each photograph. A: Interaction at rest and during
the plateau was nearly 0%. B: In
another impalement in which the cell
was injured by the two electrodes, the
resting potential was low and the degree
of interaction was high. C-D: In an-
other impalement in which the cell was injured, the resting potential was low, and the
degree of interaction was high (channel 1 deflection not shown because of bridge
imbalance). The contraction accompanying the action potential caused one of the
electrodes to become dislodged from that cell and penetrate into a neighboring cell
which had a normal resting potential; the degree of interaction then became nearly
zero.

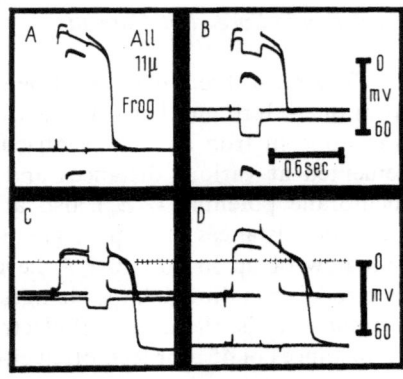

that while the two electrodes were in the same cell, the percent interaction is
large (C). However, the contraction accompanying the action potential caused
one of the microelectrodes to leave this cell and to impale an immediately
neighboring cell (right-hand portion of C). Then in D, while the electrodes were
in neighboring cells, the original cell having a low resting potential and the
new cell having a normal resting potential, the percent interaction became zero.

Therefore, there is a step change in degree of interaction which appears to
depend on whether the electrodes impale the same cell or neighboring cells.
Polarizing current applied in one cell does not have a substantial effect on the
membrane potentials of adjacent cells. The apparent longitudinal and transverse space constants are one cell length (about 150 µ) and one cell width (about
10–16 µ), respectively, where a sharp discontinuity in potential occurs. These
data indicate that cardiac muscle is not a two- or three-dimensional electrical
syncytium and suggest that the intercalated discs are high resistance.

Since there is transverse propagation in cardiac muscle[11], there must be
functional junctions effectively oriented in this direction. Thus, there still should
be substantial electrotonic interaction between neighboring cells not in the same
fiber tract if the mechanism of spread of excitation is purely electrical. We
stated[10] that '…the results presented indicate that the resistance of the intercalated disc is high if crossover junctions between laterally contiguous cells are
not very infrequent (e.g. not more than about four cell lengths apart) and if the
true space constant is not much less than one cell length'. Recently, JOHNSON
and SOMMER[12] reported that some fiber tracts lying side by side are not interconnected by tight junctions for distances of many microns, although many of
the adjacent fibers have very frequent tight-junction connections over a part of

their length. However, their observations cannot explain our findings of no electrical coupling between contiguous cells lying side by side because in 24 successful double impalements made at distances of 60, 154, and 390 μ, not one gave a substantial interaction. The probability is that only a few of these impalements would have been several length constants away from a transverse junction.

Our fundamental observations described above have been confirmed in rabbit ventricle by TILLE[13] who found that '...no electrotonic interaction could be observed in the majority of V (ventricular) fiber pairs', and who concluded that '...the rabbit ventricle cannot be regarded as a single freely interconnected syncytium'. Similarly, in dog papillary muscle, KAMIYAMA and MATSUDA[14] stated that '...contrary to expectation, the routine procedure for polarizing the single fiber by means of an intracellular microelectrode could not produce electrotonic potentials of reasonable size and spatial spread in the adjoining fibers'.

Interelectrode resistance and electrotonus measurements were also made simultaneously with double microelectrodes at distances of less than 30 μ in order to help resolve why intermediate degrees of interaction are sometimes obtained[15]. When the electrotonic interaction is nearly 100%, the interelectrode resistance is nearly zero. When the interaction is nearly 0%, the interelectrode resistance is large; however, the percent interaction often increases simultaneously with spontaneous depolarization due to injury, and the interelectrode resistance decreases concomitantly. There is an inverse correlation of percent interaction with resting potential (correlation coefficient of -0.75). Thus, the variable and intermediate degrees of interaction often found can be accounted for by injury produced by impalement with two closely spaced electrodes which leads to decreased resistance of the discs. The injury also causes an increase in R_m of the surface membrane; hence this type of injury depolarization occurs due to increase in R_m (decrease in g_K) and not a decrease. The phenomenon is the opposite of 'healing-over', in which injury is presumed to increase the resistance of intercalated discs[16].

Additional support for the hypothesis that there is no electrical coupling between contiguous cardiac cells comes from the fact that similar data were obtained using cat intestinal smooth muscle (circular muscle layer)[17]. With the electrodes oriented longitudinally, at interelectrode distances of 100 μ or greater, current pulses up to 20×10^{-9} A applied through one intracellular electrode produce almost no change in potential at the second intracellular electrode, although a large ΔE_m is produced at the first electrode (Table 4). At shorter distances of 40 μ or less, similar currents produce, depending on the impalement, either an insignificant or a substantial ΔE_m at the second electrode, the degree of interaction approaching 100%. With transverse electrode orientation, substantial interactions never occur at distances beyond 10 μ. Since there is transverse propagation of impulses[3], there must be effectively transversely oriented junctions which occur frequently; therefore, one cannot account for the fact

Table 4. Incidence of impalements in cardiac and smooth muscles giving substantial and weak interactions as a function of interelectrode distance and electrode orientation

Preparation	Electrode orientation	Interelectrode distance (μ)	Number of impalements			Percent substantial interactions
			Substantial interaction (>2 or 5%)	Weak or no interaction (<2 or 5%)	Total	
Frog and cat cardiac muscle	Parallel	7	38 (>1 mV/8 nA or >2%)	21 (<1 mV/8 nA or <2%)	59	64
		10, 11, 12	23	53	76	30
		17, 21	4	6	10	40
		28, 35	2	4	6	33
		45	4	4	8	50
		60	0	4	4	0
		154	0	13	13	0
		390	0	7	7	0
Cat intestinal muscle	Parallel	4, 7	45 (>7.5 mV/20 nA or >5%)	7 (<7.5 mV/20 nA or <5%)	52	92
		10, 14	15	12	27	56
		17, 18	20	28	48	42
		25, 28	24	52	76	32
		40	3	18	21	14
		100	0	24	24	0
		200	0	15	15	0
Cat intestinal muscle	Transverse	4	7	17	24	29
		10	5	22	27	18
		17, 18	0	18	18	0
		25, 28	0	22	22	0
		40	0	6	6	0
		200	0	14	14	0

that cells lying side by side are not electrically coupled on the basis of low-resistance junctions being several length constants away. Our fundamental observations have been confirmed in cat intestinal muscle by KOBAYASHI, PROSSER and NAGAI[18], and in guinea-pig *Taenia coli* by KURIYAMA and TO-MITA[19].

Probability Arguments

As stated above, it is clear that contiguous cells lying side by side are not electrically coupled, but the tentative conclusion that cells lying end to end also

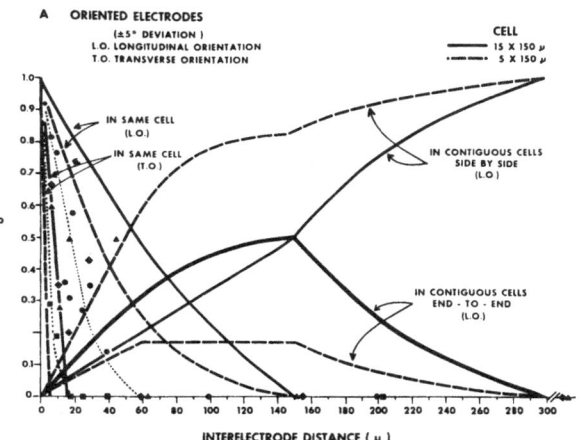

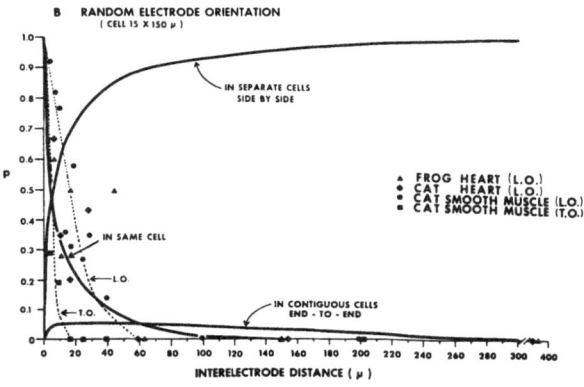

Fig. 3.
Probability calculations for the intracellular location of pairs of microelectrodes at various fixed interelectrode distances. Calculations made both for electrodes oriented parallel (L.O.) and transverse (T.O.) to the long axis of the cells, allowing a ± 5° deviation (part A), and for random electrode orientation (part B). The four possible locations of the electrodes are that they are: (a) in the same cell, (b) in contiguous cells end to end across the myo-myo junction, (c) in contiguous cells side by side in two adjacent fiber tracts, and (d) in separate cells transversely in non-adjacent fiber tracts. Calculations made for two sizes of cylindrical cells: 15 by 150 μ (continuous lines) and 5 by 150 μ (dashed lines). The points plotted (duplicated in parts A and B) represent the summary of previously published data for the incidence of substantial electrotonic interactions as a function of interelectrode distances in frog and cat cardiac muscle and in cat intestinal muscle. Two dotted lines are drawn through these data for T.O. and L.O.

are not coupled rests on the following probability arguments. Fig. 3 gives the probability calculations for the position of the two intracellular microelectrodes as a function of interelectrode distance. The probability calculations were made for oriented electrodes ($\pm$ 5° deviation), both longitudinal and transverse with respect to the long axis of the cells (Fig. 3A), and for two sizes of cells: $15 \times 150\ \mu$ (for cardiac muscle) and $5 \times 150\ \mu$ (for intestinal muscle). The probabilities were also calculated for random electrode orientation in cells $15 \times 150\ \mu$ (Fig. 3B). The calculations for oriented electrodes should be reasonably reliable for the longer interelectrode distances, such as those greater than $50\ \mu$, and those for random electrode orientation should be more realistic for the shorter distances. At distances less than $50\ \mu$ it is difficult to be certain that the electrodes remain oriented during the actual impalement because of flexibility of the tips.

These calculations show that the probability of both electrodes impaling the same cell falls off very rapidly as a function of distance, being for example only about 7% at $50\ \mu$ for random electrode orientation (Fig. 3B). At $50\ \mu$, p is 5% for the two electrodes being in contiguous cells end to end across an intercalated disc, and 88% for their being in separate cells of different fiber tracts lying side by side. With longitudinally oriented electrodes at, say, $120\ \mu$, p is about 48% that the two electrodes would be in contiguous cells end to end for $15\ \mu$ cells (about 17% for $5\ \mu$ cells) and 40% that they would be in contiguous cells lying side by side (81% for $5\ \mu$ cells) (Fig. 3A). Thus, for longitudinal orientation at distances less than $150\ \mu$, when the electrodes are in contiguous cells, the probability is actually higher that they will be in cells end to end across an intercalated disc than that they will be in cells side by side (for $15\ \mu$ cells). With electrodes oriented transversely, there is a very steep fall-off in the probability that the two electrodes will impale the same cell, such that the probability becomes zero at lengths greater than one cell width (Fig. 3A).

Thus, these probability predictions come close to fitting our published experimental data which are superimposed on the theoretical curves of Fig. 3A and 3B (see Table 4). For example, in frog and cat cardiac muscle, no significant interactions were found at interelectrode distances of $60\ \mu$ and greater out of a total of 24 successful impalements (longitudinal electrode orientation). It is clear from Fig. 3 that, with longitudinal electrode orientation at distances close to one cell length, there is a large probability of up to 0.5 that the two electrodes would impale contiguous cells end to end across an intercalated disc. Yet in 17 impalements at distances of 60 and $154\ \mu$ none of the impalements gave interactions. Similarly, no interactions were found at 100 and $200\ \mu$ in cat intestinal muscle out of a total of 39 successful impalements (longitudinal electrode orientation); with transverse orientation, no interactions were found in 60 impalements at distances of 17–$200\ \mu$. It must be concluded, therefore, that the resistance of the intercalated disc is very high.

Studies on Cultured Heart Cells

Electrotonic Spread of Current

Monolayer cultures of heart cells were prepared in order to reduce cardiac muscle to a one- or two-dimensional system and thereby simplify the interpretation of electrotonus measurements. As is found in intact cardiac muscle, the degree of interaction is not dependent on interelectrode distance per se, but appears to depend on whether both electrodes impale the same cell[20]. In many preparations, however, it was not possible to determine the location of the cell junctions. At short distances, the electrotonic interaction is nearly 100% in some impalements, whereas in other impalements the interaction is nearly 0%. Generally there is only little fall-off of electrotonus within one cell (Fig.4). Some-

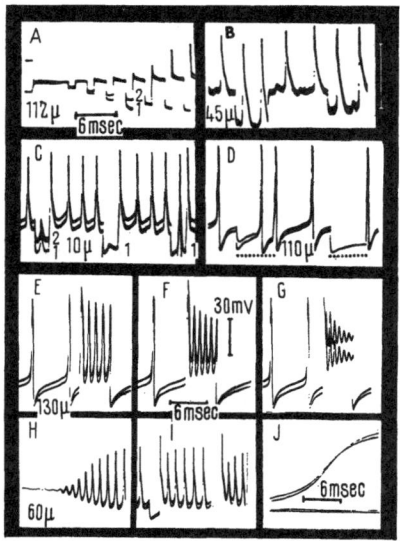

Fig.4.
Strong electrotonic spread in cultured heart cells illustrating some criteria used to help determine when both electrodes were in the same cell, and showing that the bridge technique can be used to obtain accurate measurements of r_{in}. Nearly identical changes in potential were recorded by both electrodes when current was injected through one of two electrodes impaling a cell. A–D: Four separate impalements in strand preparations at the interelectrode distances marked. Time calibration and voltage calibration (50 mV) apply to all photos. A: 1.2, 2.4, 3.6, 4.8, 6.0, 7.6, 9.0 nA hyperpolarizing pulses. B: 2.2 and 1.4 nA hyperpolarizing pulses. C: hyperpolarization of a driven nonpacemaker cell with 1.4 nA pulses. D: hyperpolarizing pulses of 2.4 and 4.8 nA in a pacemaker cell suppressed frequency of firing at both electrodes; dotted lines mark application of pulses. E–J: Two impalements (E–G and H–J) at the interelectrode distances indicated. Voltage calibration applies to all photos; time calibration in F applies to E–I. E–G: Sequential photos from an impalement in a strand; depolarizing pulses of 9.0 (E), 10.2 (F), and 11.6 nA (G) applied through one electrode. H–J: Sequential photos from another impalement. I: Hyperpolarizing pulses of 3.6 and 6.0 nA applied through one electrode suppressed firing at both electrodes. J: Simultaneity of the rising phases of the action potentials.

times, significant interactions were found at long distances of over 500 μ, but these occurred only in large strand preparations in which the cells were abnormally thick and long. Because of the varied sizes and arrangements of cells in culture, the electrotonus data from intact muscle are more meaningful.

Additional criteria which help to determine when both electrodes impale the same cell include: (a) a short diminution of the resting potential to a common intermediate level following insertion of the second electrode (Fig. 4A); (b) graded anodal-break responses of equal magnitude (Fig. 4A); (c) equal changes in magnitude of action potentials produced by polarizing current (Fig. 4B–D), and (d) abolition of the spikes at both electrodes with sufficient hyperpolarization, leaving only equally sized driving junctional potentials (Fig. 4C). The fact that 100% interactions can be obtained (Fig. 4E–J) indicates that the bridge technique, if carefully used, will give faithful measurements of ΔE_m and r_{in} at electrode No. 1, the site of current injection.

Artificial Junction Experiment

Whenever cultured cells grow naturally into physical contact with one another, they usually beat synchronously. However, when independently beating cells are pushed together for several minutes they do not synchronize, but continue to beat independently[4,21]. Thus, such artificial junctions do not result in significant electrotonic interaction. Others, including MARK et al.[22], have observed that synchrony of beating upon natural contact of cells does not occur instantaneously, but requires 10–30 min. Typical intercalated disc structures are found in monolayer cultures of heart cells (ref.[21, 23–25] and unpublished observations). Time necessary for formation of intercalated discs may explain why permanent physical contact between adjacent cultured heart cells is necessary for transmission of excitation.

Synchrony of Cells Connected Only by a Long, Narrow Process

Pairs of isolated cells in sparse cultures, although separated by distances up to several hundred microns, contract synchronously if connected only by a narrow process 0.5–2 μ wide[4,21]. Hence, the two cells must functionally interact by means of the long narrow process. It is unlikely that this interaction is electrical because of impedance mismatch; that is, action current associated with propagation in a narrow process would not be sufficient to produce threshold depolarization of a relatively large cell body[26]. Others, including MARK et al.[22] and MERCER and DOWER[27], have reported similar synchronization of beating by long narrow cell processes. The problem of synchronization is further compounded by the observation[22] that two cultured rat heart cells are synchronized by a third large non-muscle cell ('endothelioid' cell) which is connected to the muscle cells only by very narrow short processes. It seems unlikely that any form of electrical transmission can account for this phenomenon.

Propagation in a Single Chain of Cells

Visual observations of synchronous contractions of chains of cells in loose nets of monolayer cells show that single chains of cells propagate excitation from cell to cell over distances of many cell lengths[21]. Similar observations have been made in cultured smooth muscle cells from chick amnion[28]. Thus, nondecrementing propagation does not need summation from many parallel fiber tracts, as has been proposed for smooth muscle on the basis of cessation of propagation when fiber bundles are dissected to diameters below 100μ[19,29]. If the junctional transmission process is purely electrical it is necessary that sufficient current pass from one active prejunctional cell to excite to threshold one inactive postjunctional cell.

Propagation at Low Resting Potentials

Synchronized beating of cultured heart cells stops only at $[K^+]_o$ levels of 50–70 mM and above, corresponding to resting potentials of -10 to -25 mV[21,30]. The action potentials at these low resting potentials just about reach the zero potential level, and their maximum rate of rise is only about 0.5 V/sec. Yet, excitability spreads from cell to cell, causing synchronized beating among groups of cells. Since the longitudinal current associated with the propagating action potential is proportional to dV/dt, it is difficult to see how sufficient action current can be developed to bring contiguous cells to threshold.

Resting Potential across Each Junctional Membrane, and Lack of Spread of Injury Depolarization

Impalement of cultured cell doublets, joined and beating synchronously, with a large-tipped microelectrode to produce injury depolarization shows that one of the cells possesses a substantial resting potential and continues to beat at a time when its adjoining neighbor is entirely depolarized and not beating[4,21]. Thus, there is sharp discontinuity of potential between the myoplasm of two adjoining cells across a functional junction, i.e. there must be a resting potential across each junctional membrane.

Injury depolarization of one cell resulting from impalement, whether in culture or in intact myocardium, does not spread to a neighboring cell impaled by the other electrode (e.g. see Fig. 2). This finding is also corroborated by visual observation of beating in cultured cells. Others have made similar observations[31].

Driving Junctional Potentials and Partial Block of Transmission

The cardiac action potential consists of three major components: a junctional prepotential, an initial rapid spike component, and a later plateau or slow-wave component. Under various experimental conditions, a separation of the spike and plateau components to various degrees occurs[32]. The plateau can be abruptly terminated by the application of relatively brief hyperpolarizing current pulses. Anodal repolarization appears to be all-or-none in nature, and propagates from cell to cell, terminating the action potential of many cells prematurely. Thus, the membrane potential appears to switch between two stable states.

The junctional potential is a small local potential which triggers the action potential. The junctional potential represents the excitatory interaction between contiguous myocardial cells, and appears as a step on the rising phase of the action potential (Fig. 5 G–I). It becomes prominent and exaggerated under

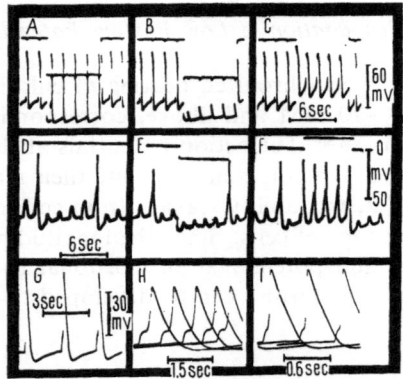

Fig. 5.
Spread of excitation to an impaled cell from a neighboring cell is evidenced by the presence of driving junctional potentials. Upper trace in A–F indicates relative intensity of current pulse; upward deflection, depolarizing current and downward deflection, hyperpolarizing current. A–C: Sequential photos from one impalement. Hyperpolarizing pulses of 4.8 (A) and 7.6 nA (B); the larger pulse abolishes firing and reveals the driving junctional potentials. Depolarizing pulse of 3.6 nA applied in C. D–F: Sequential photos from an impalement illustrating the variable magnitudes of the prepotential. D: No current applied; amplitude of prepotential increases during diastolic depolarization evident between spikes. E: Hyperpolarizing pulse of 2.0 nA diminishes the local excitatory response and produces complete transmission block. F: Depolarizing pulse of 1.8 nA causes each junctional potential to trigger a spike, thus circumventing the naturally-occurring partial block. G–I: From other impalements illustrating a prominent step on the rising phase of the action potential; several sweeps superimposed in H and I.

conditions of impeded propagation[5,33]. The presence of the junctional potential in every cardiac action potential, in which the impaled cell is not a pacemaker, is best illustrated by phase plane recording[34]. In the phase plane, membrane potential is recorded against its first time derivative and small changes in slope become very prominent. Exponential functions can be identified as linear regions and their rate constants and equilibrium potentials measured. Three linear regions occur during the rising phase of the cardiac action potential, the first two before the inflection point and the third after. Since the first exponential is absent in pacemaker cells, the threshold potential probably corresponds to the intersection of the first and second exponentials. Thus, the first exponential process corresponds to part of the rising phase of the junctional potential.

The junctional potential can be further revealed by hyperpolarizing the membrane of cultured cells sufficiently to inhibit development of spikes (Fig. 5 A–B)[4,21]. This failure of spike development occurs suddenly as a step function, thus indicating that the junctional potential remaining is not an electrotonically decremented spike due to a graded spreading of anodal block. Sometimes junctional potentials without spikes occur naturally, indicating a partial block of cell-to-cell transmission (Fig. 5 D–F). In such cases, a second electrogenic component, the local excitatory potential, is sometimes superimposed on the junctional potential; hyperpolarizing current depresses this second component (Fig. 5 E). A spike is always accompanied by a contraction of the impaled cell as well as of the neighboring cells, whereas a junctional potential is not accompanied by contraction of the impaled cell but only of the neighboring cells. Thus, transmission of excitation fails spontaneously in many cases, the impaled cell firing in response to every second or third junctional potential. Failure of transmission is increased with depression of excitability produced by hyperpolarizing current and decreased with enhancement of excitability produced by depolarizing current (Fig. 5 F). The junctional potentials become larger with hyperpolarization and smaller with depolarization in some cases, consistent with the notion that it is an EPSP with an equilibrium potential near zero potential. Some drugs impair and block transmission, and it has also been reported that low O_2 tension blocks cell-to-cell transmission in cultured heart cells[22]. Junctional potentials and failure of transmission are also observed in intact cardiac muscle[5]. These observations support the hypothesis of a rather labile junctional transmission process. Junctions activated chemically would be expected to exhibit fatigue more readily than an electrical junction. For support of the hypothesis of a labile junctional transmission process in smooth muscle, the reader is referred to ref.[35] and[36]. GOTO et al.[37,38] also have described myo-myo junction potentials in cardiac muscle and uterine smooth muscle, and MASHIBA[39] has described them for cardiac muscle.

High Input Resistance

The higher the input resistance, r_{in}, the greater the likelihood that the intercalated discs are high resistance. A high r_{in} means that most of the current injected intracellularly exits into the extracellular fluid close to the site of injection. In a freely branching three-dimensional syncytium, r_{in} should be very low. In a true cable, r_{in} is equal to $0.5\ r_i^{0.5}\ r_m^{0.5}$. Thus, for a fiber of given diameter and R_m, r_{in} is a function of r_i. The resistance of the discs (r_{md}) is in series with and is indistinguishable from r_i so that if r_{md} is large, r_{in} should be large.

The input resistance of myocardial cells, both in intact hearts[7,10,15] and in cultured heart cells[4,20,21] is very large, being about 10 MΩ. Other investigators have also reported similar high values of r_{in} for cardiac muscle[40]. Similar values have been obtained for cat intestinal muscle (7.5 MΩ)[17] and cat uterine muscle

Table 5. Summary of electrotonus studies made on cardiac muscle and smooth muscle.

Reference	Muscle	r_{in} (MΩ)	λ (μ)	Method	Comments
I. Vertebrate cardiac muscle					
SPERELAKIS, HOSHIKO and BERNE (1960)	Frog ventricle	12	≪500	Two intracellular electrodes	Either no low-resistance pathways between cells or short λ
WOODBURY and CRILL (1961)	Rat atrial trabeculae	0.030	130 longitudinal 65 transverse	Probably extracellular	Large currents of up to 1 μA applied; ΔE_m of injected cell not measured
JOHNSON and TILLE (1961)	Rabbit ventricle	0.047		Double-barrelled intracellular microelectrodes Bridge circuit	Large coupling artifact; rheobase current of ∼0.5 μA; $\tau_m < 1$ msec
VAN DER KLOOT and DANE (1964)	Frog ventricle	11			
TARR and SPERELAKIS (1964)	Frog and cat ventricular muscle	6	(cell length) (∼150)	Bridge, and second voltage electrode	
SPERELAKIS and LEHMKUHL (1964)	Cultured chick heart cells	13	(cell length)		
LEHMKUHL and SPERELAKIS (1965)	Cultured chick heart cells	7	(cell length)	Bridge, and second voltage electrode	
TILLE (1966)	Rabbit ventricle	(states low)	100–300 (large I_0 of up to 10^{-6} A had to be used) (not monotonic)	Intracellular	No electrotonic interaction in majority of fiber pairs; ventricle not syncytium
KAMIYAMA and MATSUDA (1966)	Dog papillary	Not measurable	1300	Extracellular	No electrotonic interaction between adjoining fibers by intracellular method
TANAKA and SASAKI (1966)	Mouse ventricle	0.200 to 0.600	70 (increases as a function of distance)	Pairs of intracellular microelectrodes, and pencil-type microelectrodes	Conclude low R_{md} on basis of low r_{in}

Reference	Muscle	r_{in} (MΩ)	λ (μ)	Method	Comments
Weidmann (1966)	Sheep ventricular trabeculae		1550	K⁴² diffusion (extracellular)	Bundles 0.6–0.8 mm thick
Schanne et al. (1966)	Rat atrium	0.177		Double-barrelled intracellular microelectrodes	Concluded low-resistance discs on basis of r_{in} being only 23% of expected cable value
Tarr and Sperelakis (1967)	Frog and cat ventricular muscle	4.8	cell length	Bridge, and second voltage electrode	Injury causes depolarization and large increase in r_{in} and R_m
Johnson and Sommer (1967)	Rabbit ventricular trabeculae	0.5–1.0			Their new r_{in} values are 10–20 times their previously reported value of 47 KΩ
II. Vertebrate visceral smooth muscle					
Daniel and Singh (1958)	Cat uterine muscle	10			
Barr (1961)	Cat intestinal muscle (circular)	103	90–200	Intracellular	Resistance between two cell interiors is $2 \times r_{in}$. This finding of a large r_{in} conflicts with conclusion of low-resistance connections, unless assume a high R_m of > 50 KΩ·cm² (see Fig. 6)
Shuba (1961)	Frog stomach muscle		about 1200	Extracellular	

Table 5. Summary of electrotonus studies made on cardiac muscle and smooth muscle.

Reference	Muscle	r_{in} (MΩ)	λ (μ)	Method	Comments
NAGAI and PROSSER (1963)	Cat intestinal muscle (circular)	69	187 1030 (longitudinal) 270 (transverse)	Intracellular Extracellular	Resistance between two cell interiors is $2 \times r_{in}$ (at interelectrode distances > 150 μ). This finding of a large r_{in} conflicts with conclusion of low-resistance connections, unless assume a high R_m of > 50 KΩ·cm² even for a simple cable (see Fig. 6)
SPERELAKIS and TARR (1965)	Cat intestinal muscle (circular)	7.5	($\sim$176)		
KURIYAMA and TOMITA (1965)	Guinea-pig Taenia coli	40 ($\Delta V/I_o$ curve) 30 ($\Delta AP/I_o$ curve)	250	Intracellular	Findings similar to those of SPERELAKIS and LEHMKUHL on cultured heart cells
TOMITA (1966)	Guinea-pig Taenia coli		1680	Extracellular	No electrotonic coupling between adjacent cells; τ_m of 60–100 msec
KOBAYASHI, PROSSER and NAGAI (1967)	Cat intestinal muscle (circular)	68	118 (intracellular); 1000–1500 (extracellular)	Paired microelectrodes	electrotonic coupling approached 0% at interelectrode distances > 70 μ; τ_m of 21 msec
III. *Invertebrate cardiac muscle*					
ABBOTT, LANG and PARNAS (1968)	Limulus heart			Intracellular electrodes	No electrical coupling between adjacent cells; neuromuscular junctions

$(10\ M\Omega)$[41]. Much higher values of 69 $M\Omega$[8] and 103 $M\Omega$[9] have also been reported for cat intestinal muscle (Table 5). Therefore, the high r_{in} is consistent with the junctional membranes being of high resistance. However, much lower values of 30 $K\Omega$[42], 47 $K\Omega$[43], and 177 $K\Omega$[44] have been reported for r_{in} of cardiac muscle, although the 47 $K\Omega$ value has been revised upward 10–20 times to 500–1000 $K\Omega$ in a later paper[12] (Table 5). For a typical frog sartorius fiber (radius of 40 μ), r_{in} is about 0.5 $M\Omega$[45].

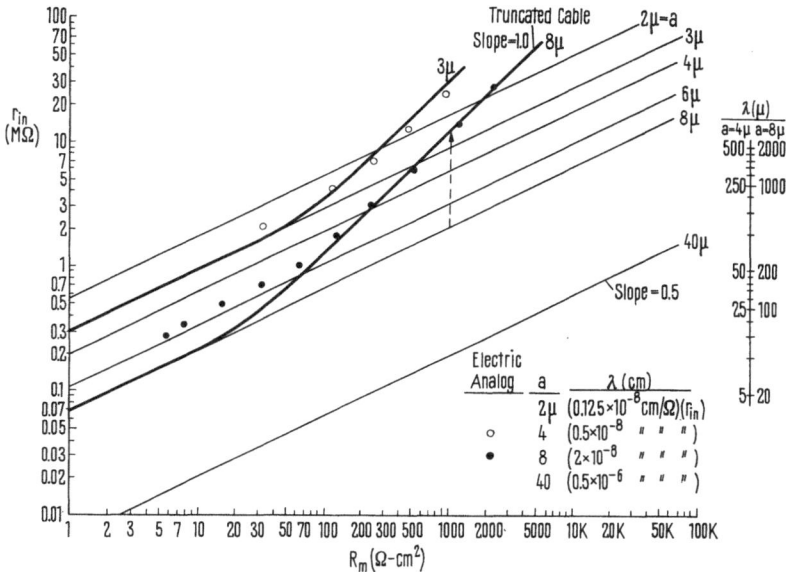

Fig. 6.
Linear relationship on a log-log scale (slope of 0.5) between input resistance, r_{in}, and membrane specific resistance, R_m, for cables of different radii (a) and infinite in length. The current is injected at the middle of each cable. The resistivity of the core conductor, R_i, was assumed to be 200 Ω-cm. The corresponding length constants (λ) for two cables having radii of 4 μ and 8 μ are given by the second ordinate at the right. Theoretical calculations were made for the effect on r_{in} produced by truncation of the cables (3 μ and 8 μ radii) by high resistance transverse membranes, assuming that the specific resistance of the junctional membranes between contiguous cells, R_{md}, was equal to R_m. The linear curve for r_{in} bends upward at higher R_m values and becomes linear again with a steeper slope of 1.0. Thus, at high R_m values, the internal resistance, r_i, is negligible and r_{in} is a simple direct function of R_m. At very low R_m values, λ is so short that the high resistance junctional membranes would have a negligible effect on r_{in}. The points plotted were obtained for cell radii of 8 μ ($\bullet$), and 4 μ ($\bigcirc$) using the electrical analog circuit shown in Fig. 7. Note that there is good agreement with the theoretical calculations, the analog curve having corresponding slopes of 0.5 and 1.0. The small absolute deviation at the lower R_m values is due to the need for greater distribution of the r_m resistors when r_i becomes more significant; lowering the relative value of r_i causes the points to fall almost exactly on the theoretical curve.

The linear relationships, on a log-log scale, between r_{in} and R_m for doubly infinite cables of different radii are given in Fig. 6. The slopes of these lines are 0.5, i.e. r_{in} is proportional to $R_m^{0.5}$. The right-hand ordinate (log scale) gives the calculated length constant values as a function of R_m for cables of 4 μ and 8 μ radii. Also plotted are two calculated curves for truncated cables (at 150 μ), one for a cell of 3 μ radius and the other for an 8 μ cell. At low R_m values, at which the corresponding λ is very short compared to a cell length (150 μ), truncation by a transverse membrane whose specific resistance is equal to that of the surface membrane ($R_{md} = R_m$) has very little effect on r_{in} because the cell is several λ long. However, as R_m approaches more realistic values, the curve gradually

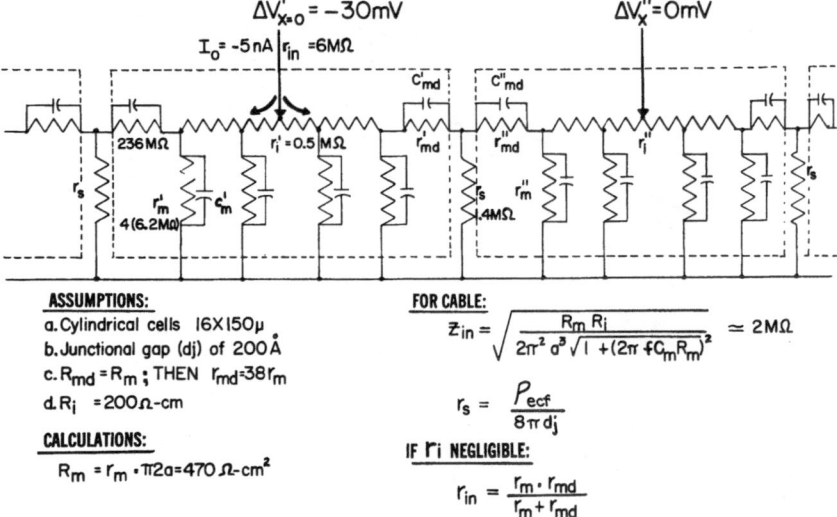

Fig. 7.
Approximate electrical circuit analog for two contiguous cardiac cells lined up end to end across an intercalated disc. Three successive intercalated discs are depicted, and the light dashed lines superimposed on the analog enclose the borders of two complete cells and ends of two other cells. The membrane resistance, r_m, has been distributed into only four resistances per cell for simplicity. The resistances of the myoplasm (r_i) and each membrane of the intercalated disc are denoted by prime and double-prime for each of the two cells. The shunting resistance of the junctional gap is given by r_s. The relative values for r_i, r_m, r_{md}, and r_s given in the figure were calculated for cylindrical cells 16 μ in diameter and 150 μ in length separated by junctional gaps of 200 Å and assuming an R_i of 200 Ω-cm and that $R_{md} = R_m$. The value for r_s was calculated from Katz's equation given in the figure. If 5 nA of hyperpolarizing current injected into the middle of the cell at the left produces a hyperpolarization of 30 mV, then r_{in} is 6 MΩ. Various values of r_{in} measured on such an analog for different values of R_m are plotted in Fig. 6. If r_i is negligible, and r_s is small compared to r_{md}, then the resistances of the surface and disc membranes are in parallel, and the simplified equation shown holds. The AC input impedance (Z_{in}) for a simple cable can be calculated from Katz's equation shown.

bends upwards and again straightens out with a new slope of 1.0; that is, now r_{in} and R_m have a direct linear relationship ($r_{in} \propto R_m^1$). Conversely, any hypothetical fiber branching would act to diminish the slope below 0.5, approaching a maximum of zero slope (r_{in} independent of R_m)[46]. Thus, Fig. 6 gives the predicted increase in r_{in} expected if a long fiber is truncated into individual cells separated by high-resistance junctions. For example, a cardiac fiber tract of 8 μ radius having an R_m of 1000 $\Omega \cdot cm^2$ should have an r_{in} of only about 2 MΩ if it forms a simple cable, compared to a value of about 12 MΩ if each intercalated disc membrane also has an R_{md} of 1000 $\Omega \cdot cm^2$. Therefore, it is obvious that the absolute value of r_{in} is critically important for helping to determine whether the intercalated discs are high- or low-resistance. To account for an r_{in} of 12 MΩ and still have low-resistance discs, R_m would have to be about 35,000 $\Omega \cdot cm^2$ (Fig. 6).

The points plotted in Fig. 6 were obtained from the electrical analog shown in Fig. 7 for two contiguous cells end to end, separated by intercalated discs having a shunt resistance (r_s) of 1.4 MΩ for a junctional gap of 200 Å (calculated from the equation shown and given in ref.[26]). The analog is an approximation because r_m is distributed into only four parallel resistances. The resistance values given in the figure are for cylindrical cells, 16×150 μ, having an r_{in} of 6 MΩ, and assuming $R_{md} = R_m$. These values were adjusted for use with cells of different radius. The points plotted in Fig. 6 are for cells of 8 μ radius (filled circles) and 4 μ (unfilled circles). It is obvious that these values obtained from the

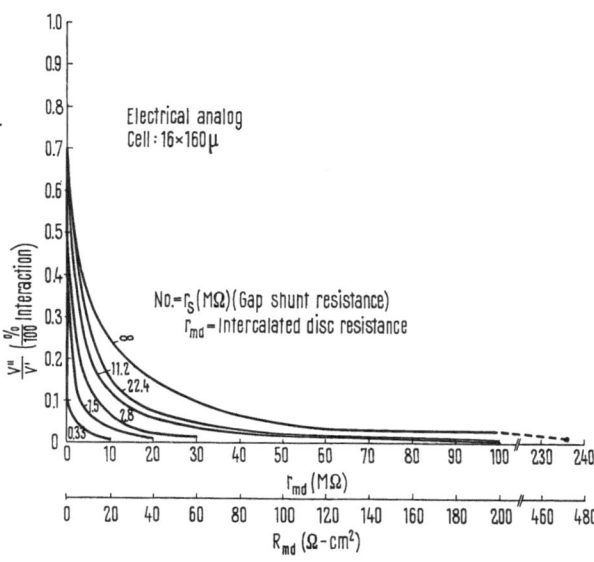

Fig. 8.
Measurements on the electrical analog shown in Fig. 7 giving percent interaction between contiguous cells end to end as a function of the resistance of the intercalated disc (r_{md}) and at different gap shunt resistances (r_s). It is seen that in order for successful electrical transmission to occur (e.g., % interaction $> 20\%$), the specific resistance of each junctional membrane of the intercalated disc (R_{md}) must be < 2 Ω-cm^2 for the calculated r_s of 1.4 MΩ. The maximum % interaction of 77% on the electrical analog at an r_s of infinity (∞) would become greater (approaching 100%) if there were an active change in E_m up to the end of the prejunctional cell. At an r_s of 1.5 MΩ, the intersection on the ordinate is 0.45.

analog give a reasonably good fit to the corresponding calculated curves. The small deviation at the lower R_m values is due to the limited distribution of the r_m resistances, this defect becoming most prominent at the larger r_i/r_m ratios.

Fig. 8 gives the relationship for percent electrotonic interaction between two contiguous cells as a function of the resistance of the intercalated discs obtained from the electrical analog shown in Fig. 7. The various curves are for different values of gap shunt resistance, varying from infinity to 0.33 MΩ. The curves have a steep hyperbolic shape, the percent interaction rapidly approaching zero for disc membranes having a specific resistance of the order of that of the cell surface membrane. They show that for substantial electrical coupling, i.e. that which is nearly sufficient for bringing the postjunctional cell to the threshold potential (e.g. $> 20\%$), the maximum specific resistance allowable for each membrane of the intercalated disc is only about 2 Ω·cm² for an r_s of 1.4 MΩ.

The evidence that r_{in} is high comes from several types of measurements: (a) slope of steady-state voltage/current curves, (b) slope of spike magnitude as a function of applied current, and (c) action potential frequency as a function of applied current. The change in membrane potential (ΔE_m) produced at the site of current injection is equal to r_{in} times the applied current (I_o): $\Delta E_m = r_{in} \cdot I_o$. The following paragraphs summarize these measurements.

(a) *Steady-state voltage/current curves* ($\Delta E_m/I_o$). Plots of the steady-state changes in membrane potential recorded from electrodes No. 1 and No. 2 at all interelectrode distances are often linear in both the depolarizing and hyperpolarizing quadrants (± 30 mV). However some rectification is apparent, especially at large hyperpolarizations. Voltage/current curves of the data obtained from electrode No. 1 give an average cell input resistance of 13 $\pm$ 1.2 MΩ for cultured heart cells[4,20], 5.5 MΩ for frog ventricle, and 6.4 MΩ for cat ventricle[10]. The degree of interaction is determined by the ratio of the slopes of the potential changes at electrode No. 2 compared to those at electrode No. 1.

(b) *Effect of current upon the magnitude of action potentials.* Spike height and frequency of discharge are affected by changes in E_m produced by small polarizing currents of less than 1 nA; hence, the resistance of one heart cell or r_{in} must be high in order to produce a sufficient ΔE_m. Polarizing current has a pronounced effect on spike magnitude in cultured heart cells[4,20] and in intact cardiac muscle[34]. Depolarizing current diminishes and hyperpolarizing current enhances the action potential (Fig. 5). The relationship between relative spike magnitude and current is linear between ± 4 nA, with a mean slope of 22% change in spike height per 1.0 nA. Thus, for a mean action potential of 72 mV, the spike height would be changed about 16 mV/nA. Allowing for an increase in overshoot, there is at least a 12 mV change in resting potential per 1.0 nA, from which an r_{in} of about 12 MΩ is obtained. Thus, data obtained by this method predict a lack of rectification and give an effective cell resistance (r_{in}) similar to that obtained by voltage/current curves in resting membrane. Similar findings have been obtained for smooth muscle[19].

(c) *Effect of current on frequency of firing of cultured heart cells.* The frequency of firing of non-pacemaker cells, driven by transmission of excitation from neighboring cells, remains unaltered with electrotonic depolarization or hyperpolarization[4,20,21]. Excessive hyperpolarization suddenly abolishes the spike responses as a step function, leaving the small driving junctional potentials at the same frequency (Fig. 5A, B). Such voltage independency of frequency and the presence of junctional potentials are the major characteristics of non-pacemaker cells.

In pacemaker cells, the frequency of firing increases during depolarizing pulses and decreases with hyperpolarization (Fig. 9). Excessive hyperpolarization abolishes firing but without the presence of junctional potentials (Fig. 9D, F). The relationship between frequency of discharge and polarizing current is

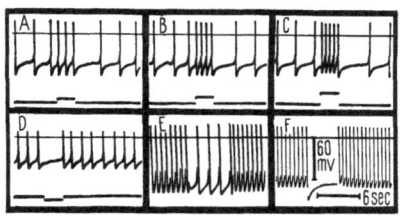

Fig. 9.
Polarizing currents of low intensity greatly modify the frequency of firing of true pacemaker cells. Lower trace in A–D is the current channel. A–D: Sequential photos from one impalement; depolarizing pulses of 0.4 (A), 0.6 (B), and 1.2 nA (C); hyperpolarizing pulse of 0.4 nA (D). E: In another cell, hyperpolarizing pulse of 1.2 nA decreased action potential frequency from 88 to 39 impulses per minute. F: Electrical activity in another pacemaker cell abolished by a hyperpolarizing pulse of 1.2 nA; note absence of junctional potentials.

sigmoid, with a linear steep portion (between 0 and 1.2 nA of depolarizing current) having a slope of about 58 impulses/min/nA. This indicates that r_{in} is very high. The effect of polarizing current appears to be mediated mainly by a large change in the slope of the pacemaker potential for a small ΔE_m.

Pacemaker and non-pacemaker cells can be interconverted[33], therefore supporting the conclusion of a high r_{in} for all cardiac cells. Ba^{++} or Sr^{++} (5–10 mM) rapidly converts driven or quiescent non-pacemaker cells into spontaneously firing pacemaker cells. Automaticity of a pacemaker cell is rapidly and reversibly abolished at elevated $(K^+)_o$ levels (< 20 mM) which do not produce a significant ΔE_m.

Although pacemaker cells always respond to very small depolarizing currents, for some unknown reason non-pacemaker cells, whether in culture or in intact myocardium, respond to intracellularly applied depolarizing current pulses either not at all or only with exceptionally large currents[21]. Thus, although these cells readily respond to anodal-break stimulation and fire action potentials during propagation, they fail to respond to large electrotonic depolarizations. Excessive current applied intracellularly may behave effectively as though applied extracellularly. It has been reported that intact ventricular cells require much more intracellularly applied depolarizing current than do Purkinje fibers[47], although a given current should depolarize a ventricular cell more than a Purkinje fiber since ventricular cells are smaller (higher r_{in}). Intracellular stimulation of smooth muscle by depolarizing pulses is also very difficult or does not occur[18,19].

Measurement of λ by Extracellular Methods

Measurements of λ by extracellular methods give values much larger than those obtained by intracellular injection of current (Table 5). Thus, these data lead to opposite conclusions with respect to the resistivity of the intercalated discs. Hence it is important that the possible reasons for these discrepancies be examined.

Glass Rod Model

In parallel-fibered muscle bundles, the electrical resistance or resistance to diffusion should be several times greater in the radial compared to the axial directions relative to fiber orientation. Therefore, the extracellular fluid of a *muscle bundle itself acts as a cable*, and one must distinguish between length constants of the tissue and of the cells (or fibers). To examine this point, a physical model to represent the extracellular fluid of a bundle of parallel cardiac muscle fibers was constructed by using glass rods and Tyrode solution. With the packing of rods used, the electrical resistivity of the 'tissue' in the transverse direction was seven times that in the longitudinal direction (Table 3). In actual cardiac muscle, this ratio can be much larger because of the smaller ECF space and the more flattened profile of cell cross-sections. The ratio can theoretically approach infinity with ideal packing. Similar data are not available for actual cardiac tissue, but somewhat analogous measurements made on human skeletal muscle give a transverse resistivity 2–3 times the longitudinal one[48].

Electrical Analog

An electrical analog circuit representing diffusion in the extracellular fluid of a bundle of cardiac muscle fibers was also constructed (Fig. 10). If the 'length constant' (λ') is arbitrarily taken as the point where the potential falls to $1/e$ (37%), this analog gives an average tissue λ' for a bundle of 500 μ radius of about 0.53 mm (λ' at the core of over 0.9 mm) if the ratio of diffusion coefficients in the longitudinal/radial directions is taken as 8 (Fig. 11). WEIDMANN[49] actually found that λ_{Br} for Br$^-$ diffusion in 0.4 mm radius cardiac bundles, Br$^-$ being used as an extracellular marker, was approximately 0.5 mm. Fig. 12 shows the bullet-shaped potential profiles at $1/e$ for different resistance ratios that are obtained from the electric analog for any longitudinal section passing through the core of the 'bundle' of 0.5 mm radius. Two additional profiles (dashed lines) are also given at a ratio of 8:1 for $1/e^2$ ($2\lambda'$) and $1/e^3$ ($3\lambda'$).

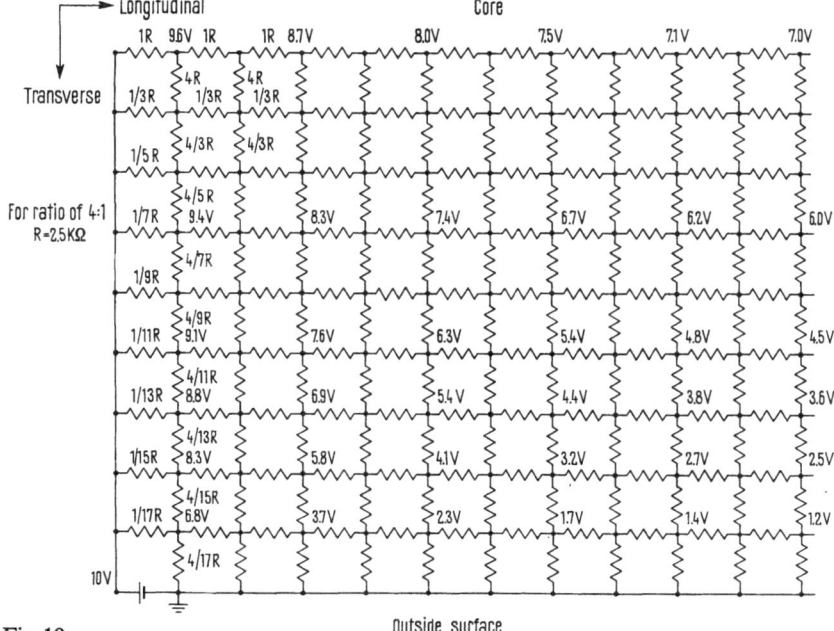

Fig. 10.

Schematic diagram of the electrical analog circuit used to study the equivalent three-dimensional profile for longitudinal diffusion in a parallel-fibered muscle bundle. The analog illustrated is for a ratio of 4:1 for longitudinal/transverse conductivities $(R_l/R_1 = D_1/D_t = 4)$; other ratios were obtained by changing the resistance ratios. The various R_1 values, going from the core to the surface, were determined by dividing the cylindrical bundle radius into 9 equidistant segments, and assuming their respective resistances to be inversely proportional to their areas. The voltage (+ 10 V) was applied in a transverse plane across the cross-sectional area of the bundle (left side of diagram). The outside surface of the bundle (bottom of diagram) was held at zero potential (ground). Maximum potential is at the core of the bundle (top of diagram). The analog shown may be arbitrarily scaled to any relative units of bundle radius and longitudinal distance. The analog was lengthened in the longitudinal dimension by use of appropriate longitudinal compression ratios. This analog circuit is similar to that used by WEIDMANN[49].

In addition, theoretical calculations were made for diffusion in parallel-fibered muscle bundles using the diffusion equations (Fig. 13). Calculations were made for bundles having radii of 0.4 mm and 1.0 mm. These curves are sums of exponentials, but approach a single exponential at the larger ratios. The points plotted are the corresponding approximate average bundle λ' values obtained from the electrical analog. It is apparent that there is reasonably close agreement between the theoretical calculations and the electrical analog measurements. Therefore, the long apparent 'length constants' measured by any extracellular means can be due entirely to ion movements through the extracellular fluid and need not have any direct bearing on the resistance of the intercalated discs.

Fig. 11.
Dependence on the resistance ratios
$(R_r/R_1 = D_1/D_r)$ of the apparent λ'
(decay in potential to $1/e$) of a bundle
of 0.5 mm radius. Data measured on an
electrical analog circuit described in
Fig. 10. Apparent λ' given for the core
and for the average of any transverse
plane of the bundle. The average λ' at
any distance along the bundle is assumed to be approximately equal to
that at $0.707 \times a$ from the core (at
which 50% of the volume is central and
50% is peripheral). The average values
obtained for ratios of 4, 8, and 16 were
adjusted for bundle radius and plotted
in Fig. 13. There is good agreement
between these values measured on the
electrical analog and those calculated
from the diffusion equations.

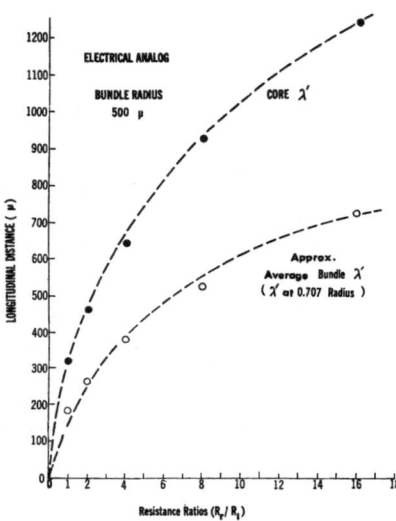

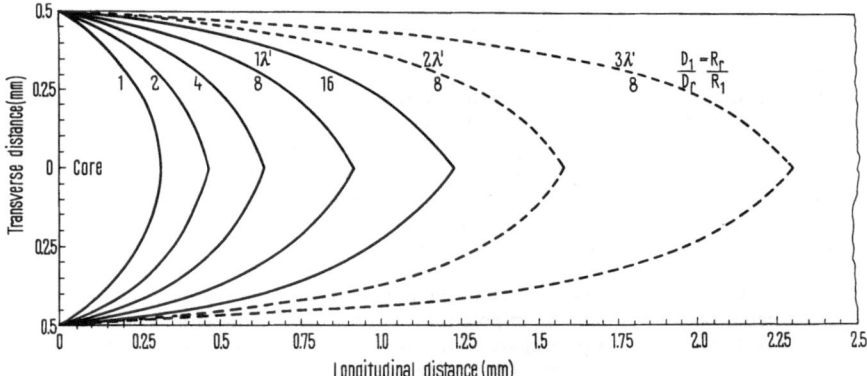

Fig. 12.
Two-dimensional profile for decay of potential to $1/e$ (apparent λ') in an electrical
analog circuit of the extracellular fluid of a bundle of muscle fibers all oriented
parallel to one another (longitudinal orientation); potential plotted as a function of
longitudinal distance from a sharp transverse boundary at the plane of voltage injection at left end of diagram. The bullet-shaped profiles give the relative potential in
the longitudinal and transverse directions with distance along the bundle for any
longitudinal plane that passes through the longitudinal axis (core) of the bundle.
Maximum potential is always at the core and zero potential is at the surface of the
bundle. The profiles were measured for different ratios of resistances, radial to longitudinal (R_r/R_1) corresponding to the inverse ratios of diffusion coefficients (D_1/D_r).
The apparent λ' is greatest at the core and is zero at the surface. The dashed lines
give the profiles for $2 \times \lambda'$ ($1/e^2$) and $3 \times \lambda'$ ($1/e^3$) for a ratio of 8:1. The diagram shown
is for a bundle of 0.5 mm radius, but the curves may be adjusted to apply to a bundle
of any radius. The core λ' is greater than 1.2 mm for a ratio of 16:1.

Fig. 13.
Theoretical decay of a
labelled substance (e.g.,
Br[82]) in a bundle of
muscle fibers all oriented
parallel to one another
(longitudinal orienta-
tion) as a function of
longitudinal distance
from a sharp boundary
with an infinite reser-
voir, and assuming the
substance remains extra-
cellular only. The curves
were calculated from
the diffusion equations
using seven terms, and
with the stipulation that
any tracer reaching the
radial surface of the
bundle is instantane-

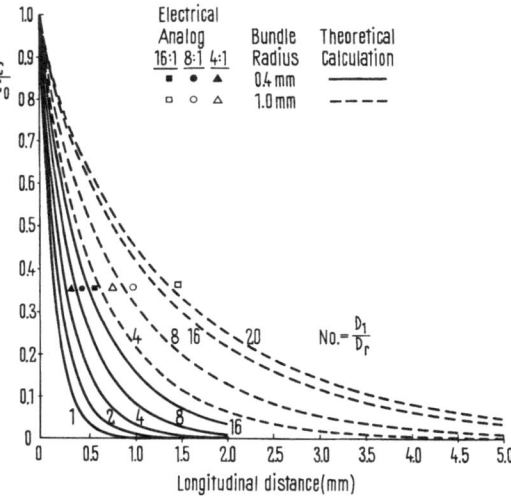

ously washed out and discarded. The curves give the relative concentrations of the
substance with distance along the bundle under steady-state conditions. The curves
were calculated for various ratios of diffusion coefficients, longitudinal to radial
(D_l/D_r), from 1 to 20 as labelled in the figure, and for two size bundles: 0.4 mm and
1.0 mm radii. These curves are a sum of several exponential terms, and approach
a single exponential at the larger ratios. Although there is no real length constant,
decay to $1/e$ occurs at 1.3 mm for a ratio of 20 in a bundle of 1 mm radius; hence,
the apparent 'λ' would be 1.3 mm. The six points plotted are from independent
measurements on an electrical analog of this diffusion problem for decay in potential
to $1/e$, as given in Fig. 10–12. Thus, there is a remarkable agreement between the
analog measurements and the theoretical calculations.

Literature Values

Table 5 summarizes some of the available data for cardiac muscle and
smooth muscle comparing the apparent λ values obtained by intracellular and
extracellular methods. It is obvious that extracellular methods give apparent
long λ and intracellular methods give short values. By intracellular methods,
short λ of the order of one cell-length have been measured in cardiac muscle[4,7],
[10,13,15,20,42,50], and in smooth muscle[8,17–19,29]. By extracellular methods, long λ
of the order of 1 mm have been measured in cardiac muscle[14,49] and in smooth
muscle[18,29,51,52].

WEIDMANN[49] obtained a long length constant of 1.6 mm for sheep ventricular trabeculae by measuring the exponential profile for steady-state K[42] diffusion ($\lambda_K = 1.6$ mm). These experiments would be excellent if done on an ideal
single fiber. But in a muscle bundle, it is difficult to adequately distinguish
between longitudinal K[42] movement through the myocardial cells and across
the intercalated discs with that through capillaries, other blood vessels, and

neurons. Also because of the problem of the ratio of diffusion coefficients in the longitudinal versus radial directions, any K^{42} ion which exits from one cell has a greater possibility of moving longitudinally down the tissue than moving in a radial direction and getting washed out of the tissue.

WOODBURY and CRILL[42] measured the electrotonic potential surrounding a current electrode in rat atrial trabeculae and obtained an elliptical isopotential contour with a λ of 130 μ in the longitudinal direction and 65 μ in the transverse direction. From these data, they concluded that the myocardium formed a two-dimensional or pancake syncytium. The elliptical contour may be similarly explained by the problem of differential current flow through the extracellular fluid. Furthermore, they injected extremely large currents (about 1 μA) through a supposedly intracellular microelectrode. In my experience, this tremendous amount of current causes the electrode to become dislodged from the cell and perhaps the microelectrode tip to break. In addition, this amount of current should produce sufficient heat to damage the impaled cell. It seems likely that the behaviour when large currents are applied is such that the current is effectively applied extracellularly. Their λ measurements then are probably due to IR drops in the interstitial fluid, and this can account for their very low apparent r_{in} of 30 KΩ (extrapolated value). As shown, λ_{ecf} can be expected to vary anywhere from about 100 to 600 μ or more, depending on the geometry of the tissue and other factors. In the electrical analog shown in Fig. 10, the current is injected uniformly through a cross-sectional plane of the bundle and not from a point source as in WOODBURY and CRILL's experiments. The 1 μA of current applied compares with values of 0.1–4 nA applied in our experiments, which are sufficient to substantially alter resting E_m, frequency of firing, and spike magnitude, duration, and maximum rate of rise. The important parameter for measuring the electrical coupling between cells is the ratio of potential change in the second cell to that of the injected cell. The ΔE_m in the injected cell is not known in the experiments of WOODBURY and CRILL. The fact that a potential change can be produced in the second cell is not sufficient because if large currents are applied, a ΔE_m would be produced even in cells known not to be electrically coupled.

BARR, DEWEY and BERGER[53] showed that the action potential in cardiac and smooth muscles can jump across a sucrose gap if current is allowed to flow through the external circuit. For small bundles, the maximum width of the sucrose gap permissible is about 400 μ. These experiments show that the functional length constant is short, since the maximum gap width should represent about 2 X λ. However, these experiments cannot easily distinguish between electrical and non-electrical junctions because longitudinal current must flow in both cases. In chemical transmission, the source of the depolarizing current is the postjunctional cell itself when the postsynaptic membrane is activated by the chemical mediator from the prejunctional cell. Hence, failure of propagation by itself does not reveal the mechanism of the junctional transmission process. In addition, because it is impossible to get sharp borders with the sucrose gap, the shorter the gap and the thicker the muscle bundle, the less sharp the gap borders will be. The fact that the ends of the individual cells are not lined up side by side, but are staggered, creates extra problems of interpretation. Since efflux of Na^+, Ca^{++}, K^+, etc. can occur from cells in the gap, there is a possibility of membrane excitability remaining, especially near the peripheral regions of the gap; action potentials can occur in low $[Na^+]_o$, or Ca^{++} spikes can occur. Therefore, the behavior of the cells in the gap should be determined. If an ideally sharp gap of inexcitability could be obtained, its minimum width should be at least two cell lengths.

Summary

From the general facts concerning the nature of cell-to-cell interactions in cardiac muscle, several very general summarizing statements can be made: (1) Many of the electrophysiological properties of single isolated cultured heart cells are similar to those of cells in intact adult hearts; thus, the individual myocardial cell is the smallest viable unit for electrical and mechanical activity characteristic of cardiac muscle. (2) Propagation can occur in a single chain of cells. (3) Injury depolarization does not spread from one cell to another. (4) The intercalated discs have a high capacitance. (5) The input resistance of myocardium is very high, consistent with high-resistance junctional membranes. (6) Cardiac muscle is not a two- or three-dimensional branching electrical syncytium. (7) The cable properties of myocardial cells must be distinguished from those of parallel-fibered muscle bundles. (8) A labile process of junctional transmission, with junctional potentials, occurs for transmission of excitation from cell to cell.

A total of 36 cylindrical glass rods were used in 6 rows of 6 per row (giving 6 columns of 6 per column) in a lucite tank. Each rod was 10.3 cm long and 0.32 cm in diameter. The tank was 10.3 cm long, 1.9 cm wide, and 2 cm deep. All the rods were orderly lined up in one direction parallel to one another. The volume of the rods occupied 76.5% of the tank volume, leaving an 'extracellular space' of 23.5%. (The maximum packing possible for cylindrical rods is $\pi/4$ or 78.6%.)

Measurements were made in normal Tyrode solution (1 X), in Tyrode solution diluted tenfold (0.1 X), and in Tyrode solution concentrated tenfold (10 X). Resistivity measurements were made in the longitudinal direction (ϱ_l) of the tank and in the transverse direction (ϱ_t) both with and without the rods. The resistance measurements were made with a bridge circuit using a 1 V peak-to-peak sine wave. All resistivity measurements were made at 10 Kc to minimize electrode polarization of the copper plate electrodes.

With any of the solutions, the dielectric constant decreased linearly with frequency on a log-log plot, at frequencies above 100 cps. Addition of rods decreased the dielectric constant.

References

1   N. SPERELAKIS, T. HOSHIKO, R. F. KELLER JR. and R. M. BERNE, Am. J. Physiol. *198*, 135 (1960).
2   N. SPERELAKIS and T.HOSHIKO, Circulation Res. *9*, 1280 (1961).
3   C. L. PROSSER and N. SPERELAKIS, Am. J. Physiol. *187*, 536 (1956).
4   N. SPERELAKIS and D. LEHMKUHL, J. gen. Physiol. *47*, 895 (1964).
5   T. HOSHIKO and N. SPERELAKIS, Am. J. Physiol. *201*, 873 (1961).
6   N. SPERELAKIS, Circulation Res. *12*, 676 (1963).
7   N. SPERELAKIS, T. HOSHIKO and R. M. BERNE, Am. J. Physiol. *198*, 531 (1960).
8   L. BARR, Am. J. Physiol. *200*, 1251 (1961).
9   T. NAGAI and C. L. PROSSER, Am. J. Physiol. *204*, 910 (1963).
10 M. TARR and N. SPERELAKIS, Am. J. Physiol. *207*, 691 (1964).
11 T. SANO, H. TSUCHIHASHI and T. SHIMAMOTO, *Electrical Activity of Single Cells* (Ed. Y. KATSUKI; Igaku Shoin, Tokyo 1960), p. 261.
12 E. A. JOHNSON and J. R. SOMMER, J. Cell Biol. *33*, C103 (1967).
13 J. TILLE, J. gen. Physiol. *50*, 189 (1966).
14 A. KAMIYAMA and K. MATSUDA, Jap. J. Physiol. *16*, 407 (1966).
15 M. TARR and N. SPERELAKIS, Am. J. Physiol. *212*, 1503 (1967).
16 J. DÉLÈZE, *Electrophysiology of the Heart* (Ed. B. TACCARDI and G. MARCHETTI; Pergamon Press, London 1965), p. 147.
17 N. SPERELAKIS and M. TARR, Am. J. Physiol. *208*, 737 (1965).
18 M. KOBAYASHI, C. L. PROSSER and T. NAGAI, Am. J. Physiol. *213*, 275 (1967).
19 H. KURIYAMA and T. TOMITA, J. Physiol. *178*, 270 (1965).
20 D. LEHMKUHL and N. SPERELAKIS, J. cell. comp. Physiol. *66*, 119 (1965).
21 N. SPERELAKIS, *Electrophysiology and Ultrastructure of the Heart* (Ed. by T. SANO, V. MIZUHIRA and K. MATSUDA; Bunkodo Co., Tokyo 1967), p. 81.
22 G. E. MARK, J. D. HACKNEY and F. F. STRASSER, *Factors Influencing Myocardial Contractility* (Ed. R. D. TANZ, F. KAVALER and J. ROBERTS; Academic Press, New York 1967), p. 301.
23 M. JELLINEK, N. SPERELAKIS, L. M. NAPOLITANO and T. COOPER, J. Neurochem. *15*, 959 (1968).
24 I. HARARY and B. FARLEY, Expl. Cell Res. *29*, 466 (1963).
25 M. J. HOGUE, Anat. Rec. *99*, 157 (1947).
26 B. KATZ, *Nerve, Muscle and Synapse* (McGraw-Hill, New York 1966).
27 E. N. MERCER and G. E. DOWER, J. Pharm. Exp. Therap. *153*, 203 (1966).
28 A. WOLLENBERGER, personal communications.
29 T. NAGAI and C. L. PROSSER, Am. J. Physiol. *204*, 915 (1963).
30 A. J. PAPPANO and N. SPERELAKIS, Expl. Cell Res. *54*, 58 (1969).
31 K. E. ROTHSCHUH, Arch. ges Physiol. *253*, 238 (1951).
32 T. HOSHIKO and N. SPERELAKIS, Am. J. Physiol. *203*, 258 (1962).
33 N. SPERELAKIS and D. LEHMKUHL, J. gen. Physiol. *49*, 867 (1966).
34 N. SPERELAKIS and H. K. SHUMAKER, J. Electrocardiol. *1*, 31 (1968).
35 H. F. BRUNE and H. KOTOWSKI, Pflügers Arch. *262*, 484 (1956).
36 T. SUZUKI, A. NISHIYAMA and K. WADA, Tohoku J. expl. Med. *77*, 394 (1962).
37 M. GOTO, T. TAMAI, Y. ABE and T. YANAGA, Kyushu J. Med. Sci. *12*, 177 (1961).
38 M. GOTO, H. KURIYAMA and Y. ABE, Proc. Jap. Acad. *36*, 509 (1960).
39 H. MASHIBA, Japan. Heart J. *2*, 487 (1961).
40 W. G. VAN DER KLOOT and B. DANE, Science *146*, 74 (1964).
41 E. E. DANIEL and H. SINGH, Can. J. Biochem. Physiol. *36*, 959 (1958).
42 J. W. WOODBURY and W. E. CRILL, *Nervous Inhibition* (Ed. E. FLOREY; Pergamon Press, London 1961), p. 124.
43 E. A. JOHNSON and J. TILLE, J. gen. Physiol. *44*, 443 (1961).

44   O. F. SCHANNE, L. J. THOMAS JR. and E. CERETTI, Federation Proc. *25*, 635 (1966).
45   N. SPERELAKIS, M. F. SCHNEIDER and E. J. HARRIS, J. gen. Physiol. *50*, 1565 (1967).
46   E. P. GEORGE, Australian J. expl. Biol. *39*, 267 (1961).
47   J. USHIYAMA and C. McC. BROOKS, Am. J. Cardiol. *10*, 688 (1962).
48   H. C. BURGER and R. VAN DONGEN, Physics Med. Biol. *5*, 431 (1961).
49   S. WEIDMANN, J. Physiol., London *187*, 323 (1966).
50   I. TANAKA and Y. SASAKI, J. gen. Physiol. *49*, 1089 (1966).
51   M. F. SCHUBA, Biophysics *6*, 56 (1961). (Trans. from Biofizika *6*, 52.)
52   T. TOMITA, J. Physiol. *183*, 450 (1966).
53   L. BARR, M. M. DEWEY and W. BERGER, J. gen. Physiol. *48*, 797 (1965).

Neuromuscular Transmission in the Heart of the Lobster
Homarus americanus

by MARGARET ANDERSON and I. M. COOKE
The Biological Laboratories, Harvard University, Cambridge,
Massachusetts 02138

Contractions of the single-chambered neurogenic lobster heart are produced by spontaneous bursts of impulses generated by neurons in the cardiac ganglion[1,2]*. In this study dangling microelectrodes were used to record intracellularly from single, striated cardiac muscle fibers of isolated lobster hearts.

The muscle fibers exhibit resting potentials of -50 to -60 mV. In the spontaneously beating heart the individual fibers show regular complex depolarizations. Each depolarization is accompanied, following a short latency, by a contraction. By recording extracellularly from a cut proximal nerve branch of the ganglion while simultaneously recording intracellularly from a muscle fiber, it can be seen that each muscle fiber depolarization is associated with a burst of nerve impulses exiting from the ganglion. The intracellular potential changes consist of many depolarizing increments which we interpret to be excitatory junction potentials (ejp's) elicited by efferent nerve activity. The typical configuration of the depolarizations is illustrated in Fig. 1. The rapid

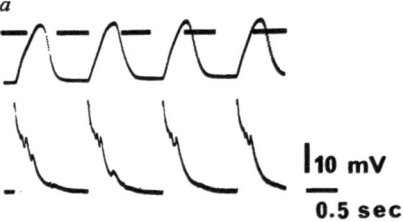

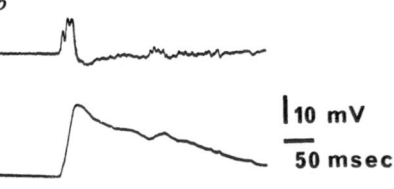

Fig. 1.
Intracellular recordings from muscle fibers of the normally beating neurogenic lobster heart.
a. The bottom trace shows spontaneously evoked activity recorded intracellularly from a heart muscle fiber. The top trace is a mechanical record of heart contraction. The dashed line indicates the zero potential. It is clear that the muscle fiber depolarizations neither reach nor overshoot the zero level.

b. The bottom trace is an intracellular recording from the same muscle fiber as in *a*, but taken at a faster sweep speed. The top trace is a record of the bottom trace which has been differentiated by a simple RC circuit (time constant = 1 msec). Note that the differentiated trace clearly distinguishes three components of the fast rising phase of the intracellular record.

* Numbers refer to References, p. 168.

rising phase is composed of several depolarizing increments occurring in rapid succession; this can be clearly seen from the differentiated trace (Fig. 1 b). The rising phase reaches a peak of 35 to 40 mV; it never depolarizes the cell beyond 10 mV negative to the zero level. The peak is followed by a plateau of lower amplitude which then declines to the resting level. The duration of the period of depolarization varies from preparation to preparation within a range of 400–700 msec. This corresponds to the duration of the regular spontaneous bursts produced by the cardiac ganglion.

Spontaneously initiated ganglion bursts produced heart contractions at approximately 1/sec in the isolated preparations. This is in close agreement with the rates observed in EKG records taken from freely moving lobsters which were fitted with electrodes implanted in the dorsal carapace. Such EKG's revealed frequencies of 1.1–1.5 contractions/sec. These data suggest that approximately the same level of spontaneous activity is maintained in the isolated heart preparations as that in intact animals.

When the ganglion is removed the heart ceases to contract. In such a quiescent state no regular spontaneous depolarizations occur. It is possible to evoke activity by stimulating electrically the cut distal nerves of these deganglionated preparations.

The nerves used in this study contain three axons; by gradually increasing the intensity of the applied stimulus it is possible to evoke responses, monitored intracellularly in a muscle fiber, which show up to three distinct threshold steps. This and other observations suggest that each muscle fiber in the heart is poly-neuronally innervated.

Series of ejp's are produced by trains of brief (0.5 msec duration), constant frequency (0.5, 1.0, 2.0, 5.0 per second) stimuli adjusted in voltage to stimulate only the lowest-threshold axon of a given nerve. The evoked ejp's recorded in a muscle fiber increase in amplitude; that is, they show a facilitation as one stimulus follows another (Fig. 2 a). At a given frequency, the amplitude of each successive response increases over a period of seconds until it reaches a maximum value above which it is no longer augmented.

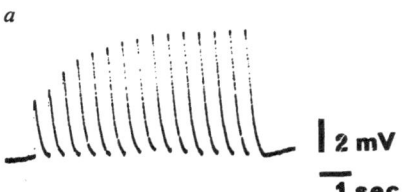
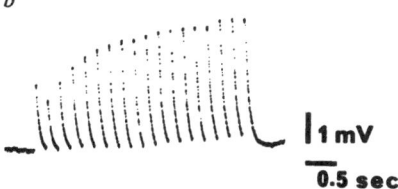

Fig. 2.

Intracellular records of ejp's evoked by electrical stimuli adjusted in voltage to stimulate a single axon in distal nerve fibers of deganglionated lobster hearts. *a*. Simple facilitation: stimuli were applied at 2/sec.

b. Complex response pattern: stimuli were applied at 5/sec. Facilitation follows after the second response. *a* and *b* are records from different preparations.

This simple facilitation phenomenon grows and decays exponentially. The time constants of growth and decay are of the order of a few seconds. The value of the plateau of facilitation (i.e. the final maintained amplitude of the junctional responses) is dependent on the stimulus frequency.

In addition to the simple facilitation phenomenon there was observed a complex response pattern in which the first response to a train of applied stimuli is greater than the second response (Fig.2b); facilitation usually follows after the second response.

References

[1]   T. H. BULLOCK and G. A. HORRIDGE, *Structure and Function in the Nervous Systems of Invertebrates. II* (W. H. Freeman and Co., San Francisco, 1965), p. 988–997.
[2]   D. M. MAYNARD, *Circulation and Heart Function*, in: *The Physiology of Crustacea. I* (Ed. T. H. WATERMAN; Academic Press, New York, 1960), p. 161–226.

This work was supported by a National Institutes of Health Postdoctoral Research Fellowship (1-F2-HE-35, 383-01) to M. ANDERSON and by a National Science Foundation Grant (GB 4315X) and a Grant-in-Aid (No. 879) from the Greater Boston Chapter of the Massachusetts Heart Association.

A Study of the Active State in Molluscan Ventricular Muscle[1]*

by Robert B. Hill
Dept. of Zoology, University of Rhode Island, Kingston, Rhode Island 02881
U.S.A.

The tension-length relationship of mammalian cardiac muscle is character-ized by the presence of resting tension on the rising limb of the active tension curve and by the high value of resting tension at the length where active tension is maximal[2]. This is in contrast to skeletal muscle in which resting tension begins to appear only beyond the length for maximal active tension.

Similar relationships are emerging for *Busycon* muscle. Strips of ventricle muscle have been studied[3]. The ring tears just as the active force approaches a plateau or begins to turn down, so that only the ascending limb of the tension-length curve can be recorded, but it differs from the radula protractor muscle[4] in which the resting tension appears only at a relatively greater length.

Entire isolated ventricles of *Busycon (Busycotypus) canaliculatum* were studied in order to confirm that the properties of the cardiac muscle were reflected in the tension-length diagram of the whole ventricle. Spontaneously beating ventricles were cannulated through the auricle and perfused with arti-ficial sea water at a pressure of 45 cm water at 25 °C. Tension was recorded across the ventricle between auricle and aortic attachment point.

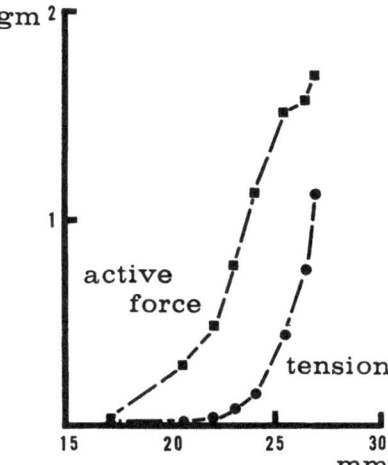

Fig. 1.
Length-tension diagram for an isolated
ventricle of *Busycon canaliculatum*.
Abscissa: length in mm between auricle
and aorta. Ordinate: grams tension.

* Numbers refer to References, p. 172.

The results of one out of seven experiments are shown in Fig. 1. The maximum tension in single spontaneous beats of the whole ventricle was only about 2 gm, in contrast to the 90 gm developed by a maximally tetanized ring of ventricular muscle[3]. The tension in single beats would understandably be less than in a tetanus, but the length-tension diagram has the shape characteristic of cardiac muscle[5,6].

Studies of the active state of vertebrate cardiac muscle are hindered by the time dependency of contractility during the beat. Since the molluscan cardiac muscle can be normally tetanized[7] it would appear to be a favorable material for the study of active state[8]. The quick stretch method[9] previously used to follow the time course of active state in vertebrate cardiac muscle[10] was used on the perfused *Busycon canaliculatum* ventricle[11]. The time course of spontaneous rise of isometric tension is compared in Fig. 2 with the tension developed as a result of an 0.75 mm stretch applied in 5 msec[11] at instants throughout the beat. Active state rises and falls rather slowly, paralleling the force developed by the ventricle.

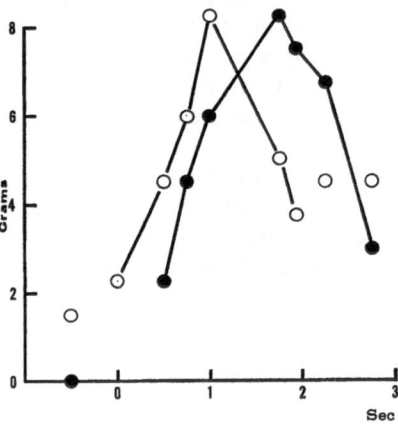

Fig. 2.
Time course of spontaneous isometric contraction (filled circles) and of resistance to stretch (unfilled circles). Abscissa: time in seconds. Ordinate: grams tension.

These experiments have been repeated with driven rings of ventricle in order to determine if this time course of active state really represents an activation pattern or whether it might be due to the complex trabecular architecture of the entire ventricle[12].

The rings used included nearly the entire ventricle. They were cut, allowed to equilibrate in natural sea water for 30 minutes, and mounted stretched in air between two glass rods wrapped with platinum foil to serve as stimulating electrodes. The rings were kept moist with natural sea water throughout each experiment and were blotted and weighed at the end. One glass rod was mounted on a Palmer screw stand and as the ring was stretched in 1 mm increments passive tension was recorded by an RCA 5734 mechano-electronic transducer attached to the other glass rod. Each ring was stretched until the resting tension, after stress-relaxation, remained at about ten grams. The rings were then driven by single shocks of 30 volts (loaded) and ten msec duration. Quick stretches were timed by a delayed relay and applied with a Levin-Wyman ergometer on whose moving beam was mounted the glass rod with 5734 transducer attached. Extent and velocity of displacement were recorded by use of a Langham Thompson PD 20 displacement transducer. Stimulus artifact, tension of the ventricle ring, and ergometer displacement were photographed from a Tektronix 4-channel oscilloscope.

Some differences were immediately obvious between the spontaneous beats of the entire ventricle (perfused but with only about 0.1 gm resting tension between auricle and aorta) and the driven contractions of the ventricular ring, stretched to about 10 gms resting tension. The time from the shock to the peak of isometric tension was about 1.75 sec for the ventricle but about 3 sec for the ring. The time from peak contraction to 3/4 relaxation was about 1 sec for the ventricle but commonly 20 sec for the ring.

In an experiment with a 0.3 gm ring (7 mm wide and stretched to 18 mm in length at 10 gms resting tension) the maximum active tetanic force at 15 shocks/sec was found to be 22 gms while force following a single shock was 14 gms. A 0.5 mm stretch, applied at the peak of the active state, produced 22 gms tension additional to the resting tension (of 10 gms). When the 0.5 mm stretch was applied at instants throughout the contraction, the active state appeared to reach its peak earlier than did the isometric tension. Furthermore, by the criterion of resistance to quick stretch, active state appeared to be over before half-relaxation was reached. Thus in the stretched ring of ventricular muscle the time course of active state is no longer parallel to that of active tension development, but it is not possible to say, on the basis of data now at hand, whether it is because of the reorientation of the stretched fibers or because of the greatly slowed time course of contraction.

Summary

The length-tension diagram for a driven ring of ventricular muscle is very similar to that for the entire spontaneously beating isolated ventricle of *Busycon canaliculatum*. Both are of the 'cardiac' type in that maximum active force is *not* developed at a length corresponding to negligible passive tension. The time course of active state in the entire spontaneously beating ventricle is also of 'cardiac' type, being roughly parallel to the time course of isometric tension. However, in the driven stretched ring of ventricular muscle, relative inextensibility rises more quickly after a shock than does isometric tension.

References

1   Research supported by NSF grant GB 6923.
2   A. F. GRIMM and W. V. WHITEHORN, Am. J. Physiol. *214*, 1378 (1968).
3   R. B. HILL, Experientia *23*, 772 (1967).
4   R. B. HILL, E. MARANTZ, B. A. BEATTIE and J. M. LOCKHART, Experientia *24*, 91 (1968).
5   J. F. LAMB and J. A. S. McGUIGAN, J. Physiol. *186*, 261 (1966).
6   S. E. DOWNING and E. H. SONNENBLICK, Am. J. Physiol. *207*, 705 (1964).
7   R. B. HILL, Comp. Biochem. Physiol. *23*, 1 (1967).
8   A. J. BRADY, Physiol. Rev. *48*, 570 (1968).
9   H. S. GASSER and A. V. HILL, Proc. Roy. Soc. (London) B*96*, 398 (1924).
10   L. J. O'BRIEN and J. W. REMINGTON, Am. J. Physiol. *211*, 770 (1966).
11   R. B. HILL and P. J. SCHUNKE, Experientia *23*, 570 (1967).
12   R. BRUNET and A. JULLIEN, Arch. Zool. Exp. Gén. *78*, 375 (1937).

III
Initiation of Rhythmicity

Introduction
by ERNST FLOREY

Apart from the fact that all hearts are hollow muscles, the most characteristic property of hearts is their rhythmicity, i.e. their ability to become electrically and mechanically excited at regular intervals. This rhythmicity can be myogenic, arising from events that occur within the tissue of the heart itself, or it can be neurogenic, initiated by nerve impulses arising within a cardiac ganglion.

Myogenic rhythmicity usually arises in specialized regions of the myocardium, which are commonly referred to as pacemaker regions. In some cases the specialization is morphologically defined, in others it can be recognized only as a functional differentiation, in that many or all regions of the myocardium have pacemaker properties and can assume pacemaker functions once they are liberated from the influence of the previously dominant pacemaker region.

Where the heart beat is neurogenic, the pacemaker activity is shifted to the pacemaker neurons of the heart ganglion. Cardiac muscle cells do not function independently but in co-ordination with the other muscle cells of the heart. Special structural features, such as are described elsewhere in this volume, facilitate the co-operative functioning of heart muscle. A similar interaction occurs among the nerve cells within the cardiac ganglion. The patterns of the spread of excitation through the myocardium and through the cardiac ganglion are topics of prime importance for our understanding of heart physiology.

The requirements of rhythmicity can be particularly well studied through the approach of comparative physiology. Interaction and co-operation of the individual cells of the myocardium and of cardiac ganglia are particularly amenable to experimental analysis. Comparative physiology here involves the study of the functioning of hearts, not only of different species, but also of different stages of development. The following papers range through several animal phyla and different developmental stages. Each paper illuminates the problem of initiation of rhythmicity from a different angle.

Electrical Recording from Embryonic Heart Cells Isolated in Tissue Culture

by Robert L. DeHaan

Carnegie Institution of Washington, Department of Embryology, Baltimore, Maryland, 21210

When the heart of a 7-day-old chick embryo is dissociated into its component cells, a majority of those cells (50–60%), completely isolated from contact with neighbors, are capable of spontaneous rhythmic activity when maintained under suitable in vitro culture conditions[1]*. Any such single isolated cell which is seen to beat must be initiating its own activity, and may thus be defined as a pacemaker[2]. Moreover, such a cell, within 5–30 min after coming into contact with a more slowly beating neighbor, may cause that neighbor to beat in synchrony with it.

What are the essential characteristics of a functional contact between two cells, which permit conduction of a contractile stimulus between them? With the advent of techniques for the dissociation and cultivation of heart tissue in vitro, several investigators have succeeded in recording with intracellular electrodes from sheets or clumps of cells beating in synchrony. Gaining an answer to the above question, however, depends upon being able to record at will either from an isolated single pacemaker, or from such a cell in electrical contact with a small group of neighbors under circumstances in which the number of cells in the group may be determined and the points of contact examined at the ultrastructural level. We have recently described conditions which satisfy these requirements[3].

Success in recording from heart cells in small groups or those completely isolated from neighbors appears to depend upon at least four conditions being met: (a) the use of ultrafine microelectrodes (100–150 MΩ), (b) the use of a nutrient medium which contains a growth-promoting agent such as embryo extract, (c) the utilization of a low-oxygen (10%) ambient atmosphere, and (d) the maintenance of other parameters such as pH, pCO_2, temperature, and osmolarity at suitable levels during the recording period.

Even under these circumstances, however, we found approximately a tenfold difference in frequency of success in recording from single isolated cells as compared with cells in contact with two or more neighbors. In cultures plated at a density designed to yield cells randomly attached to the dish singly and in groups of 2–50, out of a total of 189 attempts, isolated single cells could be successfully impaled only 8% of the time, whereas in the same dish and with

* Numbers refer to References, p. 175.

the same electrode good recordings could be obtained from 75 % of the cells in contact with neighbors. To attempt to explain this differential sensitivity to impalement, a hypothesis was put forward, based upon a consideration of the ultrastructure of intercellular connections[3].

The cells of cardiac muscle exhibit 'tight junctions' or 'nexuses' at regions of cell-cell contact, where the usual intercellular space of 70–150 Å disappears and the outer leaflets of the apposed cell membranes appear to fuse. These junctions have been associated with regions of low intercellular impedance and the transcellular flow of ions and other substances. When a microelectrode penetrates a cell membrane, it seems reasonable to assume that some tearing of the membrane would occur. Although repair of the membrane and sealing around the electrode must be rapid processes, they can hardly be instantaneous. Therefore, during these reparative events some intracellular materials must be lost to the external medium, and ionic and electrochemical gradients will tend to decay. In order for an isolated single cell to restore its transmembrane differentials, it must actively pump materials across its high-resistance plasma membrane. For cells already damaged and possibly substantially depolarized, such a task may require too great an expenditure of energy, and therefore (it is postulated) a majority succumb.

A cell in electrical contact with neighbors, on the other hand, while repairing its membrane, may act as a sink for ions and other intracellular components which will flow into it from neighboring cells across the low-resistance points of intercellular contact. A much smaller expenditure of energy would therefore be required to re-establish transmembrane gradients. And, therefore, presumably, most cells would survive. This hypothesis has the advantage of suggesting an experimental test of its validity. It is predictable that the frequency of success of impalements of isolated single cells might be substantially increased by placing the cells in a medium designed to mimic the intracellular milieu, while the electrode was being inserted. Experiments to test this prediction are now in progress.

References

[1]   R. L. DE HAAN, Devel. Biol. *16*, 216 (1967).
[2]   R. L. DE HAAN, *Factors Influencing Myocardial Contractility* (Ed. R. D. TANZ, F. KAVALER and J. ROBERTS; Academic Press, New York 1967), p. 217.
[3]   R. L. DE HAAN and S. H. GOTTLIEB, J. gen. Physiol. *52*, 643 (1968).

Effects of Na^+ and Ca^{++} on the Spontaneous Excitation of the Bivalve Heart Muscle

by H. Irisawa, A. Irisawa and N. Shigeto
Department of Physiology, School of Medicine, Hiroshima University,
Hiroshima, Japan

This work was aided by a research grant from the Ministry of Education of Japan.

If one were to attempt a study of the invertebrate heart muscle, the preparation which would immediately come to mind would be the classical clam heart preparation. Almost 30 years ago Prosser[1]* discovered that the venus heart was extremely sensitive to acetylcholine. Subsequently Welsh[2,3] used the heart of the *Venus mercenaria* as a tool for the bioassay of acetylcholine and its analogs and also for the study of the pharmacology and mechanisms of cardioregulation. Recently Greenberg[4] examined more than 40 different bivalve hearts for their response to acetylcholine. Many of the physiological studies on these muscles have been summarized by Krijgsman and Divaris[5] and more recently by Hill and Welsh[6].

The heart muscles of oyster and mytilus can continue their spontaneous activities even in Na^+-free solution. Although the principal role of Na^+ for the generation of action potentials has been well established in many excitable tissues, there are several exceptional muscles which apparently do not require Na^+ in the extracellular fluid for the generation of action potentials. In these muscles the inward current of the action potential is not dependent on Na^+ but on Ca^{++}. Examples of these muscles are found in the barnacle[7,8], crayfish and lobster[9], and mammal[10,11]. These muscles are known to be insensitive to tetrodotoxin and to changes in $[Na^+]_0$, but are sensitive to changes in $[Ca^{++}]_0$ and $[Mn^{++}]_0$. Is the insensitivity of the bivalve heart muscle to $[Na^+]_0$ related to the mechanism of the Ca^{++} spike of the crustacean muscles? What is the functional role of Na^+ in these muscles? In order to answer these questions, extensive structural and electrophysiological studies are necessary.

The present paper describes some findings of the fine structure and the electrophysiology of these heart muscles.

Classification of Bivalve Heart Muscle

The bivalve heart has been classically regarded as the most primitive chambered heart. Striations in this heart muscle can be observed when viewed under the light microscope but these striations are not due to the structural orientation

---

* Numbers refer to References, p. 190.

of the myofilaments as in the case of the vertebrate myocardium. The array of myofilaments in this myocardium is very different from that of the cross-striated myocardium in that they do not show a regular band pattern, A, I and Z. Under the light microscope a dark line between each sarcomere can be observed. This is not a typical Z-line, but rather an aggregation of dense substances. The sarcoplasmic reticulum is not so well differentiated as that in other cross-striated myocardia for there is neither a mesh-like organization of longitudinal sarcoplasmic reticulum nor T-tube invaginations at the level of the Z-line. Many prominent invaginations of the plasma membrane are seen together with dyad formations. Thus, these bivalve myocardia cannot be classified and treated as a typical cardiac muscle in the classical sense.

A similar discussion has been presented on the classification of the smooth muscles of invertebrates[12]. Traditionally, smooth muscle has been defined according to the morphological findings obtained by the light microscope using mammalian visceral muscles. After the development of the electron microscope, the mammalian smooth muscles were found to show a homogeneous arrangement of one set of myofilaments with scattered dense substances dispersed within the cytoplasm. If one were to extend these criteria set by examination of mammalian smooth muscle to the invertebrate visceral muscle, one would immediately find the classification impractical.

The same is true in the case of the cardiac muscle. Therefore it would be desirable to classify the cardiac muscle and smooth muscle of invertebrates by different criteria. It is thus considered appropriate to review the classification of muscles reported in the literature.

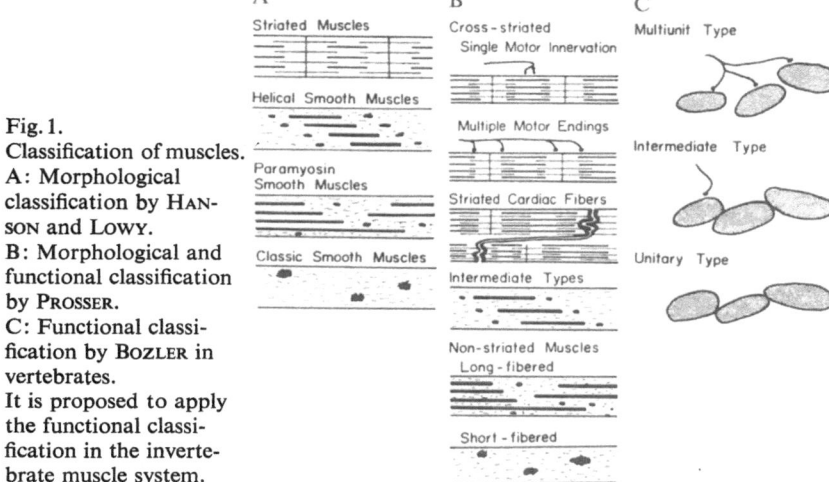

Fig. 1.
Classification of muscles.
A: Morphological classification by HANSON and LOWY.
B: Morphological and functional classification by PROSSER.
C: Functional classification by BOZLER in vertebrates.
It is proposed to apply the functional classification in the invertebrate muscle system.

Fig. 1 summarizes the three kinds of classifications of muscles. Column A is rearranged from the concept of HANSON and LOWY[13], who classified the muscles solely from the morphological findings of the array of myofilaments. Between the striated muscles and the smooth muscles they added two different types of invertebrate muscles: helical smooth muscles and paramyosin muscles (Fig. 1 A). Fig. 1 B, after PROSSER[14], classifies muscles on the basis of the array of myofilaments (cross-striated or non-striated), on the pattern of innervation (single or multi-motor endings), on junctional apparatus between the adjacent cells (presence of the intercalated disks) and on the length of the fibers (long-fibered or short-fibered). This classification provides a more general clue for the evaluation of muscles from the viewpoints of both morphology and physiology. The third scheme, shown in Fig. 1 C and initially presented by BOZLER[15] in 1948, was classified from the functional viewpoint only.

The multi-unit type muscle is typified by the vertebrate skeletal muscle. The individual muscle cell is independent of the adjacent muscle fiber, being surrounded by a complete plasma membrane and a basement membrane. Excitation in this type of muscle is by way of the motor nerve. Examples of multi-unit invertebrate muscles are found in Mytilus (anterior byssus retractor)[16,17], golfingia (buccal retractor)[18], pecten (adductor)[19], Busycon (radula protractor)[20], and earthworm (body wall)[21].

Unitary muscle is typified by vertebrate myocardial striated and visceral smooth muscles; such muscles are spontaneously active and the conduction of excitation occurs from cell to cell. Examples of the unitary type in the invertebrate muscle are seen in the intestines of chitons[22] and echinoderms[23].

The intermediate type demonstrates both motor innervation and cell-to-cell conduction, as can be seen in the vas deferens of guinea-pig. Few intermediate types have been described in invertebrates because of the paucity of both functional and histological data. One example is the heart muscle of a stomatopod, which has motor innervation[24], shows electrical connections between the adjacent fibers[25] and also has a nexus structure[26].

The following histological and functional observations demonstrate that the bivalve myocardium shows a typical unitary characteristic in many respects.

Fig. 2.   ▷
An electron micrograph of a longitudinal section of oyster muscle fiber.
Dense bodies (DB) are shown by an arrow. Two kinds of myofilaments (MF),
dyads (D) and glycogen granules (G) are seen within the cytoplasm. This and
Fig. 3 were obtained through s-collidin buffered osmium fixation, potassium
permanganate block staining in 100% acetone solution during dehydration
modified from Parson's method[19]. Section staining was made with lead hydroxide
solution. SR, sarcoplasmic reticulum; E, extracellular space. Magnification 63,000×.

Fig. 3.   ▷
An electron micrograph of a cross-section of a part of oyster myofibril.
A large thick filament is surrounded by several thin filaments. Very large irregular
shaped mitochondria are located in the middle of the picture. The extracellular
space filled with many collagen fibers is also noted. E, extracellular space;
M, mitochondria; MF, myofilaments. Magnification 77,000×.
Inset: Higher magnification of two sets of myofilaments in mytilus heart muscle.
s-collidin-buffered OsO₄-fixation, lead hydroxide staining. Magnification 150,000×.

Fine Structure of the Oyster Myocardium

Fig. 2 illustrates a longitudinal section of the heart muscle of the oyster, and Fig. 3 a transverse section. An aggregate of large dense substances (Z-body)

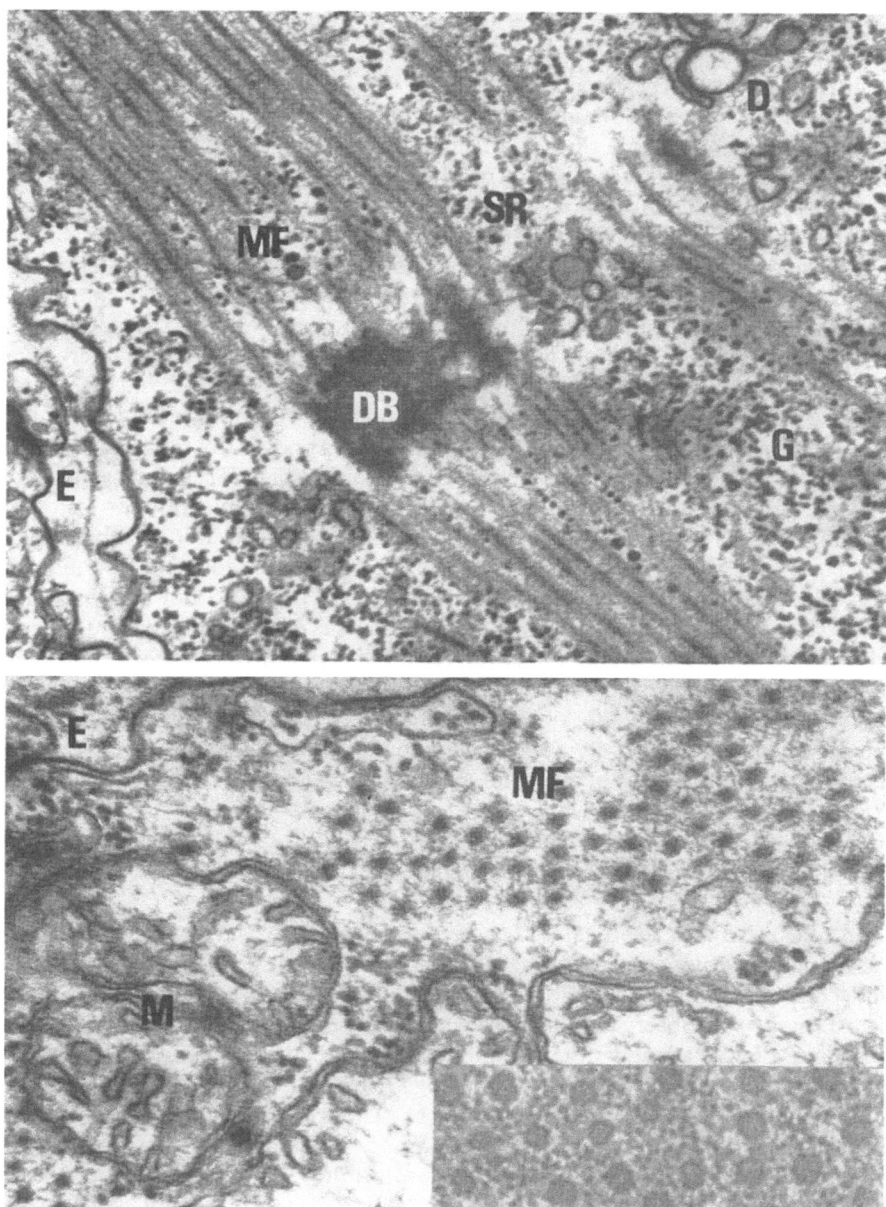

Fig. 4.   ▷
A cross-section of a portion of a cell process and the adjoining cell. Arrow
indicates an electron dense junctional membrane (*CJ*). *DB*, dense bodies;
M, mitochondria; *MF*, myofibrils; *G*, glycogen granules. The inset shows an
enlargement of the closely apposed membranes. There is no obvious fusion of two
outer leaflets of each plasma membrane (arrow). Magnification 110,000×.
Inset picture 1,540,000×.

is situated between the two sarcomeres in Fig. 2. The thick filament is about
200 Å in diameter. Thin filaments about 50 Å in diameter can be seen between
the thick filaments. These thick filaments appear to taper at both ends into the
dense substances. It is not certain from Fig. 2 whether the thin filaments penetrate into the dense substances or are independent of them. The band pattern
is not regular as in the mammalian myocardium. Glycogen granules are abundantly scattered throughout the cytoplasmic matrix. Sarcoplasmic reticulum is
not as highly organized as in the vertebrate cardiac muscle. Vesicular configura-
tions of the sarcoplasmic reticulum are seen on both sides of the dense substances and dyad formation is observed (Fig. 2 and 3). However, there was
neither T-tube invagination nor triad organization in this cardiac muscle. Fig. 4
shows a transverse section of the peripheral portion of the oyster myocardium.
The membrane of a process of a cell is closely apposed to the adjacent cell
membrane and makes a nexus connection between them[27]. The inset picture
shows a higher magnification of the nexus in which the intercellular space of
extremely narrow width is observed. Outer leaflets of the unit membranes of
these two cells are closely apposed at this area in a dense ladder-like configuration. The outer leaflets of the apposed unit membranes do not appear to be
fused. The overall thickness of the junctional part is estimated to be 160–170 Å,
while the thickness of a unit membrane in other parts of the cell measures 65 to
75 Å. The space between two unit membranes at the area of contact is therefore
calculated to be about 20 Å with a range from 10 to 30 Å. The statistical
measurements of the closely apposed membranes indicated that the area of the
nexus occupied $1/40 - 1/20$ of the total surface area of a cell.

Na+ Requirement for Spontaneous Excitation

When Na+ in artificial sea water was removed and replaced by Tris,
spontaneous activity frequently ceased temporarily, but within 5 min it resumed
and continued to beat for an extended period (over 6 h). Contraction of the
heart muscle was observed but not recorded. The resting potential of the mytilus
heart muscle was around − 45 mV in normal artificial sea water. The amplitude
of the spike was about 25 mV with a range from 16.5 to 29.0 mV. None of the
action potentials displayed any overshoot. The maximum rate of rise of the
action potential ranged from 0.1 to 0.6 V/sec. The action potential continued
after the heart was immersed in Na+-free solution[28, 29] without any appreciable

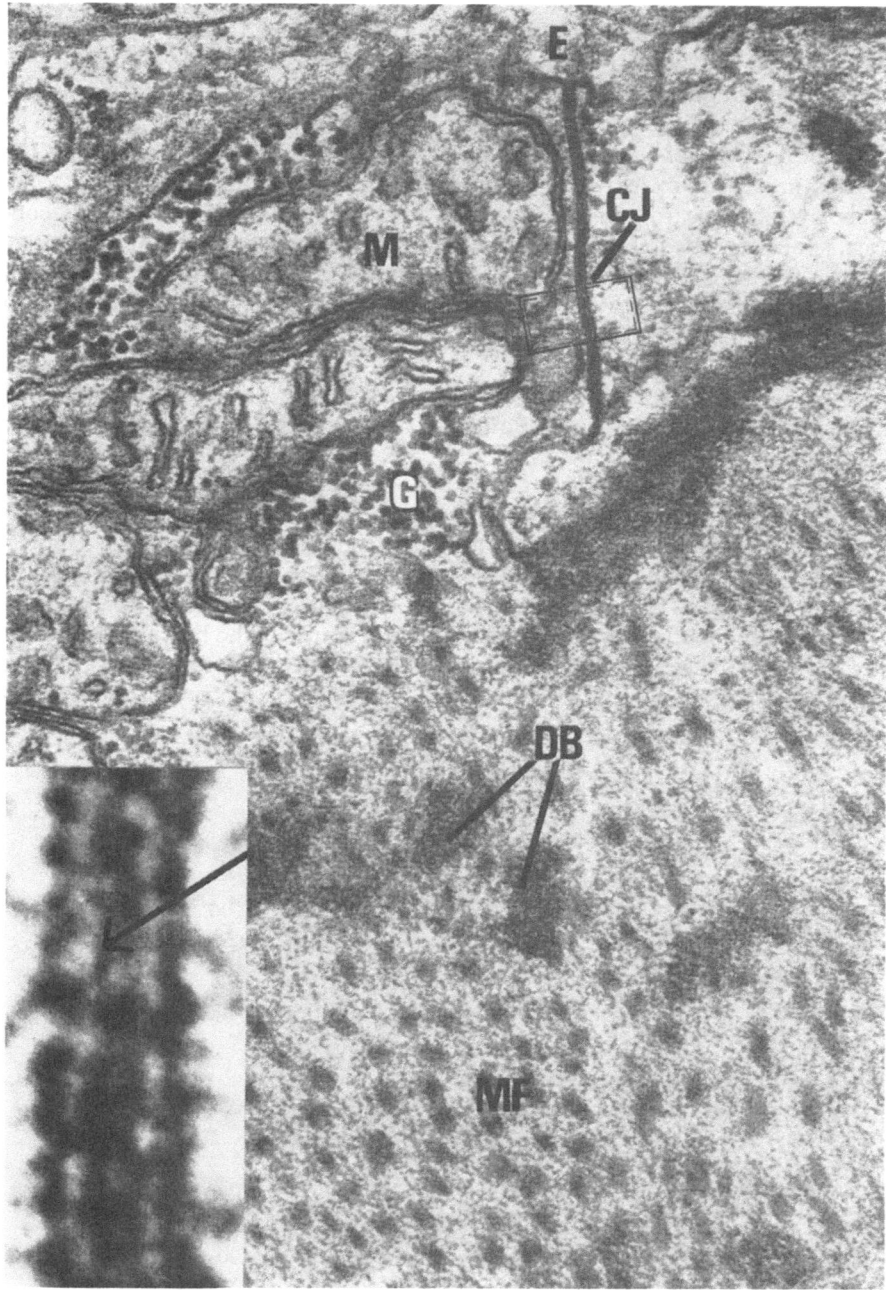

change in the amplitude or the maximum rate of rise of potential. The membrane was not sensitive to tetrodotoxin but was sensitive to $[Mn^{++}]_0$. Administration of 10 mM $MnCl_2$ reduced the amplitude of the action potential and increased its duration but the spontaneous activity ceased. When $[Ca^{++}]_0$ was increased in Na^+-free solution, the maximum rate of rise and the amplitude of the action potential increased. The amplitude of the action potential in the mytilus myocardium increased by 13 mV for each decade increase in $[Ca^{++}]_0$ in the oyster myocardium[28] and by 22 mV in the mytilus myocardium. By contrast, spontaneous excitation ceased when $[Ca^{++}]_0$ was reduced from the control value of 9 mM to 2.3 mM Ca^{++} in Na^+-free solution. These results are consistent with the findings that Ca^{++} is partially responsible for the generation of action potential in various excitable tissues.

Effect of Changes in Membrane Potentials on the Action Potentials

The current applied through a suction electrode was used to hold the membrane potential at various levels. Hyperpolarization of the membrane was obtained when the polarity of the suction electrode was cathodal, while depolarization of the membrane was obtained when the electrode was anodal. In the oyster myocardium the pattern of action potential varied from specimen to specimen. Frequently after isolation of the heart muscle, the heart did not show any spontaneous activity. These differences in activity were found to be due to the value of the membrane potential. In Fig. 5, four examples of action poten-

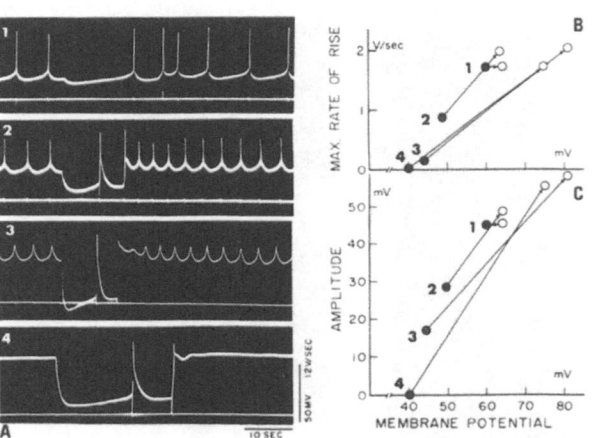

Fig. 5.
Effect of hyperpolarizing current on both membrane potentials and action potentials.
A: The original action potential (upper trace) and the first derivative (lower trace) indicating the maximum rate of rise of the action potential. 1 and 2 are obtained from one cell, while 3 and 4 are recorded from another cell. From 1 to 4, the membrane potential is gradually hyperpolarized.
B: Relationship between the maximum rate of rise of action potential and the membrane potential. C: Relationship between the amplitude and the membrane potentials. Numerals in B and C correspond to those in A. Closed circles indicate the values obtained before and the open circles illustrate the values taken during the application of the hyperpolarizing current.

tials at various resting potentials ranging from − 40 to − 60 mV are shown. During the control resting state, the amplitude and the maximum rate of rise of the action potential is largest in A-1 and smallest in A-3. In A-4, the muscle did not show any spontaneous activity. As a result of hyperpolarization, the amplitude and the maximum rate of rise of the action potentials were increased in all four examples to approximately the same magnitude as can be seen in Fig. 5 B and C.

These observations suggest that the magnitude of the action potential is dependent upon the membrane potential, as in other excitable tissues[30−32].

A series of experiments was conducted by applying the polarizing current through the suction electrode and recording the potential from the same single heart cell (Fig. 6 A). When the membrane was depolarized, the rate of rise and

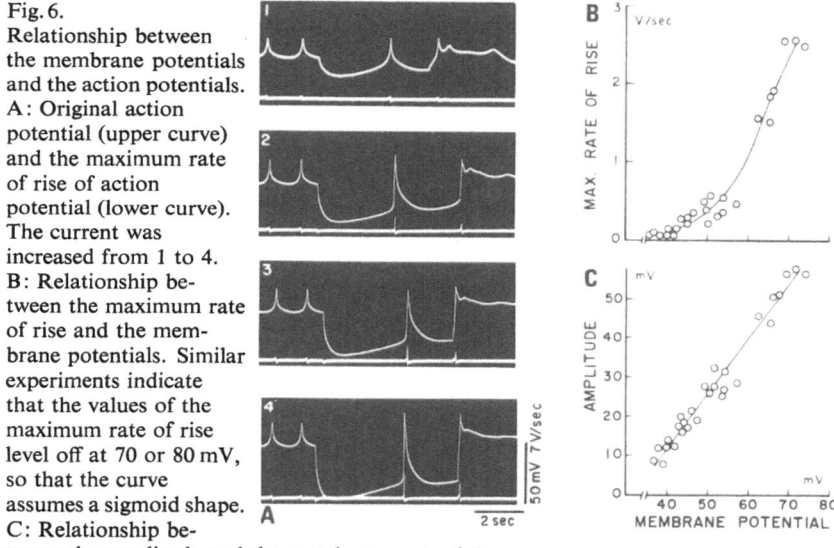

Fig. 6.
Relationship between the membrane potentials and the action potentials.
A: Original action potential (upper curve) and the maximum rate of rise of action potential (lower curve). The current was increased from 1 to 4.
B: Relationship between the maximum rate of rise and the membrane potentials. Similar experiments indicate that the values of the maximum rate of rise level off at 70 or 80 mV, so that the curve assumes a sigmoid shape.
C: Relationship between the amplitude and the membrane potential.

the amplitude of action potential decreased and when the membrane was hyperpolarized, the reverse was obtained. The amplitude of the action potential was directly proportional to the membrane potential (Fig. 6 C), but the rate of rise of depolarization was non-linear (Fig. 6 B). Since the maximum rate of rise of depolarization reflects the total current flow according to the analysis made by HODGKIN and KATZ[33], it can be stated that the total current was 1×10^{-7} A/cm^2 and was 10^{-4} times that of the squid axon. When the membrane potential was hyperpolarized, the current increased to 2×10^{-6} A/cm^2. At the normal resting potential of about − 50 mV, only $1/_{20}$ of the maximum carrier of the net inward current is in a state of activation.

The magnitude of the net inward current varies in different muscle fibers. In the Purkinje fiber of the sheep, the control resting potential was -90 mV and most of the carrier system was activated in the resting state[31]. In the skate heart muscle, the resting potential was found to be -60 mV and about $4/5$ of the carrier was inactivated[34,35]. In the smooth muscle of the uterus, it has been reported that a part of Na^+-carrier was in a state of inactivation[36].

Effect of Changes of $[Na^+]_0$ on the Action Potential

When $[Na^+]_0$ was increased, the spontaneous action potential usually ceased, but when hyperpolarizing current was applied, spontaneous action potentials were observed. When the two action potentials were compared under -60 mV membrane potential, no obvious difference in the action potential was observed between normal and 924 mM $[Na^+]_0$. When the membrane was hyperpolarized further (-78 mV), the amplitude of action potential in normal $[Na^+]_0$ solution was 53 mV while in 924 mM $[Na^+]_0$ it was 68 mV. No appreciable difference in the maximum rate of rise of the action potentials measured under -60 mV was observed between normal $[Na^+]_0$ and 924 mM $[Na^+]_0$, but at -78 mV it was twice as large in 924 mM $[Na^+]_0$ than in normal $[Na^+]_0$. Thus both the maximum rate of rise and the amplitude of action potentials were larger in the higher Na^+ solution at higher membrane potentials. These results indicate that the sensitivity of the myocardium to changes in $[Na^+]_0$ can only be seen at higher membrane potential where most of the carrier of the inward current is available. At low membrane potential the carrier of the net inward current is mostly in an inactive state.

Effect of Reduction of $[Na^+]_0$ on the Action Potential

When external Na^+ was removed and replaced by Tris, the action potential could be elicited only at lower membrane potentials between -40 and -53 mV (Fig. 7A), but at a higher membrane potential, no action potential was observed (Fig. 7A-3). Spontaneous action potentials were observed at low resting potentials (-40 mV) in Na^+-free solution, but not at higher membrane potentials. In Fig. 7B, when 45 mM Ca^{++} was added to Na^+-free test solution, the amplitude (Fig. 7D) and the maximum rate of rise (Fig. 7C) increased considerably in higher Ca^{++} solution, but spontaneous action potentials were never observed during the application of the hyperpolarizing current. At the end of polarization, anodal break excitation was recorded. The rate of rise of the action potential at break excitation was very small, suggesting that in Na^+-free solution the maximum rate of rise of the action potential did not increase at a higher membrane potential (Fig. 7A-3 and Fig. 7B-3). The maximum rate of rise of the action potentials was perhaps saturated at the membrane potential of about

Fig. 7.
Effect of $[Na^+]_0$ and
$[Ca^{++}]_0$ on the action
potential. A: Action
potentials were recorded
in Na^+-free artificial sea
water. Action potentials
during the application
of current were elicited
only by electrical stim-
ulation. Note at 3,
no response was
observed in spite of
repetitive stimulations.
B: Action potentials
in Na^+-free + 45 mM
Ca^{++}. Even when the
Ca^{++} concentration was
increased five times

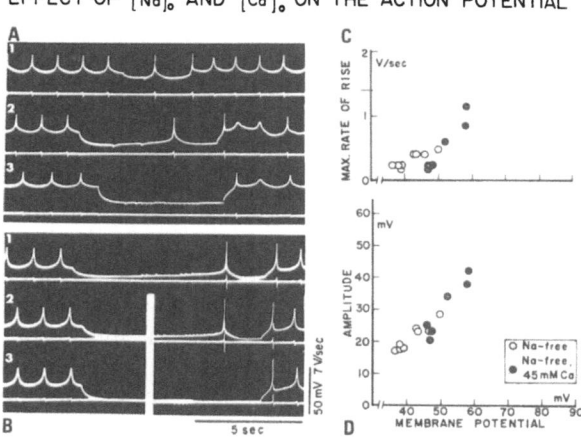

EFFECT OF $[Na^+]_0$ AND $[Ca^{++}]_0$ ON THE ACTION POTENTIAL

greater than normal, action potentials could not be elicited by external stimulation
even when the membrane potential was over -60 mV as illustrated in B3.
C: Relationship between the maximum rate of rise and the membrane potential in
Na^+-free (open circles) and in Na^+-free, $5 \times Ca^{++}$ (closed circles). It was not possible
to observe the action potential at higher membrane potentials. D: Relationship
between the amplitude and the membrane potential. Symbols are the same as in C.

50 mV in Na^+-free solutions. These observations indicate that at higher membrane potential, the action potential is likely to depend upon $[Na^+]_0$, but it is
not dependent simply upon the change in $[Na^+]_0$ at a lower membrane potential
where the carrier of the net inward current is in an inactive state.

Effect of Changes of $[Ca^{++}]_0$ on the Action Potential

When higher Ca^{++} was applied with $[Na^+]_0$ kept constant, the maximum
rate of rise of the action potential was higher than in normal $[Ca^{++}]_0$ at a
membrane potential of about -50 mV. At membrane potentials higher than
-60 mV no spontaneous action potentials were observed, possibly because
the threshold potential became elevated. However, the fact that the S-shaped
relationship between the maximum rate of rise of the action potential and the
membrane potential shifted to the left in higher Ca^{++} solution coincided well with
the similar findings of WEIDMANN[32] in mammalian Purkinje fibers. These results
indicate that although the muscle exhibits the characteristics of a Ca^{++} spike
when the membrane potential is low, it also shows some Na^+ spike dependency
at higher membrane potential.

NOMA and UEDA in unpublished observations found that when $[Ca^{++}]_0$ was
omitted from the artificial sea water but not removed by chelation with EDTA,
the action potential persisted over 14 h after perfusion with Ca^{++}-free solution,

as long as $[Na^+]_0$ remained normal. The action potential showed a prolonged plateau phase and a remarkable after-potential, which was followed by a diastolic depolarization. Hyperpolarization of the membrane caused a large action potential. The duration of the action potential decreased and the positive afterpotential was reduced when the hyperpolarizing current was applied.

If the spike-genesis of this muscle is ascribable simply to the inward current of Na^+ and not affected by Ca^{++}, then the removal of Ca^{++} from the extracellular fluid will not necessarily cause any difference to the S-shaped relationship between the maximum rate of rise of action potential and the membrane potentials. However, NOMA and UEDA's observations demonstrate that the rate of rise of the action potential was reduced even though $[Na^+]_0$ was maintained at a normal concentration. If the muscle spike is purely a Ca^{++} one, and Na^+ is not required at all, the removal of Ca^{++} from the extracellular fluid should cause the spontaneous activity to cease. The complete removal of $[Ca^{++}]_0$ was not performed in this experiment, but with a very small amount of $[Ca^{++}]_0$ (0.2 ppm) the membrane was able to continue rhythmic activity. These findings suggest that action potentials of the muscle may be neither a pure Na^+ spike nor a pure Ca^{++} spike. Instead it can be said that these ions have an intricate link with the membrane.

Comparison of Bivalve Heart Muscles to Other Ca^{++} Spike Tissues

Several mechanisms have been considered to account for the phenomenon of the genesis of action potentials in the absence of Na^+. Although BRADY[37] pointed out the inadequacy of diffusion in Na^+-free solutions in general, muscle layers of the bivalve heart are thinner and the heart size is smaller than those of the vertebrate, thus the possibility of inadequate perfusion would be very small.

KOKETSU and NISHI[38], reporting on the Ca^{++} spike of the bullfrog sympathetic ganglion cells, concluded that in a normal Ringer solution containing 1.8 mM Ca^{++}, the Ca^{++} current during the action potential is very small compared to Na^+ current. JUNGE and GEDULDIG[39,40] found in the Aplysia giant neuron that both Na^+ and Ca^{++} ions play a part in the production of action potentials. These findings appear to coincide with the present experiments on the bivalve heart muscle, where the action potential could be recorded in either a Na^+-free, normal Ca^{++}, sea water solution or Ca^{++}-free, normal Na^+, sea water solution. The role of Ca^{++} in action potential generation has been reported in various excitable tissues. The appended table presents a comparison of various excitable tissues where the Ca^{++} spike has been confirmed. It can be seen that the tissues which show Ca^{++} spikes range from mammals to molluscs[7-12,28,29,38-43].

When the various responses were carefully compared, a subtle difference was noted in the participation of Ca^{++} for the generation of action potentials among these excitable tissues in Na^+-free solution. In the mammalian stomach muscle, the initial rapid component is a Na^+ spike while the second slow potential is known to be a Ca^{++} response[44]. Similarly, in the frog myocardium, the first rising phase corresponded to the Na^+ spike, but the subsequent slow rising phase to the peak of the action potential was attributed to the Ca^{++} inflow[45]. Therefore in these muscles, the Ca^{++} spike is partially responsible for the generation of normal action potentials.

On the contrary, in the Aplysia giant neuron[39,40], the ureter smooth muscle[43] and the skate heart muscle[34,35], Na^+ is responsible for the generation of action potentials in the normal milieu, but when Na^+ is removed, the role of Na^+ can be replaced by Ca^{++}. Thus the action potentials of these muscles are the multi-ionic spikes. The oyster myocardium also displays multi-ionic spikes and these are dependent upon the membrane potentials. Whenever the membrane potential is low, the membrane shows the characteristics of a Ca^{++} spike, but when it is high, the membrane no longer exhibits the characteristics of the Ca^{++} spike but shows the features also of a Na^+ dependency. As tetrodotoxin is known to inhibit specifically the inward Na current, it cannot be used as a crucial criterion of a Ca^{++} spike[46]. The fact that the oyster muscle exhibited multi-ionic spikes also coincided with the finding that in isotonic sucrose containing a small amount of Ba^{++}, the spontaneous action potential continued for more than 2 h although the action potential was very prolonged[10,47].

Conclusion

In view of the apparent confusion that prevails in the classification of invertebrate muscles, it is here proposed to classify muscles on the basis of a functional criterion as multi-unit, unitary, and intermediate. Multi-unit muscles have the characteristics of isolated conduction, motor innervation and non-spontaneity, while unitary muscles show cell-to-cell conduction, spontaneity, autonomic innervation and nexus structures. The intermediate type shows motor innervation and nexus structures.

The bivalve heart muscle is a typical unitary muscle. Nexus structures are seen between the adjacent muscle fibers, and the heart muscle is spontaneously active in a Na^+-free solution. Though not sensitive to tetrodotoxin, it is sensitive to $[Mn^{++}]_0$ and $[Ca^{++}]_0$. When $[Ca^{++}]_0$ was increased, the amplitude of action potential increased 13 mV for each decade increase in $[Ca^{++}]_0$ in the oyster heart muscle and 22 mV in the mytilus heart muscle. Such findings suggest that these muscles are typical Ca^{++} spike muscles.

However, insensitivity of the action potentials to changes in $[Na^+]_0$ was observed only at the control membrane potential where most of the carrier of the net inward current was in a state of inactivation. When the membrane potential was high, action potentials became larger in high $[Na^+]_0$ than in low $[Na^+]_0$. When the membrane potential was high, no spontaneous action potentials were observed in Na^+-free solutions.

Species	Excitable in Na+-free normal Ca++ solution	Excitable in Na+-free Hi-Ca++ solution	Response to tetrodotoxin	Response to Mn++	Ca ion as a charge carrier	Remarks and references
Mammalia **GUINEA-PIG** Smooth muscles *Taenia coli*	+	+	−	+	+	11
Vas deferens	...	...	−	...	+	partly Ca spike[41]
CAT Smooth muscles Ureter	−	+	−	+	+	Na spike in normal Ca spike in Na+-free[43]
Stomach	initial phase + second slow −	...	...	...	+	multi-ionic[44]
Amphibia **BULLFROG** Sympathetic ganglia	−	+	+*	...	+	*when Na+-free high Ca Na spike in normal Ca spike in Na+-free[38]
FROG Myocardium	−	...	maximal rate + amplitude −	+	partly +	rapid-rising phase Na slow-rising phase Ca[45]

Elasmobranchii SKATE myocardium	−	−	...	partly +	Ca inflow when Na reduced[35]
Crustacea BARNACLE Adductor muscle	+	+	+	+	solely Ca spike[7,8]
CRAYFISH Abdominal muscle	+	+	...	+	solely Ca spike[9,42]
LOBSTER Abdominal muscle	+	+	...	+	[9]
Insecta MOTH myocardium	+	+	...	−	undetermined[48]
Mollusca APLYSIA Giant neuron	amplitude + rate of rise reduced ...	+	amplitude and rate of rise reduced	amplitude + rate of rise reduced +	multi-ionic[39,40]
MYTILUS and OYSTER Myocardium	+	−	+	+	higher membrane potential Na spike, in lower membrane potential Ca spike[12,28,29]

Five-fold increase in Ca^{++} concentration in the extracellular fluid shifted the S-shaped curve of the relationship between the rate of rise of the action potential and the membrane potential toward the left, suggesting a decrease in Na^+ inactivation. This phenomenon coincided with that of the mammalian Purkinje fiber. The above two findings indicate that the muscle shows both the characteristics of Na^+ spike and Ca^{++} spike muscles.

References

[1]   C. L. PROSSER, Biol. Bull. *78*, 92 (1940).

[2]   J. H. WELSH and R. TAUB, Biol. Bull. *95*, 346 (1948).

[3]   J. H. WELSH and R. TAUB, Science *112*, 467 (1950).

[4]   M. J. GREENBERG, Comp. Biochem. Physiol. *14*, 513 (1965).

[5]   B. J. KRIJGSMAN and G. A. DIVARIS, Biol. Rev. *30*, 1 (1955).

[6]   R. B. HILL and J. H. WELSH, *Physiology of Mollusca* (Ed. K. M. WILBUR and C. M. YONGE; Academic Press, New York 1966), vol. 2, p. 125.

[7]   S. HAGIWARA and S. NAKAJIMA, J. gen. Physiol. *49*, 793 (1966).

[8]   S. HAGIWARA, S. CHICHIBU and K. NAKA, J. gen. Physiol. *48*, 163 (1964).

[9]   B. C. ABBOTT and I. PARNAS, J. gen. Physiol. *48*, 919 (1965).

[10]   A. F. BRADING and T. TOMITA, J. Physiol. *197*, 30 p (1968).

[11]   E. BULBRING and H. KURIYAMA, J. Physiol. *166*, 29 (1963).

[12]   H. IRISAWA, *A Symposium on Comparative Smooth Muscle*, Proc. Int. Union of Physical Sciences, Washington D.C. (1968), in press.

[13]   J. HANSON and J. LOWY, *The Structure and Function of Muscle* (Ed. G. H. BOURNE; Academic Press, New York 1960), vol. 1, p. 265.

[14]   C. L. PROSSER, *Invertebrate Nervous Systems* (Ed. C. A. G. WIERSMA; University of Chicago Press 1967), p. 133.

[15]   E. BOZLER, Experientia *4*, 213 (1948).

[16]   B. M. TWAROG, J. gen. Physiol. *50*, 157 (1967).

[17]   B. M. TWAROG, J. Physiol. *192*, 857 (1967).

[18]   C. L. PROSSER and C. B. MELTON, J. cell. comp. Physiol. *44*, 255 (1954).

[19]   DE F. MELLON JR., Science *160*, 1018 (1968).

[20]   R. B. HILL, Biol. Bull. *155*, 471 (1958).

[21]   H. KURIYAMA, T. HIDAKA and H. ITO, J. exp. Biol. (1969), in press.

[22]   G. BURNSTOCK, M. J. GREENBERG, S. KIRBY and A. G. WILLIS, Comp. Biochem. Physiol. *23*, 407 (1967).

[23]   C. L. PROSSER, R. A. NYSTROM and T. NAGAI, Comp. Biochem. Physiol. *14*, 53 (1965).

[24]   H. IRISAWA, A. IRISAWA, T. MATSUBAYASHI and M. KOBAYASHI, J. cell. comp. Physiol. *59*, 55 (1962).

[25]   I. SEYAMA and H. IRISAWA, *Membrane Characteristics of the Myocardium of Stomatopod Heart Muscle*, J. physiol. Soc. Japan *30*, Abstract, in Japanese (1968).

[26]   A. IRISAWA and K. HAMA, Z. Zellforsch. *68*, 674 (1965).

[27]   M. M. DEWEY and L. BARR, J. Cell Biol. *23*, 553 (1964).

[28]   H. IRISAWA, A. NOMA and R. UEDA, Jap. J. Physiol. *18*, 157 (1968).

[29]   H. IRISAWA, N. SHIGETO and M. OTANI, Comp. Biochem. Physiol. *23*, 199 (1967).

[30]   A. L. HODGKIN and A. F. HUXLEY, J. Physiol. *116*, 497 (1952).

[31]   S. WEIDMANN, J. Physiol. *127*, 213 (1955).

[32]   S. WEIDMANN, J. Physiol. *129*, 568 (1955).

[33]   A. L. HODGKIN and B. KATZ, J. Physiol. *108*, 37 (1949).

[34]   I. SEYAMA and H. IRISAWA, J. gen. Physiol. *50*, 505 (1967).

[35] I. SEYAMA, Am. J. Physiol. *216*, 687 (1969).
[36] J. M. MARSHALL, Am. J. Physiol. *204*, 732 (1963).
[37] A. J. BRADY, Ann. Rev. Physiol. *26*, 341 (1964).
[38] K. KOKETSU and S. NISHI, Nature *217*, 468 (1968).
[39] D. JUNGE, Nature *215*, 546 (1967).
[40] D. JUNGE and D. GEDULDIG, Physiologist *10*, 215 (1967).
[41] M. R. BENNETT, J. Physiol. *190*, 465 (1967).
[42] P. FATT and B. L. GINSBORG, J. Physiol. *142*, 512 (1958).
[43] M. KOBAYASHI, Am. J. Physiol. *208*, 715 (1965).
[44] M. P. PAPASOVA, T. NAGAI and C. L. PROSSER, Am. J. Physiol. *214*, 695 (1968).
[45] R. NIEDERGERKE and R. K. ORKAND, J. Physiol. *184*, 312 (1966).
[46] A. WATANABE, I. TASAKI, I. SINGER and L. LERMAN, Science *155*, 95 (1967).
[47] H. IRISAWA and M. KOBAYASHI, Jap. J. Physiol. *14*, 165 (1964).
[48] F. V. McCANN, J. gen. Physiol. *46*, 803 (1963).

Patterns of Reversal in the Heart of *Ciona Intestinalis* L.

by H. Mislin
Institute of Physiological Zoology, University of Mainz, Germany

Introduction

The reversal of the heart beat is not exclusive to tunicates, for Marcello Malpighi in 1660 described periodic reversal in *Bombyx mori* and in some Orthoptera. In 1821 Kuhl and van Hasselt reported the phenomenon in *Ciona intestinalis* L. We have observed reversal in many active pulsating blood vessels, both in vertebrates and in invertebrates, and even in embryonic heart tubes. Using four or five suction electrodes simultaneously, we were able to record electrical activity from multiple sites. Normally, the contraction wave originates from a localized site at one end of the tube and then traverses the whole length of the heart. When a wave of excitation does not propagate the full length, the blocked area may initiate a wave that then travels opposite to the initial direction (contrawave). This pattern, where activity arises in two separate areas such that the one contractile wave is orthograde and the other retrograde, may persist for a prolonged period. Each different pattern seems to result from some disturbance of the normal pacemaker in the myoepithelial cells.

The following changes in activity of the pacemaker in *Ciona intestinalis* L. have been observed and are reported here: 1. Pauses of the heartbeat without reversal, 2. Total reversals of the heartbeat, 3. Continuous progressive reversal, 4. Partial reversal and contrawaves.

Methods

The animal selected for study was the tunicate *Ciona intestinalis* L. Only those animals which had recently been removed from the sea were used (90 experiments). Investigations were carried out on intact heart and pericardium, or on the completely isolated heart tube. The recording of electrical activity by suction electrodes was as in previous publications[1]*.

Results

1. *Pauses of the heartbeat without reversal.*

In some cases pauses were not followed by reversal (Fig. 1). The time period T, measured as the interval between the initial rise of monophasic action potentials, was equal both before and after the pause. The frequency of beats immediately before and immediately after a pause was lower than the normally observed frequency (T + a). Data on interelectrode distances and quantitative values of

*Numbers refer to References, page 198.

Fig. 1.
Manifestation of pauses without reversal.

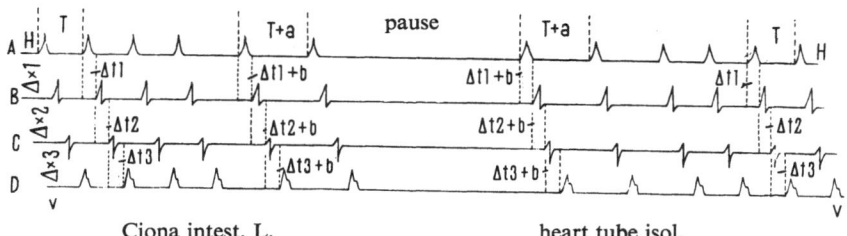

Ciona intest. L.   heart tube isol.

$\Delta x1$ Distance between electrodes A and B = constant, $\Delta2$ between B and C, etc.
$\Delta t1$ Conduction of the wave from A to B, $\Delta t2$ conduction from B to C, etc.
$\Delta t1 + b$ Conduction of the wave from shortly before or after pause A to B, $\Delta t2 + b$
conduction before or after pause B to C, etc.
T Duration of the period
$T + a$ Duration of the period shortly before or after pause
H Hypobranchial center (active)
v Visceral center (passive)

the conduction velocity will be published later. The conduction velocity $CV = \Delta x/\Delta t$ was unchanged after a pause. Immediately before and immediately after a pause some waves progressed at diminished velocity, as evidenced by the longer time interval between two different electrodes $\Delta t(\Delta t + b)$ with Δx unchanged. The duration of a pause was usually 20–40 sec (Fig. 2) and therefore longer than

Fig. 2.
Pause duration without reversal.

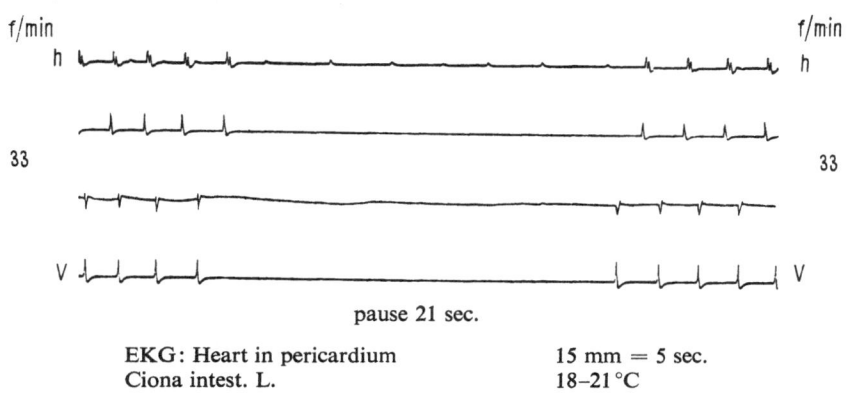

pause 21 sec.

EKG: Heart in pericardium   15 mm = 5 sec.
Ciona intest. L.   18–21 °C

V = Visceral active end
h = Hypobranchial passive end with rhythmical active potentials during the pause.

$T + a$ and longer also than the duration of the pause when total reversal occurred. For any one experiment the pause duration as well as the interval between action potentials remained relatively constant, the latter persisting longer than a pause. The shape and amplitude of the action potentials were unchanged, only the first action potential of the leading end being changed in form. Hearts which tended to exhibit pauses without reversal showed this tendency for more than 30 seconds (Table 1). We observed in this type of heart on many occasions that the passive end during the duration of a pause produced rhythmic action potentials of smaller amplitude and similar shape (Fig. 2h). We found only one case in which arrest of the heart was preceded by an extrasystole.

2. Total reversal
a) Sustained unidirectional wave propagation

While a total reversal could be precipitated by a pause or an extrasystole, these disturbances were not always observed prior to reversal (Fig. 3). The

Fig. 3.
Total reversal.

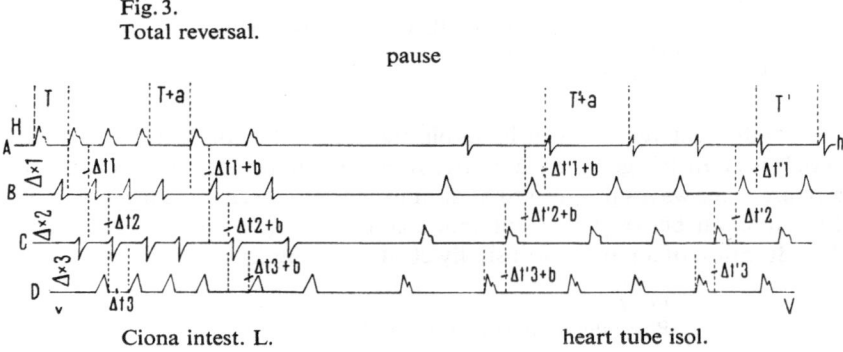

Ciona intest. L.   heart tube isol.

$\Delta x1$ Distance between electrodes A to B (constant), etc.
$\Delta t1$ Conduction time from A to B, etc.
$\Delta t'1$ Conduction time from B to A, etc.
$\Delta t1 + b$ Conduction time from A to B shortly before reversal, etc.
$\Delta t'2 + b$ Conduction time from B to A shortly after reversal, etc.
T Duration of period before reversal
$T + a$ Duration of period shortly before reversal
T' Duration of period after reversal
$T' + a$ Duration of period shortly after reversal
H, h Hypobranchial end, active, passive
V, v Visceral end, active, passive

duration of the period before reversal (T) was always of greater or lesser value following reversal (T'). Hearts exhibiting total reversal resumed their original rate of contraction even after repeated reversals. The direction of pulsation of a particular heart had a specific period, the frequency being direction specific and reproduced over long periods of time.

The duration of the period before and after reversal was in most cases prolonged, T becoming $T + a$ or $T' + a$. Only a few contractions reverted to

T' following reversal. The velocity of wave propagation ($\Delta x/\Delta t$) both before and after reversal was equal and is, therefore, independent of direction. When Δt was reduced before reversal, it was likewise diminished following reversal so that Δx and Δt remained unchanged, the conduction time before reversal in these instances being described by $\Delta t + b$. Occasionally there was a small increase in the length of conduction time following reversal indicated by $\Delta t' + b$. During total reversal, the maximum duration of the pause was ten seconds. In most cases of total reversal the pause at the active end was not longer than $T + a$. Only rarely did we observe a longer pause and then it never equalled the duration which occurred with reversal.

b) Short periods of unidirectional wave propagation

This type of pattern was frequently observed and is illustrated in Fig. 4.

Fig. 4.
Total reversal for short periods of time.

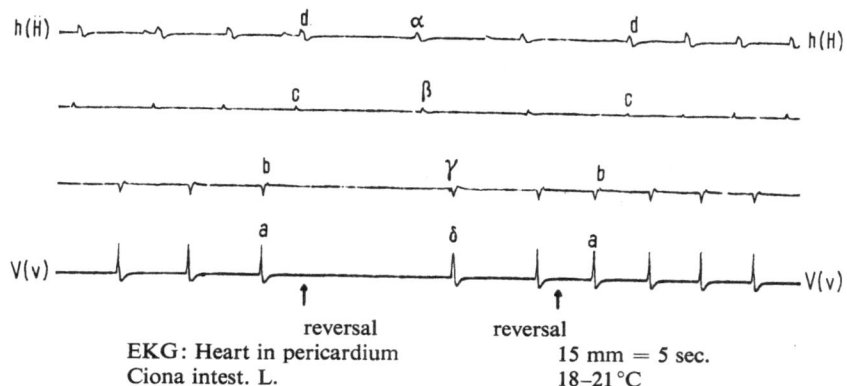

EKG: Heart in pericardium
Ciona intest. L.

15 mm = 5 sec.
18–21 °C

a-b-c-d = wave moving in the direction visceral-hypobranchial, ending in d
α-β-γ-δ = total contrawave (*H-v*)
a-b-c-d = contrawave (*V-h*)

On some occasions pauses and extrasystoles occurred before reversal. The form of the action potential was direction specific, the velocity of the wave v in these cases was the same before and after reversal, t remained unchanged.

3. *Continuous progressive reversals*

The phenomenon of total reversal has been observed as a continuous progressive form, which implies that the contractions in the heart tube normally initiated at the active end were initiated at the passive end, these latter contractions now becoming dominant (Fig. 5). Thus two contrawaves were precipitated.

Fig. 5.
Continuous progressive reversal.

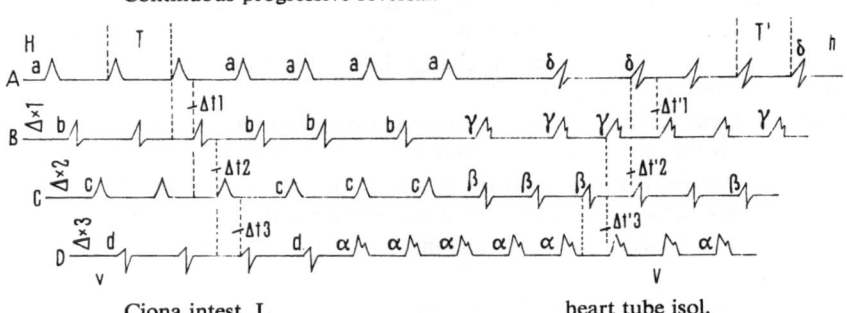

Ciona intest. L.   heart tube isol.

$\Delta x1$ Distance between location of electrodes A-B (constant), etc.
$\Delta t1$ Conduction time from A to B, etc.
$\Delta t'1$ Conduction time from B to A, etc.
T Duration of period before reversal
T' Duration of period after reversal
H, h Hypobranchial end center active, passive
V, v Visceral end center active, passive
$\alpha, \beta, \gamma, \delta$ Origination and progression of contrawaves

The progressive continuous reversal as recorded between two electrodes could result in a progressive or a total reversal. Both types were present during the continuous reversal, step for step, until the whole heart tube was completely reversed.

Preceding extrasystoles or pauses were seldom observed. The wave velocity was the same before and after reversal, due to a constant Δx and Δt before and after reversal. When progressive reversal was observed using two electrodes, it was found that after completion of reversal, t of $(t + b)$ became t'.

Recordings taken from 4 separate points A, B, C, D, indicated that waves originated at the passive end near D when the assumption was made that wave direction proceeded from A to D. If the contrawaves which initiated reversal went past D, then a change in the form of the action potential was observed at this point. If reversal between C and D was total, then C showed, even after a following contraction, a change in action potential form. This indicates that for waves going further than C, the original wave from A did not go much further than B. If the reversal between C and D was progressive then the wave originating from the passive end moved during contraction in the opposite direction, until C was passed, and at this point a new form of action potential was manifested. If the contrawave passed B and a reversal occurred between B and C, then B showed a new form of action potential. The reversal was complete when the contrawave had passed point A, and thus the original wave was erased and point A exhibited a new action potential (Fig. 5).

4. *Partial reversal and contrawaves*

Partial reversal often originated from incomplete progressive reversals and could, like total reversals, persist for a short time only (Fig. 6). An example of

Fig. 6.
Partial reversal for short periods of time.
Partial reversals

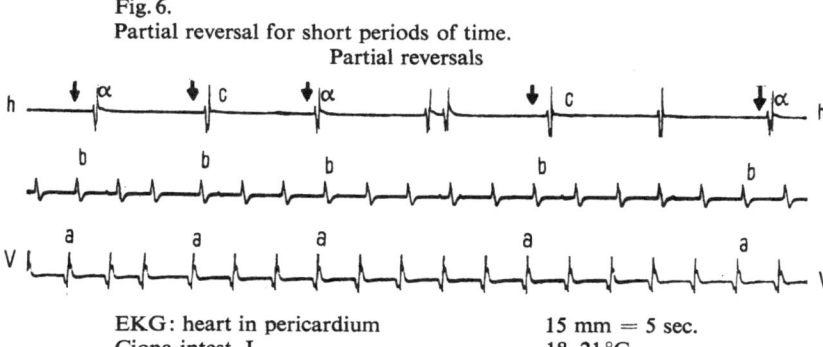

EKG: heart in pericardium   15 mm = 5 sec.
Ciona intest. L.   18–21 °C

Figure 6 shows two total waves and three partial waves.
a-b = partial wave (*V-h*)
α = contrawave
a-b-c = total wave (*V-h*)

partial reversal of long duration is shown in Fig. 7. The frequencies of the contrawaves were not identical. If the frequency of a wave of this type became synchronized with another wave, then the forms of the action potentials were cumulative, which resulted in greater amplitude and a very characteristic form of action potential. If two waves which had been travelling in opposite directions collided, then the frequencies of the resulting partial waves were not equal. Also, two successive waves that moved in the same direction could result. For example, the first wave propagated from one end to the middle part of the heart, while the second wave proceeded after a short delay from the middle to the other end. Two partial waves traveling in the same direction often resulted from two contrawaves, one of which had become reversed. It was also noted that two new contrawaves could arise when there were few contraction waves. Most partial reversals did not have an extrasystole before reversal but extrasystoles in the central part of the heart could produce contrawaves (Fig. 7).

Conclusions

As has been shown previously in studies on ligatured and transverse sections of the heart[1,2], we have concluded that the reversal of the heartbeat does not depend on rivalry between two specialized automatic centers at opposite ends of the heart. The myoepithelial cells of the one-cell-thick heart wall are potential pacemakers (multiple pacemakers[3]).

Fig. 7.

Partial reversal II and III

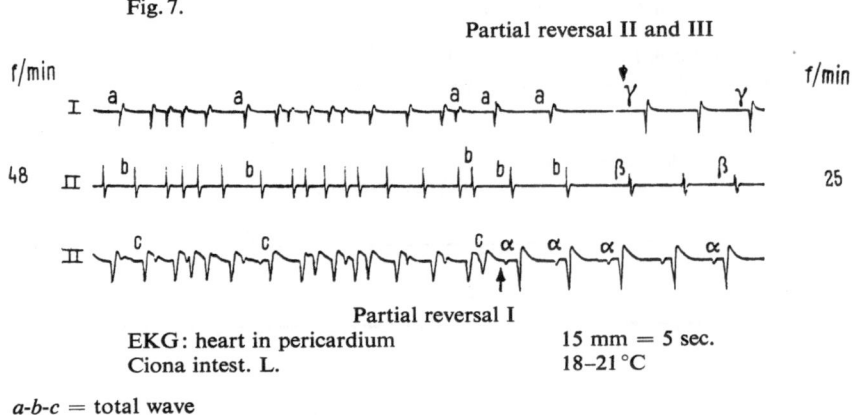

Partial reversal I

EKG: heart in pericardium   15 mm = 5 sec.
Ciona intest. L.   18–21 °C

a-b-c = total wave
a-b = partial wave
α = partial contrawave
$\alpha, \beta, \gamma, \delta$ = total contrawave

Reversal does not result from the interplay of two terminal centers, but rather from a few foci of varying location. These are usually situated at the poles of the heart, or they may be situated at an artificial end produced by cutting the heart tube. Prior to a reversal the normal frequency slows for 2 or 3 contractions and the heart pauses, and after reversal the frequency remains at the lower rate for a few contractions and then reverts to the faster initial rate. The various patterns of reversal observed in this preparation are summarized in Table 2. The shape of the action potentials associated with each direction of conduction may be different, as was shown also by KRIEBEL[4, 5].

At this time we are unable to formulate a model of the mechanism[6, 7] that effects heartbeat reversal. However, we propose that the phenomenon of total, partial, and continuous progressive reversal with contrawaves results from a disturbance of the automatic activity of multiple pacemakers in the heart tube and from interaction between two different foci.

The author wishes to express gratitude to Mrs. Bach for her technical assistance.

References

1   H. MISLIN and R. KRAUSE, Revue Suisse de Zoologie *71*, 510 (1964).
2   H. MISLIN, Revue Suisse de Zoologie *72*, 865 (1965).
3 · H. MISLIN, Helv. Physiol. Acta *24*, 41 (1966).
4   M. E. KRIEBEL, J. gen. Physiol. *50*, 2097 (1967).
5   M. E. KRIEBEL, this volume.
6   M. ANDERSON, American Zoologist *5*, 104 (1965).
7   M. ANDERSON, J. exper. biol. *49*, 363 (1968).

Table 1. Pause duration, without reversal (sec).

Pause with extrasystole (sec)	Pauses	without	extrasystoles (sec)
11,0	8,0	41,4×	52,7×
	31,0	41,4×	53,3×
	16,7×	42,3×	54,0×
	22,3	43,3×	53,3×
	21,3	44,0×	53,3×
	24,0×	44,7×	40,0×
	29,0×	44,7×	40,0×
	19,7	21,3×	40,7×
	29,7	32,0×	26,7×
	9,3	5,3×	17,7×
	8,0	52,7×	19,3×
	40,0×		20,0×

The table shows that pauses without reversal generally occurred without extrasystoles. The passive end (Fig. 2: h) shows rhythmical action potentials during pauses as indicated by values succeeded by ×.

Table 2
The different patterns of reversal found in the heart of *Ciona intestinalis* L.

Disturbance of the automatic activity	Reversals	Pauses	Duration of period $T+a, T'+a$	Wave velocity Δx constant Δt	$\Delta t+b$	$\Delta t' + b$
pause without reversal	—	+	+	—	—	+	+
total reversal long duration	+	+ or −	+	—	+	+	+
total reversal short duration	+	+ or −	—	—	+	—	—
continuous progressive reversal	+	+	+	+	+	+	+
partial reversal	+	+ or −	—	—	+	—	—

Ciona intest.   18–21 °C
heart, isol.
+indicates presence, − indicates absence.
For the definitions $T+a$, $T'+a$, Δt ... etc. vide
Fig. 1 and Fig. 3.

Observations on the Neural Rhythmicity in the Cockroach Cardiac Ganglion

by Ned A. Smith

Department of Biology, Tufts University, Medford, Massachusetts 02155 U.S.A.

Introduction

The term cardiac ganglion is used here to refer to one of the paired neuro-endocrine structures intimately associated with the heart and segmental vessels of the cockroach. It seems appropriate to adopt this new terminology, since the ganglia have been previously referred to in the literature as the lateral heart nerves[1]*, aortal nerves[2], lateral cardiac nerves[3], and lateral cardiac nerve cords[4]. These diffuse ganglia, each containing 15–20 intrinsic neurons[5], have an elongate shape. They arise from the paired corpora cardiaca in the head[2] and run parallel along each side of the heart to the posterior part of the abdomen, where they fuse on the ventral side of the heart in the ninth abdominal segment[1] (Fig. 1). The somata of the intrinsic neurons are distributed somewhat regularly,

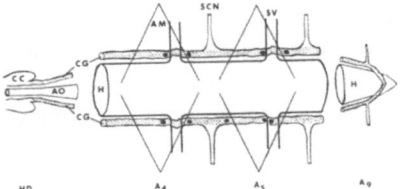

Fig. 1.
Ventral view of association of cardiac ganglia (*CG*, stippled) and heart (*H*) of cockroach, showing connection of the *CG* to corpora cardiaca (*CC*) in the head (*HD*), arrangement in abdominal segments A_4 and A_5, and the anastomosis of ganglia in the ninth abdominal segment (A_9). Dark circles mark common locations of cell bodies of intrinsic neurons. *AM*, alary muscles; *AO*, aorta; *SCN*, segmental cardiac nerve; *SV*, segmental vessel.

usually occurring where the segmental vessels connect with the heart, and occasionally where the segmental cardiac nerve joins the cardiac ganglion in each segment[1,6]. Methylene-blue staining of whole ganglia reveals not only the position of intrinsic cell bodies, but also that there are two distinct populations of axons, as was shown in electron micrographs by Johnson[3]. A group of ordinary axons is almost surrounded by a group of neurosecretory axons. The relationship of intrinsic neurons to these two axon groups is unknown.

In this short paper, I present the results of some experiments examining the neural activity in the cardiac ganglia of the American cockroach. Isolated ganglia, those of semi-isolated heart preparations, and those in the intact animal were observed and compared.

*Numbers refer to References, page 205.

Materials and Methods

Semi-isolated cockroach hearts were prepared by decapitating adult male *Periplaneta americana* and removing the thoracic and abdominal dorsum, with the heart and associated structures attached. With the dorsum pinned dorsal side down, the heart structures were easily observed and maintained in an insect saline[7].

A small length of the cardiac ganglion was freed from the heart with a fine glass needle. By drawing this free loop of the ganglion into a saline-filled suction electrode, extra-cellular recordings of ganglionic activity were made. The electrode consisted of a glass capillary tube with an inside diameter of about 60 μ, which housed a silver wire leading to a Tektronix 122 pre-amplifier, and then to a Tektronix 502A oscilloscope for display. The electrodes were placed at various positions on the ganglion within any given body segment without cutting or grossly damaging the tissue.

Recordings were also made from the ganglia of restrained intact animals. These records were obtained by carefully cutting away a 2 mm square of the exoskeleton from a point on the abdominal dorsal midline of an animal restrained by means of clay. This cut exposed a small portion of the heart, and a loop of one of the cardiac ganglia was freed and drawn into a suction electrode as before. The indifferent electrode, which was a silver wire, was placed on fat body or connective tissue in all recordings.

Records of neural activity from an isolated ganglion were made by carefully dissecting the ganglion free from the heart in all thoracic and abdominal segments, and removing it to a dish of saline. A small loop of the ganglion was drawn into a suction electrode.

Records from single neurons were made by isolating a portion of a ganglion and cutting off small pieces until the retained piece had only one spontaneously active unit remaining. The electrode served to record activity and provide a physical means of holding the tissue. Since the electrode was moved by a micromanipulator, the tissue it held could be lowered into a test extract in order to record the effect on spontaneous activity.

The extracts of the corpora cardiaca which were tested were supernatants obtained by centrifuging heat-treated homogenates of the glands at $25,000 \times G$ for 40 min at 2°C in a Sorvall RC2B centrifuge. These extracts had a concentration of 0.3 pairs of glands/ml saline, and have been shown to have a heart-accelerating action both in vitro and in vivo[8].

Results

The intrinsic cells of the ganglia were spontaneously active. In the ganglia of semi-isolated hearts and intact animals, activity was organized into a discrete pattern of spike bursts followed by a shorter inter-burst period (Fig. 2). The

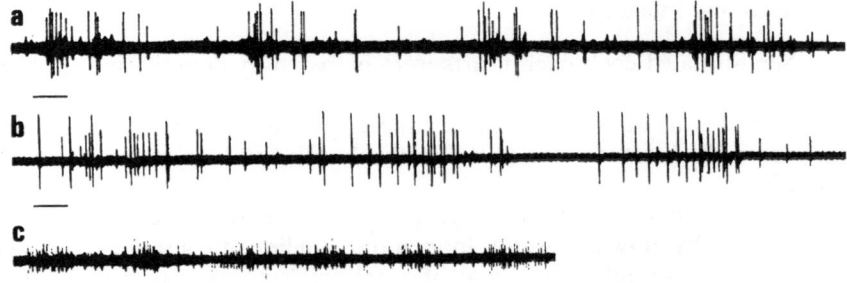

Fig. 2.
Spike bursts recorded from a cardiac ganglion (*CG*) of *P. americana*. Traces a and b show activity from the *CG* of two different semi-isolated heart preparations. In a, the firing pattern of units within the burst shifts, while in b, it remains relatively constant. Trace c, recorded with difficulty from a *CG* in the whole animal, shows the same characteristic burst activity. Time markers indicate 0.1 sec.

firing order of units in a burst did not appear to be rigidly fixed (Fig. 2a), although in a few records it is evident that the pattern did remain remarkably constant, even over a period of several hours, suggesting the existence of pace-maker and follower neurons (Fig. 2b).

Data from at least 10 bursts from 10 different heart preparations were analyzed. These particular recordings were all from the fourth abdominal segment of the ganglion, and were made while the temperature was $23 \pm 1\,°C$. Bursts were found to contain from 18 to 42 spikes, with a mean of 27. The duration of the bursts ranged from 0.36 to 1.06 sec, with a mean value of 0.66 sec, while the mean duration of the inter-burst period was found to be 0.21 sec, the range in this case being 0.14–0.29 sec.

No differences between burst activity in ganglia of semi-isolated hearts and burst activity in ganglia of intact animals could be found. However, when 10 completely isolated ganglia were examined, no characteristic burst activity was evident. Individual neurons were seen to fire repetitively at fairly constant rates for long periods of time.

Records from the ganglia of semi-isolated hearts which were made with two electrodes revealed that most of the action potentials in the abdominal portion of the ganglion were diphasic and were traveling anteriorly, the direction of the heart contractile wave. Conduction velocities of most of the axons were found to be 0.3–0.5 m/sec.

Observation of heart contraction while listening to burst activity on an audio-monitor revealed that bursting occurred during the diastolic phase of the cardiac cycle. Chilling the heart with cold saline increased the duration of the burst period, and the duration of the corresponding diastole.

The neurons appeared to fire when the diameter of the heart was increased. An artificial diastole was produced by inflating one to several chambers of the heart with air or saline. Systole was prevented, and the synchronized burst activity in the ganglion was replaced by repetitive firing of units (Fig. 3). In addi-

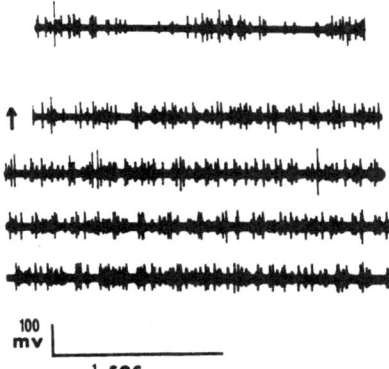

Fig. 3.
Effect on burst activity in a cardiac ganglion, of increasing the diameter of the heart by pumping in saline solution. Normal burst activity at top. Arrow indicates start of saline inflow, which results in a loss of the inter-burst period. Record is continuous.

100
mv

1 sec

tion, simply stretching the ganglion, so that the myoneural association was distorted, resulted in action potentials invading the ordinarily quiet inter-burst period. Sensitivity to stretch was also apparent in the isolated ganglion. In all six preparations tested, an increase in length of 25% caused an increase of at least 100% in the number of spikes per second produced by the spontaneously active ganglion cells.

When isolated neurons of the ganglion were exposed to extracts of the corpora cardiaca for 30 sec, the result was a dramatic increase in the spontaneous firing rate (Table). The response occurred almost immediately, and was easily reversed by washing in fresh saline.

Finally, tetrodotoxin, which is reported to block sodium conductance in certain axons[9], was perfused onto the semi-isolated heart while ganglionic activity was monitored. At 10^{-6} g/ml, tetrodotoxin abolished all action potentials in both ganglia, while the heart continued normal peristaltic contractions. This result was seen in all six preparations tested.

Discussion

Extracellular recordings of uncut cardiac ganglia in semi-isolated heart preparations of *Periplaneta americana* obtained with suction electrodes show that a characteristic burst pattern is produced by the synchronized activity of intrinsic neurons. Records from the ganglia of whole animals show that the same cyclic firing pattern occurs when the entire nervous system is intact, indicating that this is probably the normal condition in the animal. This burst activity has not been observed in isolated ganglia.

The burst occurs during the diastolic phase of the cardiac cycle, and varies in duration with diastole. The requirement for firing in situ appears to be a stretching or distention of the heart wall as its diameter increases. That the ganglia are stretch-sensitive and react to this distention is shown by the response of the ganglia to inflation of the heart, which causes what could be considered a prolonged burst, and also by the fact that cyclic activity is lost when the ganglion is detached from the heart, even though the units are still spontaneously active. The appearance of spikes in the inter-burst period in response to pulling on the ganglion, and the increase in the number of action potentials in the stretched isolated ganglion further suggest that some of the neurons in the ganglia are serving a stretch-receptor function.

Extracts of the corpora cardiaca with known cardio-accelerator action were assayed on neurons of the cardiac ganglia. The stimulating effect of these extracts on the neurons clearly implicates the ganglia as possible sites of action for any heart-active substance, especially that present in extracts of the corpora cardiaca. However, the ganglia were not considered as possible mediators of the response of the cockroach heart to corpora cardiaca extracts by DAVEY[10], who reported that it was the pericardial cells which the extract affected, and that they in turn released a substance which affected the heart.

Abolition of ganglionic activity by tetrodotoxin suggests that the active neurons produce action potentials dependent on sodium conductance. That the heart continues to beat in the absence of neural activity confirms the report that denervated hearts exhibit spontaneous activity[4].

Table. Effect of corpora cardiaca extracts on the spontaneous firing rate of individual neurons of the cockroach cardiac ganglion.

Preparation	Firing rate in saline (spikes/10 sec)	Firing rate in CC* (spikes/10 sec)	% Increase
1	35	130	271
2	27	90	233
3	50	128	156
4	41	135	229
5	18	60	233
6	22	56	154

* Supernatant from centrifugation of heat-treated homogenate of corpora cardiaca spun at $25,000 \times G$ for 40 min at 2 °C. Final concentration, 0.3 pairs CC/ml saline.

Acknowledgments

I thank Dr. K. D. ROEDER for encouragement and use of facilities. Thanks also to Dr. D. R. A. WHARTON, U.S. Army Natick Labs, Natick, Mass., U.S.A., for supplying insects. Supported by U.S.P.H.S. fellowhip 1-F2 GM-33, 738-01.

References

1   J. S. ALEXANDROWICZ, J. comp. Neurol. *41*, 291 (1926).
2   R. B. WILLEY, J. Morph. *108*, 219 (1961).
3   B. JOHNSON, J. Insect Physiol. *12*, 645 (1966).
4   T. MILLER and R. L. METCALF, J. Insect Physiol. *14*, 383 (1968).
5   K. G. DAVEY, *Advances in Insect Physiology* (Ed. J. W. L. BEAMENT, J. E. TREHERNE and V. B. WIGGLESWORTH; Academic Press, New York 1964), vol. 2, p. 219.
6   N. E. McINDOO, J. comp. Neurol. *83*, 141 (1945).
7   J. W. S. PRINGLE, J. exp. Biol. *15*, 101 (1928).
8   N. A. SMITH and C. L. RALPH, Gen. Comp. Endocrinol., in press.
9   Y. NAKAMURA, S. NAKIJIMA and H. GRUNDFEST, J. gen. Physiol. *48*, 985 (1965).
10   K. G. DAVEY, Gen. Comp. Endocrinol. *1*, 24 (1961).

Initiation of Activity in the Cockroach Heart

by T. Miller
Department of Zoology, University of Glasgow

Permanent address: Entomology Department, University of California, Riverside, California 92502.

Introduction

When all intrinsic neurons are removed from the heart of the American cockroach, *Periplaneta americana*, normal rhythmic contractions occur[1]*. This is an unusual finding since response of the cockroach heart to drugs and the presence of ganglion cells had previously served as the basis for its classification as neurogenic[2].

Preliminary works on ultrastructural characterization of the nerve cords and neuromuscular junctions in the cockroach cardiac nervous system have been published[3,4]. The properties of the cardiac neurons have been investigated using standard electrophysiological techniques[5]. Pharmacological responses of the whole heart[1], and of the cardiac neurons[6] have been described. The results of these studies will be briefly reviewed along with preliminary studies on the effects of ions on the electrical responses of the myocardium.

Cardiac Nervous System

The cardiac nervous system is composed of intrinsic neuron cell bodies and their axonal processes which are situated in paired lateral cardiac nerve cords running the length of and on either side of the heart. In addition, paired segmental nerves join the lateral cords near each ostial valve (Fig. 1). All work

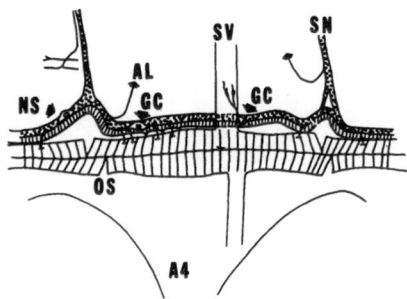

Fig. 1.
A drawing of the left lateral cardiac nerve cord in the fourth abdominal chamber (A4) of *P. americana*, dorsum down. The muscle cells are delineated on the heart tube. OS, ostial valve; SV, segmental vessel; SN, segmental nerve; NS, 'neurosecretory' neuron; GC, ganglion cell (motor neuron); AL, arrow indicating alary muscle innervation. Cardiac motor system is lined, 'neurosecretory' system is dotted.

described here pertains to the middle abdominal heart chambers, second through fifth.

*Numbers refer to References, page 218.

When examined closely in vitro (magnification c. ×100) the segmental nerve appears faintly blue, whereas the lateral cardiac cords are divided longitudinally into a blue portion (Fig. 1, dotted) and a white portion (Fig. 1, lined). The segmental nerve is comprised almost entirely of axon profiles enclosing large numbers of granules of varying electron density. Axons of similar appearance in the lateral cords were termed 'neurosecretory'[3]. There seems little doubt that the blue color of the in vitro nerve bundles represents the 'neurosecretory' portion of the cardiac nervous system.

Cardiac Axons

Four different types of axon were found in the lateral cardiac cord. Some axons contained small dense granules, others large dense granules, and a third group contained large membrane-lined 'sacks'. The respective maximum diameters of the granules were 150 and 300 μ^3, and the 'sacks' ranged up to 400 μ in diameter[4]. A fourth axon was devoid of cytoplasmic granules or sacks and was enclosed in layers of glial cells.

The fourth axon was found in the white portion of the lateral cardiac nerve cord and is considered to be motor in function.

Cardiac Neurons

Two types of cardiac neurons have been identified by ultrastructural and electrophysiological studies. Ganglion cells (GC) produced the largest potentials (motor spikes) as measured by suction electrodes, and these impulses were propagated down the cord at about 0.5 M/sec. No evidence of granule synthesis was found in the cytoplasm of the GC neuron[3].

Neurons (NS) found in the 'neurosecretory' portion of the cord were irregularly active. The NS potentials were smaller and propagated more slowly than the motor spikes, i.e. about 0.2 M/sec. The NS neurons were distinctive in ultrastructure, containing cytoplasmic evidence of synthesis such as rough endoplasmic reticulum, an ovoid nucleus and electron dense granules of varying sizes in the cytoplasm[4].

By sectioning the cord at various points along the heart while monitoring the cord activity, the NS potentials were found to originate near the junction of the segmental nerve with the lateral cardiac cord (Fig. 1). Neurons have been found in a similar position in the heart of the stick insect, *Carausius morosus*[7].

Function of the Cardiac Neurons

An increase in the rate of heartbeat caused by stimulation of the segmental nerve was always accompanied by a short-lived burst of activity from the NS neurons[8]. Since the NS neurons were considered to play a role in cardio-regulation, their position near the segmental nerve junction seems logical from the functional standpoint.

Certain of the GC neurons in each chamber in the middle abdominal portion of the heart are sensitive to stretch such that movement of the myocardium briefly exerts an inhibitory influence on the GC activity during diastole of the cardiac cycle[5]. Bursts of activity in the GC cells of the cardiac cord sometimes accompany normal heartbeat, at other times the neuronal activity is not patterned and not synchronized with the contractions of the heart.

There is evidence, then, that contractions of the myogenic cockroach heart modulate the activity of the intrinsic and spontaneously active cardiac ganglion cells. Perhaps the key to understanding the properties of this unusual heart lies in the fact that the two lateral cardiac nerve cords are electrically separate tissues. As a consequence, the active GC neurons in each cord send independent bursts of impulses to the heart with the activity from each cord co-ordinated only by myo-neural feedback.

This type of system appears essential since otherwise there is no guarantee that the impulse bursts from the two lateral cords will arrive at the myocardium in phase. Perhaps fusion of the cardiac ganglia was a prerequisite for the development of neurogenicity in other hearts. This would require the simultaneous development of a dominant pacing burst from the GC neurons and the loss of the ability of the myocardium to sustain rhythmic contractions. In addition, cardio-regulatory innervation might be expected to develop for control of the cardiac neurons rather than the myocardium. In this respect it is interesting that evidence so far has indicated a direct connection of the cardio-regulatory system both to the myocardium and to the cardiac neurons[5,8]. The cockroach heart seems to be an example of a heart which contains elements for myogenicity and most of the elements characteristic of neurogenicity.

Pharmacology

The discovery of myogenic activity in the cockroach heart offered an opportunity to establish the effects of certain neuro-active compounds both on the heart and on the cardiac neurons. Cholinergic compounds of the muscarinic type were found to act with the greatest specificity by affecting the rate of bursting of the GC neurons only. On the other hand, nicotine caused increased activity in both GC neurons and NS neurons[6], and also increased the beat rate of the denervated heart[1].

Due to the highly specific effect of acetylcholine on the GC neurons, quantitative responses of the isolated heart were dependent on the number of neurons present. A single heart chamber with its own lateral cords intact responded to one 50 μl drop of 10^{-4}M acetylcholine chloride with a brief tonus increase followed by immediate recovery (Fig.2B). The identical assay on a single chamber but with all abdominal lateral cords left attached caused prolonged systolic arrest (Fig.2A). The completely denervated heart maintained the same heartbeat before and after perfusion with 10^{-2}M acetylcholine chloride. Despite the response of the GC neurons to cholinergic compounds, no evidence was found for cholinergic synapses[6].

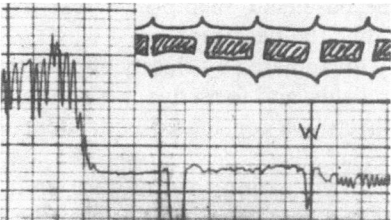

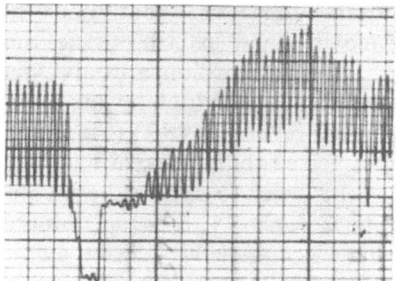

Fig. 2.
Mechanical (isotonic) responses of a single heart chamber to one 50 μl drop of 10^{-4} M acetylcholine chloride in saline. *A*. All lateral cardiac cords left intact as indicated in the inset; *W* indicates saline wash. *B*. One chamber with only its own cords. 5 seconds between vertical lines, 'down' in the record is systole.

Myocardium Ultrastructure

The myocardium is covered occasionally by connective tissue on the haemocoele side while the lumen side has no extracellular tissue (Fig. 3 and 7). Alary muscles, pericardial cells and suspensory tissues including the dorsal diaphragm all make simple connections to the myocardium, in most cases by means of

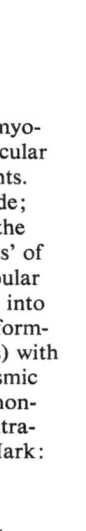

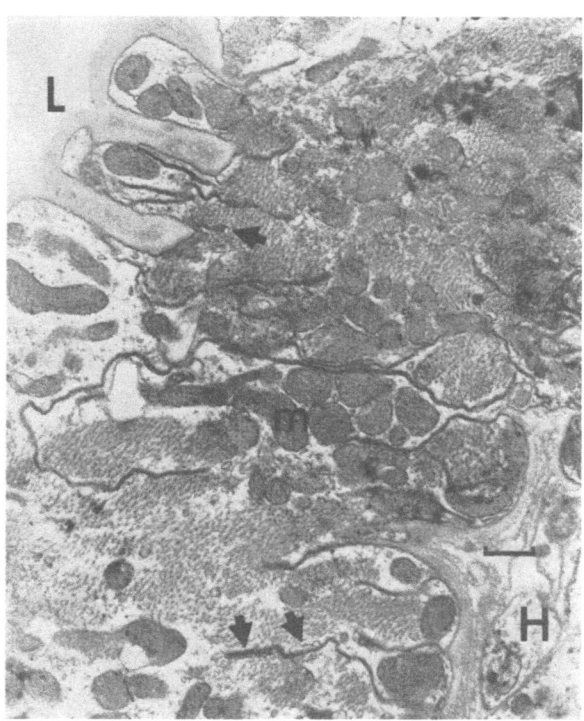

Fig. 3.
A section of the myocardium perpendicular to the myofilaments. *H*, haemocoele side; *L*, lumen side of the heart tube. 'Sheets' of the transverse tubular system invaginate into the myocardium forming dyads (arrows) with a sparse sarcoplasmic reticulum. Mitochondria, *m*, fill the intracellular spaces. Mark: 1 μ.

extracellular fibers continuous with the basement membrane. Cutting and removing these external tissues did not alter the heartbeat.

The myocardium appears extensively invaginated by closely opposed sarcolemmal 'sheets' which invade the fibrillar fields and form dyads with the sarcoplasmic reticulum (SR) (Fig. 3). These sheets are considered equivalent to the transverse tubular system.

In order to clarify the myofilament organization, hearts were pretreated with isotonic 1% 3,5-dinitro-o-cresol (DNOC) in saline. In 20 min the heart tube was completely relaxed and expanded by the elastic suspensory tissues to its largest diameter. At this time ice-cold prefixation with 2.5% glutaraldehyde caused no contractile response. Subsequent ordinary treatment and examination showed a very clean A-band of about 2 μ, an I-band with thin filaments, and a definite darkly stained Z-line (Fig. 4).

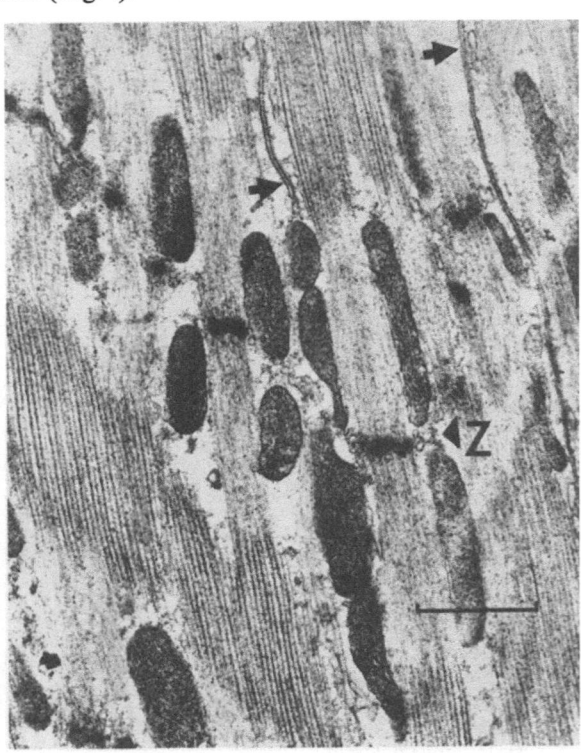

Fig. 4.
Myofibrils after 20 min. treatment in 1% 3,5-dinitro-o-cresol. Thick filament areas about 2 μ long are devoid of thin filaments. Z lines appear darkly stained (Z, arrow). TTS invaginating sheets are seen with ill-defined dyads in this plane of section (arrows). Mark: 1 μ.

The mitochondria appeared less well preserved and darkly stained. The TTS appeared unchanged and in apposition in some areas with the sparse SR.

Neuromuscular Junctions

Neuromuscular junctions of 3 distinct types were found. Small electron dense granules appear to split into synaptic vesicle structures of about 450 Å diameter in one type of presynaptic axon (Fig. 5). The vesicles were conglomerated near the synaptic membrane and the adjacent myocardium contained the undefined subsynaptic web structure (Fig. 5).

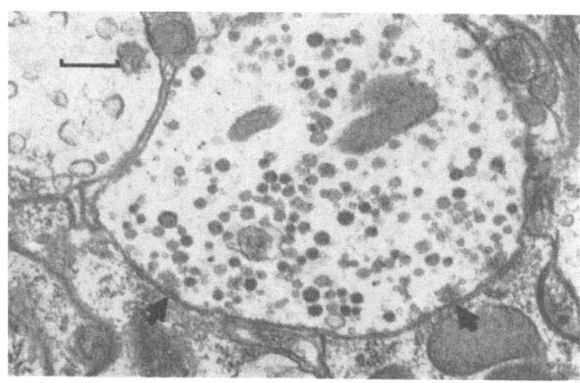

Fig. 5.
A neuromuscular junction characterized presynaptically by the presence of small electron dense granules. The granules appear to split into synaptic vesicle structures which aggregate (arrows) at the synaptic membranes. Mark: 1 μ.

A second type of neuromuscular junction was characterized presynaptically by large electron transparent vesicles or 'sacks', ranging in size downwards from 4000 Å diam. (Fig. 6). Again a subsynaptic web was clearly defined in the myocardium.

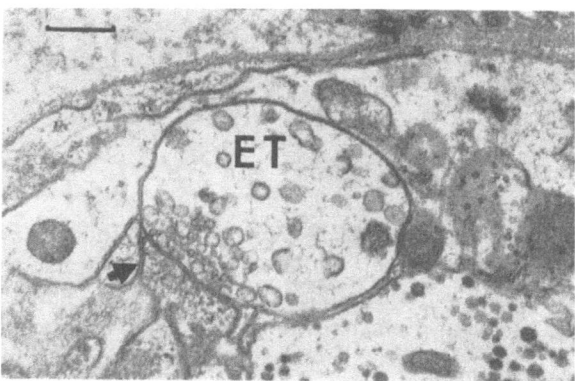

Fig. 6.
A neuromuscular junction with large electron transparent 'sacks' in the presynaptic axon, ET. Vesicles adjacent to the synaptic membranes are of non-uniform size. The postsynaptic area is characterized by a web structure, and invaginating membranes of the TTS (arrow). Mark: 1 μ.

A third type of junction was characterized by a noticeable lack of presynaptic granules with only synaptic vesicle structures present in the axoplasm and grouped near the close juxtaposition of the nerve and muscle membranes (Fig. 7). These 'ordinary' junctions also contained subsynaptic web structure and were considered to be made by motor axons from GC neurons.

The area of the myocardium near synapses was sometimes only slightly differentiated from the surrounding portions of the muscle surface (Fig. 7).

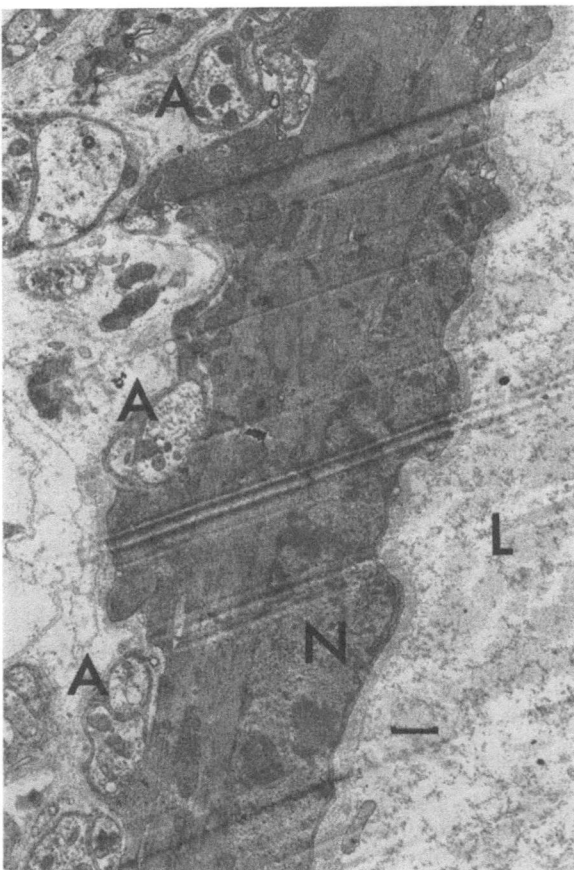

Fig. 7.
A portion of the heart
tube in cross section.
Mitochondria fill the
surface folds of the
muscle. A large elongate
nucleus, *N*, is present.
Axons, *A*, containing
synaptic vesicles join
with the myocardium
at several sites along
the haemocoele side of
the heart tube. *L*, lumen
of the heart tube.
Mark: 1 μ.

Occasionally, however, there was a pronounced evagination of the sarcolemmal membranes which enfolded the synapsing axons. Invaginating membranes of the TTS were usually evident in the postsynaptic areas of the myocardium.

The motor synapses were assumed to involve axons from the GC neurons. The nerve terminals containing small dense granules were assumed to belong either to neurons located in the ventral ganglia, or to NS neurons in the lateral cord. The origin of the nerve terminals containing electron transparent 'sacks' is as yet unknown.

Myocardial Electrogenesis

Electrically Excited Responses

For microelectrode recordings, single heart chambers were denervated by removing both lateral cardiac nerve cords. The dorsal diaphragm was also removed from the area of the heart to be impaled. $3M$ KCl-filled electrodes with an average impedance of 10–15 MΩ were used. Signals were amplified with a Bioelectric negative capacitance amplifier and displayed on a Tektronix 502 oscilloscope.

Electrical responses were similar in shape whether initiated by electrical stimulation of a small fragment of one of the lateral cords of a partially denervated heart (Fig. 8 A) or spontaneously generated from a completely denervated heart chamber (Fig. 8 B). A slow depolarizing generator phase (prepoten-

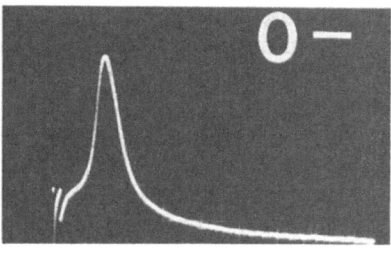

A

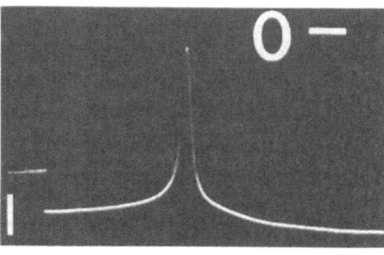

B

Fig. 8.
Intracellular responses from a myocardial cell. *A*. Neurally evoked response by stimulation of the right cardiac cord left intact on chamber A 3. 50 msec/cm sweep.

B. Spontaneous potential recorded from chamber A 3 after removal of all cardiac nerve cords. Calibration shown at left is 10 mV, 100 msec/cm sweep. Zero membrane potential is indicated.

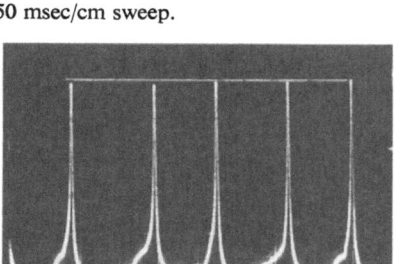

A

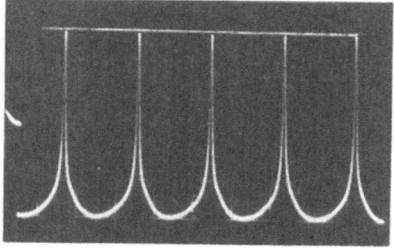

B

Fig. 9.
A. Spontaneous responses from a single cardiac cell in an innervated chamber. *B*. From a denervated chamber. *C*. From an innervated chamber. Arrows indicate junction potentials. Calibration pulse at left in *B* is 20 mv and is the same for *A* and *C*. Zero membrane potential is indicated by the straight line in each trace. *A* and *B* are 5 sec sweeps, *C* is a 2 sec sweep.

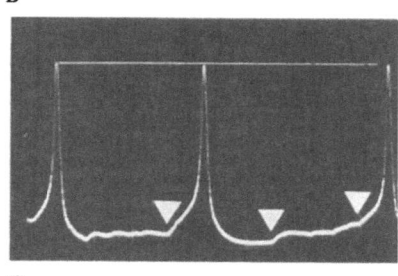

C

tial) preceded rapid depolarization at up to 1.75 V/sec and rapid repolarization at 0.87 V/sec. The term resting potential (RP) will be used here to describe the value of membrane voltage at the point of maximum repolarization.

Resting potential (RP) varied depending on the depth and placement of the microelectrode. RPs were measured from − 10 mv to about − 50 mv, and the depolarization phase fell short of, or reached zero potential, but did not overshoot the zero level. The largest potentials recorded from fibers had RPs near − 50 mv (Fig. 9).

The essential difference between the electrical activity of the innervated heart (Fig. 9 A and 9 C) and activity of the denervated heart chambers was the occurrence of junction potentials in the innervated preparation (Fig. 9 C, arrows). Note in particular that random input from the GC neurons in the lateral cardiac cord resulted in a prepotential of varying time interval (Fig. 9 A), such that the corresponding heartbeat was slightly irregular as well. The denervated heartbeat in comparison was always characterized by a smooth prepotential associated with a regular heartbeat cycle (Fig. 9 B).

Junction potentials recorded from the innervated heart probably resulted from activity of GC neurons because the GC neurons were usually active and the NS neurons usually inactive in the lateral cords in single chambers. The properties of the junctional potentials were not studied in any detail. It is possible that a variety of junctional events occur in view of the presence of at least three cytologically distinct types of neuromuscular junction (Fig. 5, 6 and 7).

Ionic Basis for Electrical Responses

The saline solution used for bathing the heart during studies of the effects of ions was based on one developed by LUDWIG, TRACEY and BURNS[9] and by YEAGER[10]:

    188 *mM* NaCl
     20 *mM* KCl
      9 *mM* CaCl$_2$
    Saline solution for *P. americana.*

The addition of 1 *mM* Mg ion either as MgCl$_2 \cdot 6 H_2O$ or as MgSO$_4$ yielded an improved heart preparation as reported by YEAGER[10], but was eliminated in the present study to make the bathing medium as simple as possible. This saline is thought to be isosmotic with, and to have an ionic composition similar to, that of cockroach blood. It is also thought to contain the best combination of ions for maintaining optimal heart activity in isolated preparations[9].

Effects of Sodium Ion Changes

Spontaneous electrical responses of the denervated heart remained unchanged on application of a saline with sucrose or choline chloride substituted for NaCl. Although initially the heartbeat remained essentially normal while the tension in the alary muscles increased, heartbeat was not maintained for more than 2 min. LiCl could be substituted for NaCl for over 1 h without any noticeable change in the character of the denervated heartbeat.

Effects of Calcium Ion Changes

When the denervated heart was perfused with saline solutions containing slightly lower or higher calcium concentrations than normal, both the electrical responses and the heartbeat rate were affected immediately (Fig. 10).

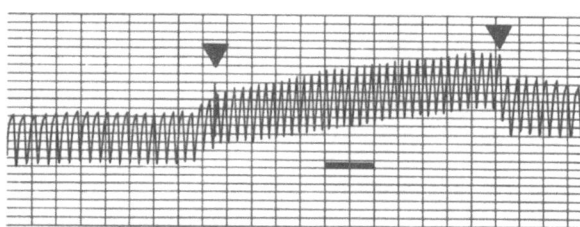

Fig. 10.
Mechanical (isotonic) responses of the denervated third abdominal chamber to application of a saline containing half normal calcium (4.5 mM), first arrow. Second arrow, return to normal saline (9.0 mM Ca). Time mark: 10 sec, systole down.

Changes in the external Ca concentration were accompanied by changes in the prepotentials. When saline containing zero Ca was applied directly to the heart preparation there was an immediate increase in the slope of the prepotential (Fig. 11, arrows) and a decrease in RP.

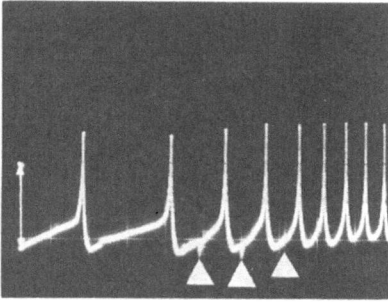

Fig. 11.
Spontaneous responses from a single myocardial cell in a denervated heart chamber. Arrows indicate drop-wise application of a zero calcium saline directly onto the heart preparation. Calibration at the left is 20 mv and the trace is a 5 sec sweep.

Cockroach cardiac muscle can be tetanized and has little or no refractory period[10]. Zero Ca causes tetany or systolic arrest during which rapidly repetitive depolarizations of the membrane maintain the myocardium in the fully contracted state.

Strontium or magnesium ions could be substituted for Ca in the saline solution without any measurable effect on the properties of the heartbeat and electrical responses. When 9 mM $MgSO_4$ was substituted for $CaCl_2$ in the saline solution, there was no immediate effect on denervated heart activity. Within several minutes, however, the heartbeat rate slowed and eventually stopped in diastole, possibly due to depletion of the intracellular Ca required for contractile function. Subsequent perfusion with normal saline resulted in full recovery of the original heartbeat in about 2 min.

The alkaline earth divalent ions, Sr, Ca, and Mg could be interchanged mole for mole in the bathing medium without immediate effect on either heartbeat character or electrical activity. This suggests that either the divalent ions do not contribute to the inward depolarizing current during the electrical response, or that all three ions can make equal contributions to the inward depolarizing current.

Effects of Potassium Ion Changes

Salines containing more than 20 mM KCl (high K) were prepared by increasing KCl and decreasing NaCl in equimolar amounts. Salines containing less than 20 mM KCl (low K) or zero KCl were prepared by a corresponding increase in the NaCl concentration.

Perfusion of the quiescent denervated heart with high K saline caused a sustained potassium contracture. The dorsal longitudinal muscles also gave a sustained contracture in high K saline.

Perfusion of the denervated heart by K-free saline resulted in elimination of the fast depolarization and repolarization phases with the muscle resting eventually in diastolic arrest. Only an irregular prepotential remained after cessation of movement (Fig. 12 B). When normal saline was reintroduced, the fast polarization phases returned in proportion to the amount of external K

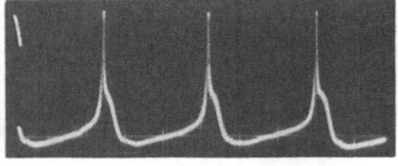

A

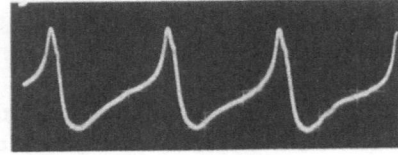

B

Fig. 12.
Spontaneous single cell responses from a denervated heart chamber. A. In normal saline, resting potential − 49 mv. B. After perfusion of K-free saline, resting potential is − 46 mv. C. During perfusion of normal saline, resting potential − 47 mv. Calibration pulses are all 20 mv, all 10 sec sweeps. Note the initial heartbeat rate (A) and recovery rate (C) are comparable.

C

(Fig. 12C). In zero K saline the resting potential remained within 3 mv of its value in 20 mM K saline.

A denervated heart bathed in K-free saline spontaneously generated small potentials less than 4 mv in height without accompanying contraction (Fig. 12B). In the absence of extracellular divalent ion the RP was found to shift towards the threshold for spontaneous responses, thereby eliminating the prepotential phase (Fig. 11). These observations suggest that a prepotential of 4 mv exists between maximum repolarization and threshold for electrically excited response. It also suggests that contraction is associated only with the 'fast' electrically excited responses and not with membrane events which determine the prepotential.

Summary

The cockroach heart is a myogenic muscle in spite of the presence of a complex and spontaneously active cardiac nervous system. Characteristic of a myogenic heart, the cockroach heart will maintain some activity after the animal is completely immobilized by ether vapors (MILLER, unpublished). In addition, the initiation of activity after removal of the lateral cardiac cords requires extreme care in dissection so that the majority of the suspensory tissues are left intact to hold the heart tube in as near the normal position as possible.

At present very little is known about the general physiology of the cockroach heart and associated musculature. For example, the alary muscles occasionally exhibit rhythmic activity[11], although the mechanisms underlying this activity have still to be examined. Segmental innervation of the heart by 'neurosecretory' axons and 'neurosecretory' neurons has been shown to play a role in cardiac acceleration. There is as yet no evidence that depression or inhibition of the heartbeat by either motoneurons or 'neurosecretory' neurons occurs in the cockroach.

The spontaneous electrical activity of the heart is apparently quite normal in an ionic environment free of extracellular sodium, although the cardiac neurons are unable to function normally in this condition. The heart muscle can be tetanized and can support graded contractions. The electrically excited responses of the cockroach heart are not generated in the absence of extracellular potassium, while the prepotential is eliminated in the absence of extracellular divalent ions.

Acknowledgement

The original work reported here was inspired by Dr. R. L. METCALF with initial support by PHS grant 5-Tl-ES-47. The electron microscopy was supported by NSF grant GB 4513 with the invaluable guidance of Dr. W. W. THOMSON. Intracellular work was supported by PHS grant PH-GM-1076 with assistance and complete co-operation of Dr. C. W. KEARNS and Dr. J. G. STERNBURG.

The invaluable comments in a critical reading of the manuscript by Dr. P. N. R. USHERWOOD are sincerely appreciated.

References

1   T. MILLER and R. L. METCALF, J. Insect Physiol. *14*, 383 (1968).
2   C. L. PROSSER and F. A. BROWN, *Comparative Animal Physiology* (W. B. Saunders, Philadelphia 1961).
3   B. JOHNSON, J. Insect Physiol. *12*, 645 (1966).
4   T. MILLER and W. W. THOMSON, J. Insect Physiol. *14*, 1099 (1968).
5   T. MILLER, J. Insect Physiol. *14*, 1265 (1968).
6   T. MILLER, J. Insect Physiol., *14*, 1713 (1968).
7   Z. OPOCZYNSKA-SEMBRATOWA, Int. Acad. Polnaisc Sci. Math. Nat [B] Sci. Nat. [II] 2, 411 (1936).
8   T. MILLER, Entom. exp. appl. 12, in press.
9   D. LUDWIG, K. M. TRACEY and M. L. BURNS, Ann. Ent. Soc. Am. *50*, 244 (1957).
10   J. F. YEAGER, J. Agric. Res. *59*, 121 (1939).
11   K. G. DAVEY, Adv. Insect Physiol. *2*, 219 (1964).

IV
Regulation of Rhythmicity

Introduction
by G. A. KERKUT

One of the most striking properties of the heart is the manner in which it goes on beating and beating. A simple calculation shows that the human heart during an average lifetime beats approximately 3×10^9 times. Though automobile manufacturers always claim good performance for their car engines, the controlled performance of the heart sets the engineers new standards, as biological engineers have realized now that they have come to design artificial hearts. Furthermore there is the added problem of regulation of heart beat. Recently the cardiac pacemakers have been fitted to younger men and it has become necessary to have some regulatory device on the pacemaker. Certain models have now been built with two speeds, 'regular' and 'fast'. The latter speed has helped the recipient in such simple tasks as running for a bus as well as the more difficult tasks of responding to complex physiological stresses.

There are two main problems involved in the regulation of the rhythmicity of the beat. There can be both neural and hormonal regulation of the heart beat. In addition, however, there is the problem of coordinating the various fibres in the heart so as to produce a controlled pumping movement. The speakers this afternoon will be concerned with these different aspects of the regulation of the heart beat. Dr. COTTRELL will discuss the action of specific chemicals on the molluscan heart and the manner in which 5-HT can bring about its observed effect, probably through the action of adenylcyclase. Dr. EBARA will describe his very elegant experiments relating to the mechanism by which activity in the different fibres in the oyster heart is coordinated through the small potentials. Dr. ANDERSON will tell us of the specific role of the nervous system in the cardiac ganglion in the regulation of crustacean heart muscle, and Dr. GREENBERG will show some of the important differences in the control systems of molluscan hearts and the manner in which these hearts show varied responses to given levels of acetylcholine and 5-HT. The afternoon session will give us insight into some of the methods by which the heart rate is controlled and altered.

Localization and Mode of Action of Cardioexcitatory Agents in Molluscan Hearts

by G. A. Cottrell and N. N. Osborne

Wellcome Laboratories of Pharmacology, Gatty Marine Laboratory, St. Andrews University, Fife, Scotland

The work described in this paper is divided into two parts. The first part deals with a histological and histochemical study of the *Helix pomatia* heart. This work was initially undertaken to determine whether the 5-hydroxytrypt-amine (5-HT) present is localized in the nerves of the heart as would be expected if this amine acts as a transmitter in this situation. The second part of the paper is concerned with the mode of action of 5-HT on the heart. We have attempted to determine whether the excitatory effect of 5-HT on the molluscan heart is dependent on an interaction of the amine with adenyl cyclase and the sub-sequent production of cyclic $3', 5'$-adenosine monophosphate.

I. Intrinsic Innervation and Localization of Biogenic Amines Within the Heart

Welsh[1]* in 1953 was one of the first to suggest a physiological role for 5-HT in an invertebrate. He observed that the amine has a pronounced excitatory effect on the isolated *Venus mercenaria* heart and suggested that it might be the chemical agent normally released from the cardio-acceleratory nerves in mol-luscs. Since 1953, various workers have obtained experimental results which support Welsh's original suggestion, for example: (a) 5-HT is found in high levels in molluscan nerve tissue. (b) Drugs which block the effect of 5-HT on the heart also block the cardio-excitatory response of nerve stimulation. (c) 5-HT has been detected in perfusates of the *Helix* heart following nerve stimulation[2]. However, although the amine has been detected in the neurones of molluscan ganglia, its presence in the nerve fibres of the molluscan heart has not been demonstrated. We decided therefore to investigate the localization of 5-HT in the heart of *Helix pomatia*, a species which has been relatively well studied with regard to the pharmacology and biochemistry of 5-HT.

Ripplinger[3] showed that the extrinsic nerve supply of the *H. pomatia* heart consists of two branches which originate in the visceral nerve. One of these branches joins the heart at the apex of the ventricle; the other enters the heart at the junction of the pulmonary vein and the auricle. He did not, how-ever, observe the distribution of nerve fibres within the heart itself.

Our first objective, therefore, was to discover the pattern of intrinsic innervation of the heart, before proceeding to study the localization of 5-HT. Hearts stained with silver or with methylene blue[4] showed only a relatively

*Numbers refer to References, page 231.

sparse distribution of nerve fibres throughout most of the heart. However, in the auricle, predominantly in an area close to the auriculo-ventricular junction, a comparatively profuse network of nerve fibres could be seen in whole mount preparations stained with methylene blue (Fig. 1). After examining a few prep-

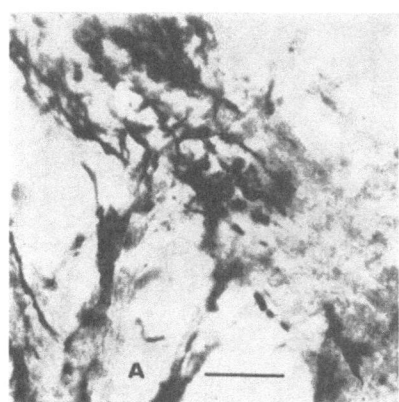
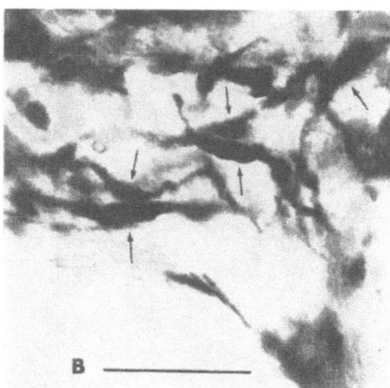

Fig. 1.
Low power (*A*) and higher power (*B*) micrographs of part of the dense network of nerve fibres in the auricle of a *Helix pomatia* heart. Note the pronounced swellings (arrows) along the fibres. Calibration, 100 μ.

arations, we reached the conclusion that this network might form part of a neurosecretory system releasing material directly into the blood stream. The observations which led us to this conclusion were as follows: (a) There was a relatively large number of nerve fibres in this particular area of the heart. (b) The fibres appeared larger than those seen elsewhere in the heart and had pronounced swellings along their lengths. (c) The system appeared to stain more intensely when the methylene blue was passed through the centre of the heart after injection into the perivisceral haemocoel, than when preparations were stained by injecting the same dye solution into the pericardial cavity. Thus most of the nerve fibres were probably situated close to, or bordering on, the heart lumen.

An electron microscopical examination of the heart[4] demonstrated the presence of large swollen axons and structures which appeared to be nerve endings in the auricle. These large axons, which contained numerous electron dense granules, were particularly abundant in that half of the auricle which adjoined the ventricle. In this area, the axons were frequently seen lying very close, or adjacent, to the heart lumen (Fig. 2). Most of the granules measured about 100–250 nm in diameter. They were very similar in appearance to the granules observed in known neurosecretory systems, e.g. the neurohypophysis[5]. Thus the ultrastructural appearance of the dense network in the auricle added support to the suggestion that the network might serve a neurosecretory role.

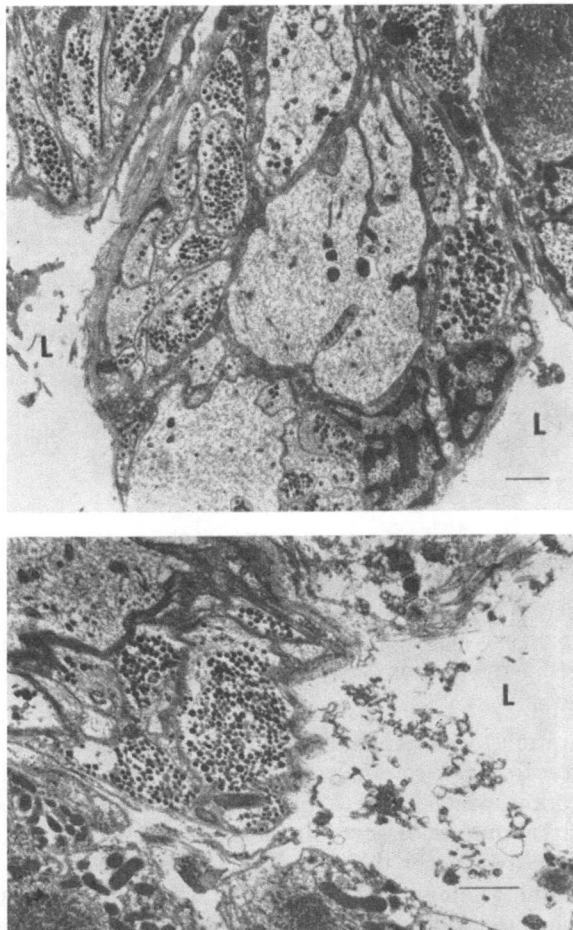

Fig. 2.
Axons, containing large
numbers of electron
dense granules, close to,
or bordering on, the
heart lumen (*L*). The
line in the bottom right
hand corner corresponds
to a length of 1 μ.

Large masses of granules in nerve fibres were seen only in the auricle. Only small axons and nerve endings were seen in the ventricle. Some of these processes contained dense centred granules and others small clear vesicles.

Having observed the pattern of intrinsic innervation, we proceeded to make a histochemical study of the distribution of 5-HT within the heart. Tissue processed and examined by means of the fluorescence histochemical method for detecting amines[6] showed the presence of yellow-green fluorescent nerve fibres throughout the heart. Such fibres were observed in sections and in whole mount preparations (Fig. 3). The primary catecholamines noradrenaline (NA) and dopamine (DA) fluoresce green and 5-HT fluoresces yellow in tissue treated and examined by this method. The observation that the fibres appeared yellow-

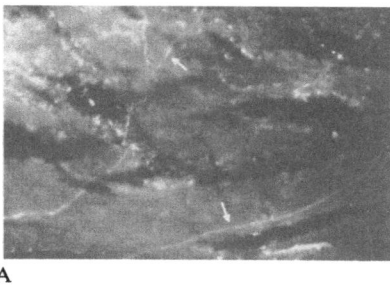

A

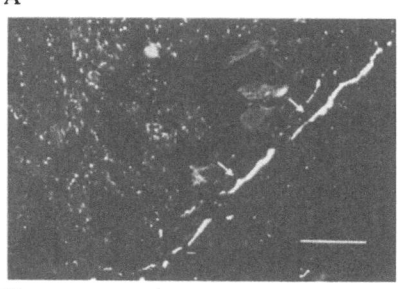

Fig. 3.
Fluorescent nerve fibres, which were
yellow-green in colour (arrows), in a
teased preparation (A) and in a section
(B) of the *Helix pomatia* heart. Much
of the fluorescence to the left in B is
not specific for amines. Calibration,
100 μ.

B

green in colour suggested to us that there is a close association of primary
catecholamines with 5-HT in the heart. Both sodium borohydride reduction of
the fluorophores, and examination of tissue not sublimated with paraformalde-
hyde indicated that the yellow-green fluorescence in the fibres specifically represented biogenic amines[6]. Furthermore, since we were able to clearly distinguish cells fluorescing yellow from those fluorescing green in the *Helix* brain,
it seems unlikely that our inability to distinguish whether the fluorescence in the
heart was mainly yellow or green was due to methodology. It would appear more
likely that the yellow-green colouration of the fibres represented the presence
of both primary catecholamines and 5-HT.

This view was substantiated with the use of drugs which are known to
influence tissue-content of 5-HT and catecholamines. The results of this study
are shown in Table 1[7-12]. Except for the experiments with NSD 1024, the effects
of which have not been extensively investigated in molluscs, the results obtained
support the view that both 5-HT and primary catecholamines are localized in
the structures resembling nerve fibres. It appears therefore that either 5-HT and
the primary catecholamines are localized in the same nerve fibres in the heart,
or that the fibres containing 5-HT and those containing primary catecholamines
are closely associated throughout the heart. The distribution of fluorescent
fibres throughout the heart appeared to parallel the 'sparse system' of nerve
fibres observed with methylene blue staining. The dense network of fibres in
the auricle did not fluoresce.

Evidence for a specialized localization of 5-HT in the heart, rather than for

Table 1. Summary of the effects of various drugs on the fluorescence in the snail heart. 2.5 mg of each drug were administered over a period of 30 h before observation.

Name of drug	Effects	Effects upon fluorescence in snail heart
Reserpine	Depletes amines from molluscan nervous tissue[7]	Almost all fluorescence eliminated
p. Chlorophenyl-alanine	Reduces 5-HT content by inhibiting the enzyme tryptophan hydroxylase in vertebrates[8]	Slight decrease total fluorescence Increase in intensity of green fluorescence
α-Methyl-m-tyrosine	Reduces CA content by inhibiting the enzyme tyrosine hydroxylase in vertebrates[9]	Slight decrease total fluorescence Increase in intensity of yellow fluorescence
5-HTP	Precursor of 5-HT in molluscs[10]	Great increase of yellow fluorescence
DOPA	Precursor of CA's in molluscs[11]	Muscle cells fluorescent yellow – impossible to observe nerves
Nialamide	Monoamine oxidase inhibitor in vertebrates	Total fluorescence slightly potentiated
NSD 1024	DOPA decarboxylase inhibitor in molluscs[12]	Total fluorescence very slightly potentiated

instance a uniform distribution of the amine in muscle cells of the heart, was obtained by estimating the content of 5-HT spectrophotofluorometrically in the auricle and the ventricle separately. It was found that there is about twice the concentration of 5-HT in the auricle ($0.29\ \mu g/g$) as in the ventricle ($0.15\ \mu g/g$). If 5-HT were localized in the muscle cells, it would be expected that the same concentration of 5-HT would be present in the auricle as in the ventricle.

The nature of the primary catecholamines in the *H. pomatia* heart has been investigated chromatographically and by bioassay[13]. Both NA and DA are present, the former being more concentrated. At present, the precise role of DA and NA in the snail heart is an enigma, since they are comparatively ineffectual in modifying the mechanical activity of the heart. (Both substances are about 10,000–100,000 times less potent than 5-HT in increasing the strength and frequency of beat of the isolated *H. pomatia* heart.)

The subcellular localization of amines in the heart has been studied histochemically by means of the technique described by WOOD[14]. This method has been reported to be specific for localizing amines with an unsubstituted amino group, such as 5-HT, DA and NA, in mammalian tissues[14,15]. In order to obtain some measure of the specificity of the method for investigations of the *H. pomatia* heart, we made a semi-quantitative estimate of all the amines and amino acids that we could detect in the heart and then tested each constituent in vitro with the histochemical reagents. The results of this study are presented in Table 2. Of all the amino acids and amines detected, only 5-HT, DA and NA

Table 2. Amine and amino acid constituents of the *H. pomatia* heart and the results of in vitro tests to determine which of these constituents yield a black precipitate when mixed with glutaraldehyde and potassium dichromate.

Constituent	μg/g wet weight *H. pomatia* heart	Reaction with 5×10^{-3} M Amino acid or Amine
Arginine	240	0
Lysine	130	0
Serine	60	0
Aspartic acid	140	0
Glutamic acid	800	0
Alanine	110	0
Histidine	120	0
Valine	10	0
Tyrosine	10	0
Leucine/Isoleucine	11	0
Phenylalanine	Trace	0
Acetylcholine	1.5	0
Histamine	<0.00025	0
5-Hydroxytryptamine	0.2–0.4	×
Dopamine	0.4–0.8	×
Noradrenaline	0.8–1.2	×

× Positive   0 Negative

gave a positive reaction when tested at a concentration of 5×10^{-3} M, i.e. produced a black deposit of some chromium oxide after the addition of glutaraldehyde and potassium dichromate solutions. Thus the method must be considered at least partly specific for the primary catecholamines and 5-HT in the *H. pomatia* heart.

Very little general cellular detail can be seen in tissues fixed and stained by WOOD's method. According to WOOD, only structures which are associated with amines are densely stained. Snail heart tissue stained by this method[4]

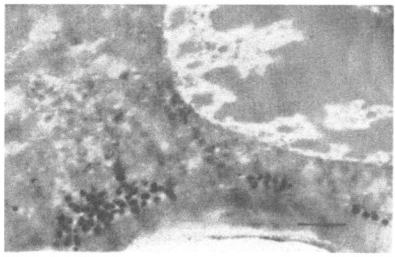 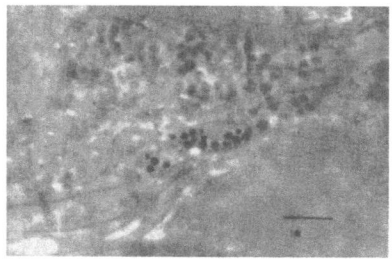

Fig. 4.
Electron dense granules, representing the sites of amine localization, in heart tissue fixed and stained according to the method of WOOD[14]. The granules are present in structures which appear to be nerve terminals. Calibration, 1 μ.

showed the presence of small clusters of densely stained granules (Fig. 4). The granules ranged in size from about 80 to 150 nm. The clusters of granules were not abundant, even in the area of the auricle where the dense network of nerve fibres occurred. The granules of the dense network did not appear particularly electron dense. Examination of tissue removed from reserpinized animals showed a marked reduction in the number of the densely stained granules, whereas the larger type associated with the dense network in the auricle were still plentiful and had exactly the same appearance as in normal tissue. In these experiments, the efficacy of reserpine in depleting 5-HT and catecholamines was monitored histochemically by examining strips of the test hearts by the fluorescence technique for amine localization.

These results, therefore, strongly suggest that the small dense centred granules contain either 5-HT and/or one or both of the primary catecholamines. The fact that only one type of particle was observed suggests that all three substances may be bound to particles with the same general appearance. Furthermore, these results support those obtained by the fluorescence method in that a positive amine reaction was obtained only in nerve fibres, and particularly in the nerve fibres of the sparse system.

Discussion

All the available data indicate that the 5-HT in the snail heart is localized in the nerves, as indeed might be expected if this amine acts as a cardioregulatory transmitter. However, the experimental results also indicate that NA and DA are also found in the nerves within the heart and that there is a close association of these catecholamines with 5-HT. The significance of this finding is yet to be determined. Neither NA nor DA is particularly effective in modifying the mechanical activity of the heart. Certainly, both amines are from 10,000 to 100,000 times less active than 5-HT in increasing the rate and strength of contraction of the isolated *H. pomatia* heart. We have also been unable so far to observe any influence of the catecholamines on the response of the heart to 5-HT.

Small amine-containing granules, which ranged in size from about 80 to 150 nm, were observed in axons and structures resembling nerve endings in the heart. Since only one type of amine-containing particle was observed, it must be considered a possible binding site for each of the biogenic amines detected in the tissue. The granules observed have the same general appearance as those which are considered to bind DA or NA in the *Spisula* ganglia[16], although the granules in the latter situation are a little smaller.

A profuse network of nerve fibres, showing thick swellings along their lengths, was observed in the auricle. Evidence has been presented which suggests that this network of fibres may form part of a neurosecretory system, releasing products directly into the blood stream. The precise physiological role of this system remains to be shown. We were unable to detect any 5-HT or catecholamines in the nerves of which it is composed. The granules in the nerves of this system were larger than those containing biogenic amine. They closely resemble the neurosecretory granules seen in the neurohypophysis[5].

II. Mode of Action of 5-HT on the Molluscan Heart

The 'Second Messenger' concept of hormone action, involving adenyl cyclase, which was developed by SUTHERLAND and co-workers[17], is now relatively well established in the control of certain biochemical processes in vertebrates. Furthermore, MANSOUR, SUTHERLAND, RALL and BUEDING[18] have shown that the adenyl cyclase of the invertebrate *Fasciola* can be activated by 5-HT. More recent work by MANSOUR[19, 20] has provided evidence that 5-HT can accelerate anaerobic glycolysis by effects, which are probably mediated by cyclic 3′, 5′-adenosine monophosphate (C-AMP), on phosphorylase and phosphofructo-kinase.

The mode of action of 5-HT in producing cardio-excitation in molluscs is unknown. We set out therefore to determine whether adenyl cyclase might in some way be involved in the response of the molluscan heart to 5-HT.

One way in which we tried to answer this question was by determining whether the enzyme adenyl cyclase is present in the molluscan heart, and if so, if it is stimulated with 5-HT. An isotope method devised by STREETO and REDDY[21] was used for this purpose. Most of our experiments were made on hearts of the clam *Spisula solida* due to the supply of animals and other factors. A membrane fraction was prepared from homogenates of the heart by the method recommended by STREETO and REDDY for tissues of the rat. Enzyme activity was measured in quantities of 100–150 mg of tissue (original weight of heart). Appreciable quantities of labelled C-AMP could only be detected when relatively large amounts of 8^{14}-C-adenosine-5-triphosphate (2 µC/100 mg original weight of heart) were employed. Since we could not be certain that any phosphodiesterase present in the tissue would be inhibited by caffeine or theophylline, 10 nmoles of unlabelled C-AMP were added to each sample in some experiments in order to dilute any labelled C-AMP formed. We routinely examined the developed chromatograms on a Nuclear Chicago Actigraph paper scanner to ensure that the portion of the chromatogram corresponding to the Rf of C-AMP was not contaminated with other labelled nucleotides. Counting of the chromatogram segments was carried out on a Nuclear Chicago Mark 1 Scintillation Counter.

Only very small amounts of C-AMP were detected in control samples. However, 5-HT in concentrations ranging from 10^{-5} to 5×10^{-4} M increased the level of labelled C-AMP formed by a factor of about one and a half to three. As with adenyl cyclase from mammalian sources[21], the *Spisula* enzyme was also activated with sodium fluoride. The degree of activation with sodium fluoride was, however, very pronounced with the *Spisula* heart enzyme (Table 3). It was sufficient, in fact, to enable us to detect the labelled C-AMP with the paper

Table 3. Percentage increase in heart adenyl cyclase activity with 5-HT and with NaF.

Experiment No.	Additive	Percentage of original activity
2	10^{-5} M 5-HT	280
3	10^{-4} M 5-HT	260
4	10^{-4} M 5-HT	160
3	5×10^{-4} M 5-HT	220
1, 2 and 3	10^{-2} M NaF	4,000–20,000

scanner and thus to ensure that the activity measured by scintillation was in fact due to C-AMP alone (Fig. 5).

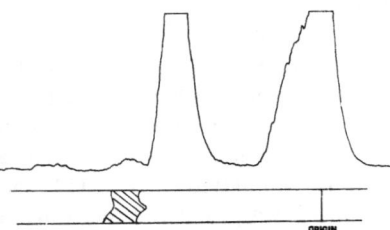

Fig. 5.
Diagram of a developed chromatogram prepared from *Spisula* heart 'membrane fraction', which had been incubated with NaF, showing the position of C-AMP. (The C-AMP was located by U.V. light after adding unlabelled C-AMP at the end of incubation.) The distribution of radioactivity along the paper strip is shown above. (Maximum upward deflection of the trace corresponds to 1500 counts.) The area of high activity at the origin is due to ATP and ADP. The next large peak represents AMP. Next to the AMP there is a small, but definite, peak which corresponds exactly with the Rf value of the C-AMP. There is another small area of activity, representing adenosine, above the C-AMP. It was only possible to detect a peak of radioactivity in the area corresponding to C-AMP in samples of heart which were incubated in the presence of NaF.

We have concluded therefore that adenyl cyclase is present in the *Spisula* heart and that the enzyme is activated with 5-HT.

We have also investigated the possible involvement of C-AMP in the response of the molluscan heart by testing the pharmacological effects of C-AMP on the mechanical activity of the isolated hearts from different species. If C-AMP is involved in the response of the heart to 5-HT, after being formed by an effect of the amine on adenyl cyclase, it would be expected that the cardio-excitatory effect of 5-HT could be mimicked with small quantities of C-AMP itself. Furthermore, if C-AMP is metabolized by a phosphodiesterase which is inactivated by caffeine and theophylline, as in the vertebrates[22], it should be possible to potentiate the effects of 5-HT and C-AMP on the heartbeat with these agents. It might also be expected that theophylline and caffeine would have an excitatory effect if applied alone. The results of this series of experiments are summarized in Table 4. High concentrations of C-AMP did in fact cause cardio-excitation in *H. aspersa*, *H. pomatia* and *E. cirrhosa*, but in the case of *M. arenaria*, even very high doses did not excite the heart. However, the difference in the sensitivity of the *Helix* and *E. cirrhosa* hearts to C-AMP compared with 5-HT was pronounced. The threshold concentration of 5-HT for excitation of both *Helix* hearts was generally of the order 10^{-10} to 10^{-9} g/ml and that for *E. cirrhosa* 10^{-9} to 10^{-8} g/ml. Thus each heart was about 50,000 times less sensitive to C-AMP than to 5-HT. Furthermore, although theophylline did have an excitatory effect on the hearts of both species of *Helix*, we were unable to obtain any clear indication that this compound consistently potentiated the responses of hearts of either species to 5-HT or to C-AMP.

Table 4. The results of pharmacological tests on the hearts of different species of molluscs to determine: (A) Whether C-AMP mimics the effect of 5-HT and if so at what concentrations. (B) Whether theophylline and caffeine produce cardio-excitation. (C) If theophylline potentiates the responses of hearts to 5-HT and C-AMP.

	Minimum concentration to mimic 5-HT effect	Comments	No. of tests
(A) C-AMP			
H. aspersa	10^{-5} to 10^{-4} g/ml (p)	(I) Occasionally caused an	30
H. pomatia	50 to 300 µg	increase in tone in high concentrations, sometimes with reduction in amplitude and frequency	
		(II) 2′ AMP + 3′ AMP gave some effect at about same concentrations	
Mya arenaria	none	In high concentrations stopped heartbeat (40 µg and up). Same effect as with 2′ AMP + 3′ AMP	3
Eledone cirrhosa	10^{-5} to 10^{-4} g/ml (p)	In these concentrations, there was an obvious excitatory effect	2
(B) Theophylline			
H. aspersa	10^{-10} (2) (p)	Wide variation in sensitivity	
	10^{-6} (1) (p)		7
	10^{-4} (3) (p)		
H. pomatia	10^{-6} (2) (p)		2
Caffeine			
H. pomatia	10^{-5} (3) (p)		4
	10^{-4} (1) (p)		
(C) *Does Theophylline Potentiate Effects of 5-HT or C-AMP?*			
H. aspersa	10^{-6} g/ml (p) Theophylline	Possibly 5-HT	1
	10^{-6} g/ml (p) Theophylline	no	3
H. pomatia	10^{-5} g/ml (p) Theophylline	no	10
	10^{-4} g/ml (p) Theophylline	no	3

p = Continuous perfusion of the drug through the heart for a period of about 2 min.

Discussion

The possibility that the ionotropic and chronotropic effects of 5-HT on the molluscan heart are mediated by C-AMP has been considered. It was shown that adenyl cyclase is present in the heart of *Spisula solida* and that the enzyme is activated by 5-HT, as might be expected if C-AMP is a mediator for 5-HT. On the other hand, all the available pharmacological data suggest that C-AMP is not directly involved in mechanical responses of the heart to 5-HT. Nevertheless, the possibility of an involvement of C-AMP cannot be entirely excluded at present, since it is possible that there are permeability barriers to the movement of C-AMP into the cells of the heart. Furthermore, it is possible that the phosphodiesterase of the molluscan heart is not sensitive to methylxanthines. CHANG[23] has recently reported that the phosphodiesterase of the slime mould, *Dictyostelium discoideum*, is not inhibited by caffeine and only partially inhibited by theophylline in high concentrations.

Summary

The observations made in this paper may be summarized as follows:

Most of the *Helix pomatia* heart is only sparsely innervated, but there is a dense network of nerves in part of the auricle.

The nerves in this dense network are positioned on the inside of the heart close to, or bordering on, the heart cavity. They have pronounced swellings along their lengths and are filled with electron dense granules similar to those seen in known neurosecretory systems. The nerves do not contain 5-HT or primary catecholamines. On the basis of this morphological data, we suggest that the network serves a neurosecretory function.

5-HT as well as dopamine and noradrenaline appears to be localized in the 'sparse nerve fibre system', which is seen throughout the *Helix pomatia* heart. No evidence was obtained to suggest that any of these amines occur in muscle cells. At least one of the amines, and perhaps all of them, is contained within granules which range in size from 80 to 150 nm.

Adenyl cyclase is present in the *Spisula solida* heart, and the enzyme is stimulated with 5-HT. It seems likely therefore that cyclic 3′,5′-adenosine monophosphate may be involved in some way in the response of the heart to 5-HT. Nevertheless, we were unable to obtain any pharmacological evidence for a direct involvement of cyclic 3′,5′-adenosine monophosphate in the ionotropic and chronotropic responses to 5-HT of hearts of various species of molluscs.

Acknowledgments

The work was supported by grants from the Medical Research Council and the Wellcome Trust. We are grateful to Mr. D. FRANK of the Biochemistry Department of Dundee for helpful discussions about adenyl cyclase estimation and Dr. J. R. T. COUTTS, Obstetrics Department, Dundee, for use of the Nuclear Chicago Actigraph paper scanner.

References

1   J. H. WELSH, Arch. exp. Path. Pharmak. *219*, 23 (1953).
2   G. A. COTTRELL and M. S. LAVERACK, Ann. Rev. Pharmac. *8*, 273 (1968).
3   J. RIPPLINGER, Ann. Sci. Univ. Besançon *2*, 3 (1957).
4   G. A. COTTRELL and N. N. OSBORNE, Comp. Biochem. Physiol., in press.
5   H. M. GERSCHENFELD, J. TRAMEZZANI and E. DE ROBERTIS, Endocrinology *66*, 741 (1960).
6   H. CORRODI and G. JONSSON, J. Histochem. Cytochem. *15*, 65 (1967).
7   M. MIROLLI and J. H. WELSH, *Comparative Neurochemistry* (Ed. D. RICHTER; Pergamon Press, New York 1964), p. 433.
8   B. K. KOE and A. WEISSMAN, J. Pharmac. exp. Ther. *154*, 499 (1966).
9   K. E. MOORE, Life Sci. *5*, 55 (1966).
10   J. H. WELSH and M. MOORHEAD, Gunma J. Med. Sci. *8*, 211 (1959).
11   J. CARDOT, Compt. Rend. *257*, 1364 (1963).
12   G. A. KERKUT, C. B. SEDDEN and R. J. WALKER, Comp. Biochem. Physiol. *23*, 159 (1967).
13   N. N. OSBORNE, unpublished observations.
14   J. G. WOOD, Nature *209*, 1131 (1966).
15   J. G. WOOD, Texas Rep. Biol. Med. *4*, 828 (1965).
16   G. A. COTTRELL, in the Report on the Symposium on Invertebrate Neurobiology, Tihany, Hungary, 1967; Annal. Biol. Tihany, in press.
17   E. W. SUTHERLAND and G. A. ROBISON, Pharmac. Rev. *18*, 145 (1966).
18   T. E. MANSOUR, E. W. SUTHERLAND, T. W. RALL and E. BUEDING, J. biol. Chem. *235*, 466 (1960).
19   T. E. MANSOUR, Adv. Pharmac. *3*, 129 (1964).
20   T. E. MANSOUR, Fed. Proc. *26*, 1179 (1967).
21   J. M. STREETO and W. J. REDDY, Anal. Biochem. *21*, 416 (1967).
22   R. W. BUTCHER and E. W. SUTHERLAND, J. biol. Chem. *237*, 1244 (1962).
23   Y.-Y. CHANG, Science *161*, 57 (1968).

Physiological and Pharmacological Properties of *Limulus* Heart

by B. C. Abbott*, F. Lang, I. Parnas, W. Parmley and E. Sonnenblick
Marine Biological Laboratory, Woods Hole, Massachusetts
Present address: Department of Biological Sciences, University of Southern California, University Park, Los Angeles, California 90007.

Perhaps the most pervasive method for classification of hearts has been that which considers the initiation of the beat—neurogenic and myogenic. Although it is known to have nervous innervation, the vertebrate heart is recognized as myogenic with the implication that the basic rhythmicity of the tissue arises from myocardium and that the nervous system acts only as a control mechanism. At the turn of the century this concept was strongly disputed and in particular Carlson[1-6]* believed the mammalian heart to be neurogenic. As models, he chose to study the physiology of various invertebrate hearts and particularly the Limulus heart where the external cardiac ganglion chain and nerve net proved to be most convenient for experimentation.

There is no longer any dispute that the vertebrate heart is myogenic but nevertheless, the Limulus heart remains of interest from the neurophysiological point of view. Although there are cardioregulatory nerves, the heart is an autonomous organ with a complex nervous center which initiates and spreads the beat in a co-ordinated fashion.

Carlson experimented with a number of different conditions affecting the heart such as drugs, ions, and temperature[1-5]. He was able to separate the effects on the ganglion from the effects on the muscle, but his preoccupation with proving the neurogenicity of the vertebrate heart led to some erroneous conclusions on his part, mainly concerning the specific action of drugs. Since these early investigations, the Limulus heart has been studied intermittently. The following is a brief review of the literature that accentuates work which has subsequently proven to be correct.

Anatomy and Histology

The Limulus heart lies on the dorsal side of the animal just under the carapace. It runs most of the length of the animal, from just between the eyes

Fig. 1.
Diagramatic representation of Limulus heart, dorsal view. g.c. ganglion chain; l.c.n. lateral cardiac nerves; s.s.n. segmental side nerves; l.v. lateral vessel; os. ostia; a.v. anterior vessel.
(Modified from Carlson[1].)

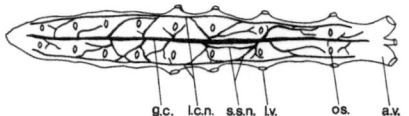

*Numbers refer to References, page 242.

in the prosoma to about the middle of the epistosoma. It is tubular in shape, up to 15–20 cm long and 2–3 cm in diameter (Fig. 1). On the outer mid-dorsal surface lies a ganglion chain which sends branches bilaterally to the muscle as well as to a pair of lateral nerves which run the length of the heart. Also on the dorsal surface are eight pairs of ostia which permit inflow of hemolymph to the heart but which have crude valves to retard backflow. Hemolymph is pumped out through four pairs of lateral vessels as well as three anterior ones to the arterial system, which, along with the venous system, comprises the most highly developed circulatory system of all arthropods[7].

Early histological studies[8–9] showed the heart to be composed of long striated fibers of 10–90 μ which branch and anastomose quite regularly. MEEK[9] reported that the fibrils likewise branch and anastomose freely within the fibers.

Electron microscopy is limited to one recent study on the Japanese horseshoe crab, *Tachypleus tridentatus* Leach[10]. Except for reporting a slightly smaller fiber diameter (20–40 μ) the author supports previous findings. In addition he describes the Z-line as being broken wavy bands and the H-zone and M-line as being absent. Sarcomere lengths vary from 2.5 μ in the contracted state to 9–13 μ in the relaxed fiber. The A-band (myosin filaments) measures 2.5 μ and remains constant. Thus in a relaxed fiber, the I-bands extend more than 3 μ on either side of the A-band. On this basis KAWAGUTI suggests that the sliding filament theory may need some revision when applied to the horseshoe crab heart[10].

The gross anatomy of ganglion and cardioregulatory nerves was described by PATTEN and REDENBAUGH[8] and later by PATTEN[11] and by CARLSON[1–2,12]. HEINBECKER described two[13] and later three[14] types of cells in the ganglion. Large unipolar cells were thought to be the pacemaker cells. Smaller bipolar and multipolar cells were also present. FEDELE[15] did an exhaustive histological study of the ganglion describing the complex interconnections of the three cell types. In contrast to earlier workers, he reported ganglion cells in the lateral nerves. This was later confirmed physiologically and will be dealt with in a later section.

Electrical Activity

The earliest electrical recordings from the Limulus heart are those of HOFFMAN[16]. He described the myogram as being similar to that of a vertebrate heart but with the distinguishing characteristic of oscillations on the plateau. This he ascribed to a tetanic character of the contraction, for when he removed the ganglion and stimulated muscle or nerve, the response was smooth. Some later workers denied the oscillatory nature (see KRIJGSMAN[17]) but later it was definitely confirmed by GARREY[18–20], HEINBECKER[13–14,21], RIJLANT[22,23], ARMSTRONG et al.[24], and PROSSER[25]. RIJLANT[22] separated the ganglion from the muscle and found ganglionic activity relatively unaffected. He also concluded that the electrical activity of the muscle was due to junctional potentials. HEIN-

BECKER, in 1936, acknowledged the difficulty in studying the activity of single cells and had the foresight to state: '...until such times when potentials can be recorded from single cell bodies with shielded microelectrodes, inferences concerning these cell potentials must be drawn from records...subject to the influence of groups of cells and fibers'[14].

The first microelectrode studies on the Limulus heart were performed by McCANN[26]. Shortly thereafter, this work was extended by ROBB and RECH[27] and TANAKA et al.[28]. The membrane resting potential is 35–45 mV. These workers described the intracellular potential as being composed of an initial upstroke and a plateau with oscillations. ABBOTT et al.[29] later showed the upstroke to be composed of a summation of junctional potentials and the plateau to be a sustained depolarization due to a tetanic volley of j.p.'s (junctional potentials) (Fig. 2).

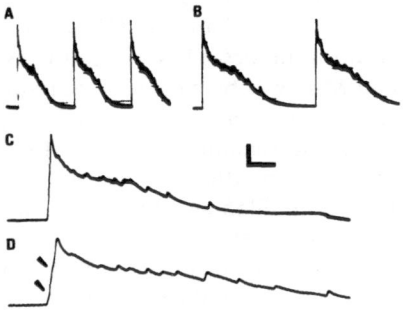

Fig. 2.
Intracellular potentials from Limulus myocardium at different sweep speeds to show that the apparent 'single upstroke' is composed of summated j.p.'s. Calibration: Vertical 10 mV; Horizontal (A) 1 sec, (B) 500 msec, (C) 200 msec, (D) 100 msec. (From PARNAS et al.[31].)

Simultaneous recording of tension, ganglionic activity and intracellular activity (Fig. 3) shows the relationship of the tetanic volley from the ganglion

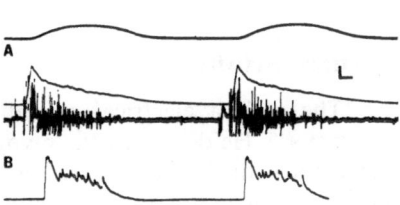

Fig. 3.
Spontaneous activity of Limulus heart. (A) Simultaneous recordings of tension (upper trace), intracellular potentials (middle trace) and ganglionic discharges (lower trace). (B) Intracellular recordings from another cell in the same heart. Calibration: Horizontal, 200 msec; Vertical, intracellular potentials 10 mV, ganglionic activity 100 μV. (From PARNAS et al.[31].)

to myocardial mechanical and electrical activity. A beat which lasts 1.0–1.5 sec starts in one of the middle segments and spreads within 100 msec to the anterior and posterior portions[2,30].

If the ganglion is removed from an isolated heart and side branch nerves are stimulated at varying intensities, up to three j.p.'s may be recorded in a nearby myocardial cell. Stimulation of another nearby side nerve can evoke up to three more j.p.'s. While no systemic count has yet been made muscle fibers have been found to be innervated by side nerves many segments away[31]. J.p.'s are recorded at all sites in the heart, indicating that the usual invertebrate pattern of multi-terminal innervation is present.

Synaptic Properties

The neuromuscular junction of arthropods and particularly that of Crustacea has been studied extensively. Similar studies on neurogenic hearts seem to be lacking. The neuromuscular junction of the Limulus heart seems to be similar to those of skeletal muscle. It is affected by drugs[1–2,5,17,22] and the synaptic delay is decreased with an increase in temperature[22]. Microelectrode studies reveal spontaneous miniature junction potentials (m.j.p.'s) of up to 1 mV[31]. Amplitude histograms show a series of broad peaks or more often a smooth curve, supporting the hypothesis of multiterminal innervation. Such circumstances make it impossible to perform quantal content calculations without recording directly from the active sites at the nerve terminals.

Lowering of Ca^{++} and raising of Mg^{++} reduces the size of evoked j.p.'s as was found in other arthropod preparations.

Peripheral Nerve Net

As mentioned above, side nerves branch off from both sides of the ganglion and run to the muscle and to the lateral nerves. CARLSON[2] showed that the side branches innervate the ipsilateral side and do not cross the ventral midline to innervate the contralateral side. GARREY[20] disagreed, stating that experiments on mechanics showed some contralateral innervation. More recent work[29] supports CARLSON.

The segmental side nerves contain a few dozen axons which go both to the muscle and to the lateral nerve[29]. GARREY[20] showed that the axons going to the lateral nerve were not merely branches of those going to the muscle. He stimulated lateral nerves repetitively until resultant contractions were very weak. He then was able to cause a nearly normal contraction by stimulating one of the segmental side nerves.

The contribution of the nerve network to the intact spontaneously beating heart was shown by a series of ablation experiments. Recording from a central segment, ipsilateral side nerves were successively cut, starting with those closest to the microelectrode. As each nerve was cut, there was a successive decline intracellularly in length, amplitude, and number of j.p.'s on the plateau. When all visible ipsilateral nerves were cut, the electrical activity was limited to the j.p.'s from one or two axons which still reached the muscle. The ganglionic activity was unaffected by this procedure and intracellular potentials from the contralateral side were the same as the control value[31].

It is well known that damage to the ganglion can result in an uncoordinated beat. If the ganglion is cut in a middle segment, each half establishes an independent beat which spreads to most of the heart through the side nerves and lateral nerves. Intracellular recording from the muscle at a level near the cut shows two independent tetanic volleys arriving at the muscle[29]. These trains establish their own frequency and remain independent for many hours, indicating that ganglionic integrity is necessary for a co-ordinated beat and that initiation of each burst is probably not dependent on peripheral events.

In many hearts segmental side nerves occasionally run parallel and close to the ganglion, meeting other side nerves and appearing to send branches back into the ganglion. This was shown to be the case[31]. The ganglion was cut between the two side nerves which appeared to join. One side nerve was cut at the point where it left the ganglion and drawn up into a suction electrode for recording. Both the other side nerve and the portion of the ganglion to which it was connected were stimulated. When the stimuli were timed to fall between spontaneous bursts, one or two axons were seen to conduct between the recording and stimulating electrodes.

Similar experiments showed that side nerves send branches directly to the lateral nerve. In a deganglionated heart, a lateral nerve was cut in segments. No. 2 and the medial end were drawn into a suction electrode for recording. Side nerves were stimulated at the proximal ends, at the point where they left the ganglion. Each side nerve evoked from one to three different size spikes in the lateral nerve, depending on stimulus intensity.

Pharmacology

In his study of the comparative physiology of hearts, the analytical tool which CARLSON put to greatest use was the action of drugs. Regarding the Limulus heart, his work was primarily aimed at proving its essential similarity to the vertebrate heart. As a result most of his studies dealt with adrenergic and cholinergic drugs and their inhibitors, mimickers, and potentiators[1-6]. Later work[32-33] (see also KRIJGSMAN[17]) led KRIJGSMAN[17] to state that the pacemaker system is cholinergic and the motor axons adrenergic.

Within the past 15 years, much work has been done on arthropod pharmacology and this has been applied to the Limulus heart by a few workers. BURGEN and KUFFLER[34] tested 5-hydroxy-tryptamine (5-HT, serotonin) and gamma-aminobutyric acid (GABA) and found both to have an inhibitory effect on the intact heart. WELSH and MOORHEAD[35] reported low concentrations of 5-HT in the heart and cardiac ganglion. Additional work has been done mainly by two groups[29,31,36-38].

The action of drugs on a relatively complex system such as the Limulus heart must be interpreted with caution. Not only are the natural transmitters unknown, but the drugs may have identical actions while working at different sites. Chronotropic effects would be presumed to be acting via the pacemaker cells, inotropic effects, however, may be exerted at follower cells in the ganglion or on the neuromuscular system in the periphery. In addition, this is the simplest model; there may well be recurrent inhibitory or excitatory loops in the ganglion and nerve network.

Gamma-Aminobutyric Acid

GABA is thought to be the transmitter at the inhibitory neuro-muscular junction of crustacean skeletal muscle[39-41]. It satisfies nearly all the criteria for identification of a synaptic transmitter[41]. Pax and Sanborn[37] tested GABA on the Limulus heart and concluded that it was not the transmitter at the cardioinhibitory nerves because application of the amino acid to the heart did not completely mimic the endogenous transmitter.

Stimulation of the cardioinhibitory nerves has both a negative inotropic and chronotropic effect. The former is caused by increase in the interval between ganglionic bursts and the latter by a decrease in the number of ganglionic units firing in a burst. Pax and Sanborn[37] reported that GABA mainly mimicked the effect on cardiac frequency, while decreasing contraction strength much less than the cardioinhibitory nerves for comparable decreases in frequency. However, they also reported that GABA did not affect the number of units firing in a burst.

Since it was shown that strength of contraction is directly related to the number of impulses reaching the muscle[29], it was desirable to further investigate the action of GABA to see if it had an inhibitory effect on the neuromuscular system, thus decreasing the strength of contraction without affecting the ganglion.

Effects of GABA were studied on the neuromuscular system of a deganglionated heart by stimulating side nerves and observing the effect on intracellular potentials. Even in high concentrations (10^{-4}M) GABA had no effect on either the electrical or mechanical activity[29]. Therefore, its effects on an intact spontaneously beating heart were tested. It was found that concentrations of 10^{-6}M always eliminated all activity in the ganglion. Lower concentrations (10^{-7}M) reduced the frequency and amplitude of the rhythm by increasing the interval between ganglionic bursts and reducing the number of units firing in a burst[29] (Fig. 4).

Fig. 4.
Effect of GABA on Limulus heart.
(A) Control of mechanical (upper trace) and ganglionic (lower trace) activity.
(B) effect of 5×10^{-6}M GABA after 10 sec. Contraction is reduced in height and duration. Ganglionic discharge has fewer units firing and is shorter in duration. Calibration: Horizontal, 200 msec; Vertical, 5 μV. (From Parnas et al.[31].)

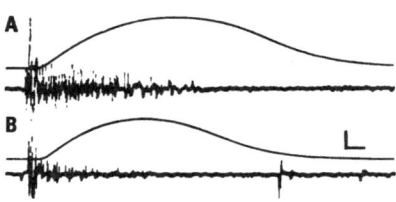

Picrotoxin, a compound known to block the inhibitory axon in the skeletal muscles of crustaceans[42] and Limulus[43], also blocks the effect of applied GABA in these systems. While picrotoxin does block the effect of the cardioinhibitory nerves in Limulus, it fails to block the effect of applied GABA[37].

Thus GABA was shown to mimic the action of the extrinsic cardio-inhibitory nerves in every way. This does not, of course, prove its status as a transmitter but does show that it cannot yet be eliminated. The effect of picro-toxin remains puzzling. There seem to be no reports in the literature where it fails to block applied GABA when GABA mimics an inhibitory axon. This in itself is insufficient evidence to discount GABA as the transmitter because the mode of action of picrotoxin is not known.

5-Hydroxy-tryptamine

5-HT has been shown to be present in the Limulus heart[35] but in contrast to its excitatory effect on the crustacean heart[44], it has an inhibitory effect on the Limulus heart[38]. The effect of 5-HT is to reduce the frequency of ganglionic bursts and to decrease the number of units firing in a burst, thus mimicking stimulation of the cardioinhibitory nerves. In addition, bromlysergic acid di-ethylamide (BOL), a specific inhibitor of 5-HT in other animals, identically blocked the action of both applied 5-HT and stimulation of the cardioinhibitory nerves[38]. 5-HT (10^{-3}M) had no effect on the electrical or mechanical activity of the neuromuscular system of the heart[29].

Glutamate

Glutamate has been proposed as the excitatory transmitter in the skeletal muscles of many invertebrates[45-48]. It was shown to cause contraction of Limulus closer muscle[43]. PAX and SANBORN[37] tested glutamate on the spon-taneously beating heart and found that 10^{-3}M reduced the strength of con-traction to barely detectable levels, while leaving the rate virtually unaffected and that 10^{-5}M had no effect on the spontaneously beating heart.

When glutamate was tested in a deganglionated heart to see its effects on the neuromuscular system, it always caused sustained contraction in concen-trations of 10^{-5}M[29]. Transient contractions were occasionally seen with much lower concentrations (10^{-7}M). In a spontaneously beating heart 10^{-5}M occa-sionally caused some increase in muscle tone; higher concentrations (10^{-4}M) caused an increase in muscle tone with a concomitant decrease in height of contraction, leaving the heart rate nearly unaffected.

Glutamate was also found to affect the firing of the ganglion cells (Fig. 5). When applied to a spontaneously beating heart, recordings from the ganglion showed repetitive firing of cells between bursts. Intracellular recordings from the muscle likewise showed j.p.'s appearing in the normally quiescent period between ganglionic bursts (Fig. 5).

Fig. 5.
Effect of glutamate on the Limulus
heart. (A) Control of tension (upper
trace) and ganglionic activity (lower
trace). (B) Effect of 10^{-5} M glutamate
after 5 sec. Note increased firing of
units between bursts. (C) Control of
intracellular activity (upper trace) and
ganglionic activity (lower trace).
(D) Effect of 10^{-5} M glutamate after
10 sec. Intracellular recording is at
double sensitivity. Calibration: Horizontal A, B, 500 msec; C, D, 200 msec.
Vertical A, B, 20 μV; C upper trace
20 mV, lower trace 50 μV; D, upper
trace 10 mV, lower trace 50 μV. (From
PARNAS et al.[31].)

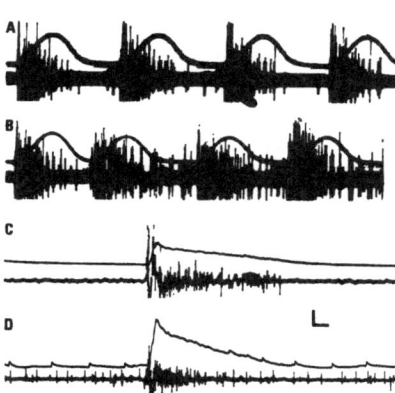

The fact that 10^{-5} M glutamate will cause contracture in a deganglionated heart yet hardly affect the spontaneously beating heart probably may be explained by the assumption that its effect in the latter is masked by the endogenous activity. The effect of 10^{-3} M in the spontaneously beating heart would be strong enough, both on the ganglion and muscle, to cause an increase in muscle tone. The quiescent heart, on the other hand, would tend to exhibit any activity that low concentrations would have on the muscle.

Cardiac Muscle Mechanics

The mechanical properties of the vertebrate cardiac muscle have received much attention[49-50] and in this regard there appear to be two basic differences between heart and skeletal muscle. First, upon stretching, skeletal muscle does not develop passive tension until the muscle has reached l_0, the point where the peak active tension can be developed. Cardiac muscle, on the other hand, develops active tension even at lengths well below l_0. The other basic difference is that skeletal muscle is tetanizable whereas cardiac muscle is not. This latter characteristic of cardiac muscle makes it difficult to study many aspects of the mechanical properties and has led to controversy as to the meaning of some aspects, particularly active state[51-52].

The mechanical aspects of the invertebrate heart, on the other hand, seem to have received little attention. Tension and pressure have been recorded from a number of hearts but for the most part invertebrate hearts are very small and difficult to work with or do not lend themselves to mechanical studies because of their structure (e.g. crustacean hearts). The Limulus heart, on the other hand, is composed mainly of a sheet of parallel circular fibers and thus lends itself readily to mechanical study.

In spite of the fact that it is tubular and the vessels are in the anterior portion, the Limulus heart contracts nearly simultaneously. The contraction starts in one of the middle segments and within 100 msec the whole heart is contracting. Systole lasts for 1.0–1.5 sec during which the ostia are closed and suspensory ligaments are stretched. Diastole consists of the heart being stretched by the suspensory ligaments and being filled through the ostia with the hemolymph[8].

Heart pressures were measured by inserting a needle directly into the heart through the hinge between the prosoma and epistosoma. The needle was attached to a pressure transducer. There was always a positive pressure in the heart (Fig. 6) of 20 cm of water. Contraction was seen to take place in two phases. The initial contraction was isovolumic, giving a quick rise to 35 cm of water, followed by an ejection phase which shows an equal (to 50 cm water) but slower rise. The measurements are probably lower than in the intact animal because the heart 'turns a corner' whenever the animal bends at its hinge thus causing the needle to tear the myocardium.

Fig. 6.
In vivo pressure recording from Limulus heart. A needle was inserted through the hinge between the prosoma and epistosoma and into the heart. (From ABBOTT et al.[53].)

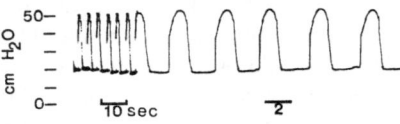

Mechanical studies were done on a 1.8 cm wide torus of heart muscle using procedures similar to those employed for mammalian papillary muscle[50]. Transverse stimulation was supplied through parallel massive gold electrodes. The muscle was mounted vertically and attached to a light afterload lever. Movement could be recorded under isometric or afterloaded isotonic conditions. Series elasticity was determined by quick release from tetanic isometric tension

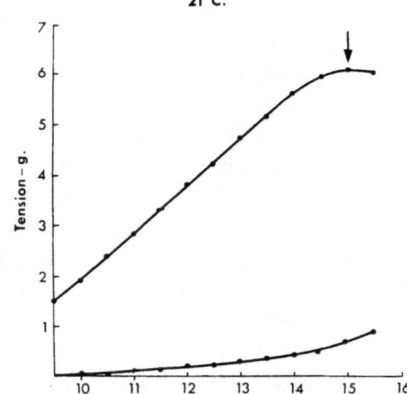

Fig. 7.
Length-tension curve from a torus of muscle of Limulus heart. (From ABBOTT et al.[53].)

to an isotonic load. Activation of the muscle was undoubtedly through the nerves: 10^{-6} M tetrodotoxin (TTX) rendered the system essentially inexcitable.

Mechanical studies and their implications will be presented in more detail elsewhere[53] but some of the results are pertinent to this discussion. Passive and developed tension varied with length as shown in Fig. 7. This shows that although the heart muscle is activated by nerves it exhibits length-tension relationships characteristic of cardiac muscle. This is not a general property of *Limulus* musculature: tail muscle shows length-tension characteristics typical of skeletal muscle[53].

Since it has been shown previously that the level of activation is dependent on the frequency and number of j.p.'s exciting the muscle, the development of isometric tension at various stimulus frequencies was studied. Fusion of the isometric response occurred at 2 stimuli/sec at room temperature. As the frequency of stimuli was increased, the tension rose. At 26–28/sec tension leveled off and at higher frequencies the tension fell.

Afterloaded isotonic contractions were measured during maximal tetanic contractions with a preload positioning the muscle length more than half way up the curve of developed tension of Fig. 7. A force-velocity curve was obtained which was replotted as $Po\text{-}P/V$ against P, where Po is the isometric tetanic tension, P is load, and V is velocity of shortening. The plot resulted in a straight line showing good approximation to the Hill characteristic equation for the contractile component.

Variations were then applied to study force-velocity relationships. Force-velocity curves were obtained at a series of different muscle lengths. As in the mammalian heart, there is usually passive tension in the muscle and hence these experiments were done under preloaded conditions. The results are shown in Fig. 8, where it is clear that maximum force varies with length. However,

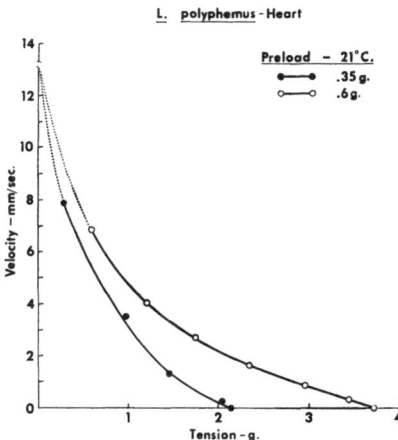

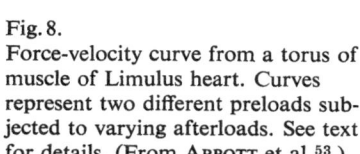

Fig. 8.
Force-velocity curve from a torus of muscle of Limulus heart. Curves represent two different preloads subjected to varying afterloads. See text for details. (From ABBOTT et al.[53].)

maximum velocity was constant over a range of lengths when normalized as speed in muscle lengths/sec.

Conclusions

Examples of analogy in evolution are manifold and they show that while the solution to similar problems has been accomplished in different ways the results may often look strikingly alike. The intracellular potentials of the Limulus heart look similar to those of vertebrate hearts, but the former is due to the summation and sustained depolarization resulting from a tetanic volley of j.p.'s while the latter is due to a propagated electrogenic spike and a delayed K^+ conductance.

Similarly, the vertebrate heart can control the strength of contraction by changes in preload extension (Starling's Law), beat frequency, and the action of hormones. The Limulus heart can likewise use preload extension but depends also on the depolarization level of the muscle, that is, the frequency and number of j.p.'s.

Synchrony of contraction in the vertebrate heart is accomplished by the myocardium acting as a functional syncytium, thus producing an all-or-none contraction. The Limulus myocardium is an anatomical syncytium but since there is no propagated spike, this seems to be of little value in the adult. Synchrony is achieved by having a dense network of multiterminal and polyneuronal innervation. This redundancy is in contrast to the usual parsimonious invertebrate pattern and provides a high safety factor.

Acknowledgments

This work was supported in part by grants from U.S.P.H.S. (HE-08218–06), the Chicago Heart Association and by a U.S.P.H.S. predoctoral traineeship, 5TI GM-619.

References

1   A. J. CARLSON, Science 20, 684 (1904).
2   A. J. CARLSON, Am. J. Physiol. 12, 67 (1904b).
3   A. J. CARLSON, Am. J. Physiol. 12, 471 (1905a).
4   A. J. CARLSON, Am. J. Physiol. 15, 99 (1906).
5   A. J. CARLSON, Am. J. Physiol. 17, 177 (1906b).
6   A. J. CARLSON, Am. J. Physiol. 18, 149 (1907).
7   A. J. CLARK, Comparative Physiology of the Heart (Cambridge Univ. Press, Cambridge 1927).
8   W. PATTEN and W. A. REDENBAUGH, J. Morph. 16, 99 (1899).
9   W. J. MEEK, J. Morph. 20, 403 (1909).
10   S. KAWAGUTI, Biol. J. Okayama Univ. 9, 11 (1963).
11   W. PATTEN, Evolution of the Vertebrates (Blakiston's Son & Co., Philadelphia 1912).
12   A. J. CARLSON, Biol. Bull. Woods Hole 8, 123 (1905).
13   P. HEINBECKER, Am. J. Physiol. 103, 104 (1933).

14 P. Heinbecker, Am. J. Physiol. *117*, 686 (1936).
15 M. Fedele, Arch. Zool. Napoli *30*, 39 (1942).
16 P. Hoffman, Arch. Anat. Physiol. *175*, (1911), quoted in [17].
17 B. J. Krijgsman, Biol. Rev. *27*, 320 (1952).
18 W. E. Garrey, Am. J. Physiol. *93*, 178 (1930).
19 W. E. Garrey, J. cell. comp. Physiol. *1*, 209 (1932).
20 W. E. Garrey, J. cell. comp. Physiol. *2*, 355 (1932b).
21 P. Heinbecker, Am. J. Physiol. *97*, 531 (1931).
22 P. Rijlant, The Collecting Net *6*, 231 (1931).
23 P. Rijlant, Arch. int. Physiol. *35*, 339 (1932).
24 F. Armstrong, M. Maxfield, C. L. Prosser and G. Schoepfle, Biol. Bull. *77*, 327 (1939).
25 C. L. Prosser, J. cell. comp. Physiol. *21*, 295 (1943).
26 F. V. McCann, Science *137*, 340 (1962).
27 J. J. Robb and R. H. Rech, J. cell comp. Physiol. *63*, 299 (1964).
28 I. Tanaka, Y. Sasaki and H. Shin-mura, Jap. J. Physiol. *16*, 142 (1966).
29 B. C. Abbott, F. Lang and I. Parnas, Comp. Biochem. Physiol. *28*, 149 (1969).
30 J. Edwards, Am. J. Physiol. *52*, 276 (1920).
31 I. Parnas, B. C. Abbott and F. Lang, Am. J. Physiol, in press.
32 W. E. Garrey, Am. J. Physiol. *136*, 182 (1942).
33 C. L. Prosser, Biol. Bull. *83*, 145 (1942).
34 A. S. V. Burgen and S. W. Kuffler, Biol. Bull. *113*, 336 (1957).
35 J. H. Welsh and M. Moorhead, J. Neurochem. *6*, 146 (1960).
36 R. A. Pax and R. C. Sanborn, Biol. Bull. *126*, 133 (1964).
37 R. A. Pax and R. C. Sanborn, Biol. Bull. *132*, 381 (1967a).
38 R. A. Pax and R. C. Sanborn, Biol. Bull. *132*, 392 (1967b).
39 E. A. Kravitz, S. W. Kuffler and D. D. Potter, J. Neurophysiol. *26*, 739 (1963).
40 M. Otsuka, L. Iverson, Z. W. Hall and E. A. Kravitz, Proc. nat. Acad. Sci. *56*, 1110 (1966).
41 E. A. Kravitz, *The Neurosciences* (Ed. G. C. Quarton, T. Melnechuk and F. O. Schmitt; Rockefeller Univ. Press, New York 1967), p. 433.
42 W. G. Van der Kloot, J. Robbins and I. M. Cooke, Science *127*, 521 (1958).
43 I. Parnas, B. C. Abbott, B. Shapiro and F. Lang, Comp. Biochem. Physiol. *26*, 467 (1968).
44 D. Maynard, Fed. Proc. *17*, 106 (1958).
45 J. Robbins, Anat. Rec. *132*, 492 (1958).
46 A. van Harreveld and M. Mendelson, J. cell. comp. Physiol. *54*, 85 (1959).
47 A. Takeuchi and N. Takeuchi, J. Physiol. *170*, 296 (1964).
48 P. Usherwood, J. exp. Biol. *49*, 34 (1968).
49 B. C. Abbott and W. F. H. M. Mommaerts, J. gen. Physiol. *42*, 533 (1959).
50 E. Sonnenblick, Am. J. Physiol. *202*, 931 (1962).
51 E. Sonnenblick, *The Myocardial Cell* (Ed. S. A. Brillger and H. L. Conn; Univ. of Pa. Press, Philadelphia 1966), p. 173.
52 A. J. Brady, Physiol. Rev. *48*, 570 (1968).
53 B. C. Abbott, E. Sonnenblick, W. Parmley and I. Parnas, in preparation.

The Role of Small Potentials in the Regulation of Rhythm in an Oyster Heart

by A. EBARA

Shimoda Marine Biological Sta., Shimoda, Shizuoka-ken, Japan

The transmembrane action potential of an oyster cardiac muscle cell displays a slowly rising initial component termed a pacemaker potential. This slow diastolic depolarization is one of the most distinguishing features of automatic cells[1,2*]. Even in a small fragment of the oyster heart, beating can be initiated under certain conditions. This paper will describe the characteristics of a 'small potential' which is obtainable under certain conditions. The role of the small potential in the regulation of rhythm will be discussed in view of the presence of numerous nexuses that are now known to be present[3] and that could serve as low-resistance intercellular pathways.

The ventricle of the oyster *Crassostrea gigas* was used in these experiments. A longitudinal incision through one side of the ventricular wall allowed the ventricle to be spread and mounted in a small Lucite box with two separate chambers. The ventricle was cut to give two half-ventricles with a small strip of muscle left as a connecting bridge. A half-ventricle was placed in each chamber so that the muscle strand bridged the partition.

Solutions could be infused into the separate chambers independently. A physiological salt solution for marine molluscs was used to bathe the tissue for control studies. Intracellular potentials of myocardial cells were recorded with conventional glass capillary microelectrodes, 3 M KCl, 20 MΩ.

The basis of these experimental procedures rests on the following assumption. When two microelectrodes are used to record electrical activity from two different cells in the endocardium of the whole ventricle, both action potentials are closely synchronized. Even when a solution of excess potassium is applied to the whole ventricle, the action potentials from the two sites are similarly affected so that the relationship between the two is not altered. Within a half-ventricle, action potentials recorded from two different cells resemble each other in their time/voltage parameters, e.g. shape and rate. The assumption is made, therefore, for the purpose of these experiments, that the action potential of a single cell in a half-ventricle is representative of the activity of all the cells of that half-ventricle. The functional relationship between two half-ventricles connected by a bridge of muscle fiber was assessed by recording from one cell in each half-ventricle. The influence of the size of the bridge was first examined.

Activity in each half-ventricle is independent of its other half if there is no connecting bridge between them, but is synchronized to a degree dependent on the size of the connecting bundle (Fig. 1 A–D). A single muscle bundle left

*Numbers refer to References, page 249.

Fig. 1.
The functional coordination between
half-ventricles was modified by the size
of the connecting bridge. A, Control.
B, Half-ventricles were connected by
several muscle bundles (total diameter
of about 500 μ). C, by a few muscle
bundles (about 300 μ). D, by a single
muscle bundle (about 150 μ). Calibration: vertical bars; 50 mV.

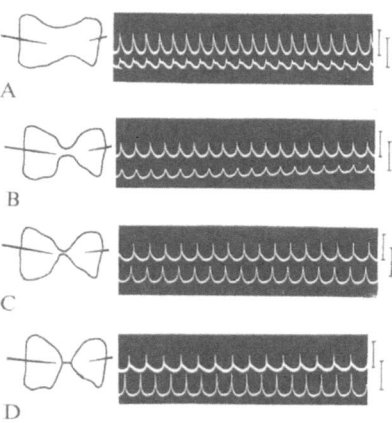

connected to both half-ventricles cannot maintain a functional coordination
between them (Fig. 1 D). When a few bundles (total diameter of about 300 μ)
remain as a bridge, a very small intracellular potential can be recorded from
one half or the other during a period of asynchrony between the two half-
ventricles and the characteristics of this so-called 'small potential' have been
reported in earlier papers[4,5]. When the half-ventricles are completely separated,
they continue to beat independently and the synchronization between them
disappears. The amplitude of the action potentials recorded in each half is
slightly decreased by the operation which separates the two halves.

In the experiments to be described, preparations were used in which the
two half-ventricles were connected by several muscle bundles adequate to maintain a complete coordination between them in the absence of treatment and
the results are summarized below in relation to the nature of the 'small potential' and to its interaction on the rhythm.

When a half-ventricle is inhibited by a treatment such as the application of
acetylcholine solution at 10^{-2} g/l the other half-ventricle continues to beat but
the rhythm changes somewhat. 'Small potentials' can be recorded intracellularly
in the inhibited half-ventricle in synchrony with the action potential of the
untreated portion.

The small potentials recorded in one half-ventricle resemble in shape, but
with greatly diminished size, the action potential which occurs in the other half
(Fig. 2). The small potential decays with distance from the active half-ventricle
with a half-decay distance of approximately 3 mm, although the distance varies
with the orientation of the fibers. Even when recorded intracellularly the potentials are only small (the upper trace in Fig. 2 is at four times the sensitivity of
the lower trace), so that this probably explains why they cannot be recorded
externally.

The small potential in either half-ventricle modifies the shape of the action
potential in the other half, as shown in Fig. 3. The action potential in the

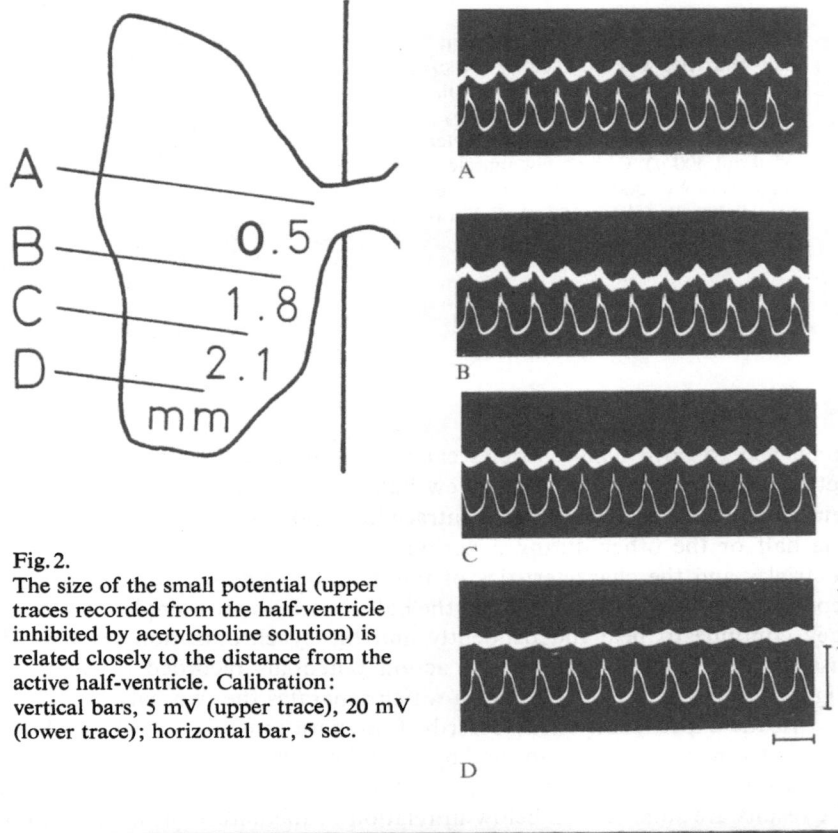

Fig. 2.
The size of the small potential (upper traces recorded from the half-ventricle inhibited by acetylcholine solution) is related closely to the distance from the active half-ventricle. Calibration: vertical bars, 5 mV (upper trace), 20 mV (lower trace); horizontal bar, 5 sec.

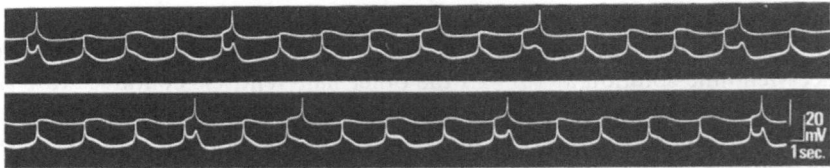

Fig. 3.
The rhythm of the dominant half-ventricle (lower trace) was changed by the small potential induced by the follower one (upper trace). A local response was often evoked in the upper trace. The half-ventricle shown by the upper trace was slightly inhibited by acetylcholine solution. Calibration: shown in the figure.

dominant half-ventricle (lower trace) induces a small potential in the half-ventricle which is behaving as a follower (upper trace). This acts as a prepotential which at times induces an action potential in the follower half. The peak of this potential is, in turn, reflected as a small potential on the falling phase of the driver, giving a plateau effect which results in a longer beat interval and so influences the rate of the dominant half.

The size of the small potential is maintained even when the membrane of the half-ventricle is partially depolarized by high potassium (20 mM/l). When the sodium chloride of the bathing medium is replaced by isotonic sucrose solution, the action potential may be diminished but the small potentials are not abolished. Neither magnesium solution at 200 mM (Fig. 4), nor acetylcholine

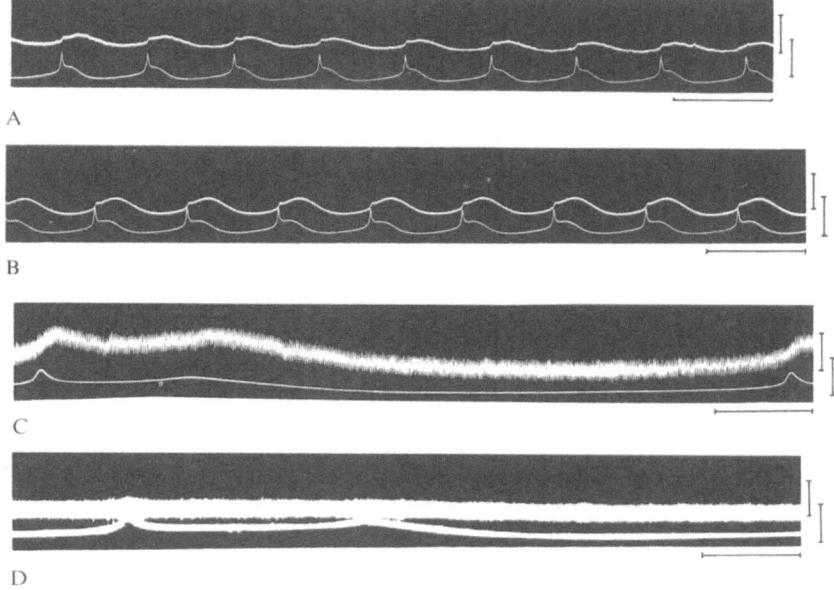

Fig. 4.
Excess magnesium solution has a minimum effect on the induction of the small potential. A, Control. The half-ventricle (upper trace) showed only the small potential without an action potential. B, Excess magnesium solution was applied to the half-ventricle (upper trace). C, The same as B. D, The upper trace was obtained by external recording. Calibration: vertical bars, 10 mV (upper traces of A and B), 50 mV (lower traces of A and B), 2 mV (upper traces of C and D), 50 mV (lower traces of C and D); horizontal bars, 5 sec (A and B), 1 sec (C and D).

solution, nor EDTA at 10^{-1} g/l, nor calcium solution which dramatically increased the resting potential of the treated half-ventricle, influenced the small potential. Therefore, it seems unlikely that the small potential is mediated by a chemical transmitter.

Hypertonic sucrose solution (700 mM/l) often abolished the small potential. When the solution was applied only to the connecting muscle bundles, the coordination of activities between the two half-ventricles disappeared, apparently as a result of the extinction of the intercellular spread of the small potential. When the solution was replaced by physiological saline, coordination between the two halves was restored.

The influence of the small potential on the rhythm formation was next studied. The beat interval was shortened or elongated according to the moment at which the small potential was superimposed on the beat cycle (Fig. 5)[8].

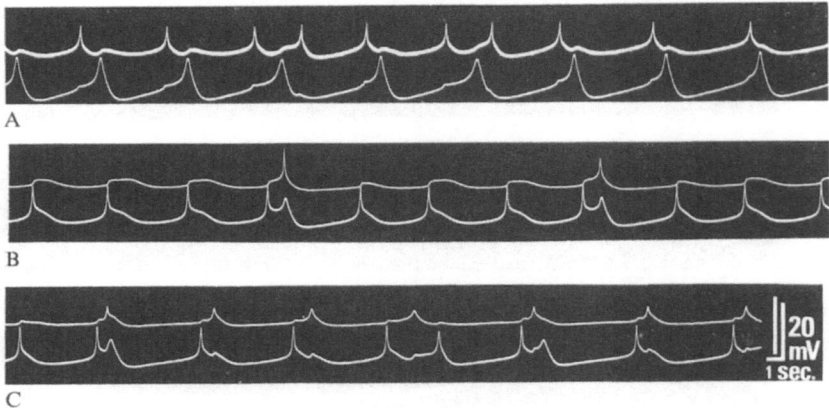

Fig. 5.
The small potential has a remarkable effect on the beat interval. A, Half-ventricles were connected by a few muscle bundles (total diameter of about 300 μ). B and C, The half-ventricle (upper trace) was weakly inhibited by acetylcholine solution. Calibration: shown in the figure.

When a half-ventricle was inhibited by treatment with an agent such as acetylcholine its beat frequency was reduced but the frequency of the beat of the other untreated half-ventricle was also markedly affected.

A modification of the beat interval similar to that produced by a 'small potential' was also produced if an artificial electrical charge was induced in the tissue by means of field stimulation.

Studies were made of the spread of current between cells of the heart from an impaled microelectrode and the results suggest that low-resistance pathways must exist between the cells. This is in conformity with the discovery of tight junctions between the cells in this ventricle[3].

Thus the 'small potential' may well represent an electrotonic spread of the action potential from an active site. The presence of the low-resistance intercellular pathway would provide for an exponential decrement of the potential; it would then play a role in modifying the beating rhythm of the whole ventricle, i.e. the 'small potential' mediates the coordination of the whole ventricle. In addition, an integrative role may be assigned to the slow potential, inasmuch as the slower follower cells can influence the rate of heart beat initiated by the fastest dominant cells.

I wish to express my sincere thanks to Professor K. MATSUI of Tokyo Kyoiku University for his criticism and encouragement.

References

1   P. F. CRANEFIELD and B. F. HOFFMAN, Physiol. Rev. *38*, 41 (1958).
2   B. F. HOFFMAN and P. F. CRANEFIELD, *Electrophysiology of the Heart* (McGraw-Hill Book Co. 1960).
3   H. IRISAWA, A. IRISAWA and N. SHIGETO, observations reported in this symposium.
4   A. EBARA, Sci. Rep. Tokyo Kyoiku Daigaku [B] *12*, 1 (1964).
5   A. EBARA, Sci. Rep. Tokyo Kyoiku Daigaku [B] *12*, 9 (1964).
6   A. EBARA, Jap. J. Physiol. *16*, 354 (1966).
7   A. EBARA, Jap. J. Physiol. *16*, 371 (1966).
8   A. EBARA, Sci. Rep. Tokyo Kyoiku Daigaku [B] *13*, 129 (1967).

The Role of Isoreceptors in the Neurohormonal Regulation of Bivalve Hearts

by MICHAEL J. GREENBERG
Department of Biological Sciences, Florida State University, Tallahassee, Florida 32306, USA

> now (more near ourselves than we)
> is a bird singing in a tree,
> who never sings the same thing twice
> and still that singing's always his[1]

The cardiac physiology and neurochemistry of bivalve molluscs have been under intensive investigation for the past 25 years or so. Hand in hand with these studies, a generalized outline of the pharmacology of four presumed neurotransmitter substances on the hearts of these animals has emerged. Most of this conceptual outline was developed in the laboratory of Professor John H. WELSH with the heart of the quahog *Mercenaria* (= *Venus*) *mercenaria* as the primary experimental object.

First, acetylcholine has been shown to be a depressor neurohumor acting at receptor sites which are, in the terms of vertebrate pharmacology, nicotinic[2]*. Curare-like drugs are the most effective ACh-blocking agents; in particular, benzoquinonium chloride (Mytolon) and its close congener ambenonium chloride (Mytelase) have become the 'standard' ACh antagonists for molluscs.

Second, 5-hydroxytryptamine (5-HT, serotonin) has been demonstrated to be an excitor transmitter in molluscan heart (evidence reviewed by COTTRELL and LAVERACK[3]). Exogenous 5-HT acts mainly to increase the amplitude of beat of isolated hearts, although higher doses also produce pronounced positive chronotropic and tonotropic effects. Lysergic acid diethylamide (LSD) mimics the action of 5-HT on many molluscan muscle preparations; on the *Mercenaria mercenaria* heart the maximally effective dose is incredibly low (10^{-16} M)[4]. The effective blocking agents for both 5-HT and LSD are 2-bromo-*d*-lysergic acid diethylamide (BOL) and methysergide (UML).

The actions of catecholamines, especially epinephrine and norepinephrine, have been studied on isolated hearts from a number of species. The effects are varied and depend on the animal and the dose; usually, tone and frequency are increased (reviewed by HILL and WELSH[5]). No effective antagonist to the actions of any of the catecholamines has been found. While dopamine, and more rarely other catecholamines, have been demonstrated in the ganglia and some other

---

[1]   1958 by E. E. CUMMINGS, reprinted from his volume, 95 poems, by permission of Harcourt, Brace & World, Inc.

*Numbers refer to References, page 264.

tissues of bivalves[3,6], a physiological role of dopamine in the regulation of bivalve hearts, while possible, seems unlikely at the present time.

Finally, a group of cardio-active substances, which may be polypeptides, have been extracted from clam ganglia and hearts, and lumped together under a nomenclatural umbrella: Substance X. These may play a role in the initiation[7] and long-term maintenance of cardiac rhythmicity in bivalves and other molluscs[7,8].

From the study of drug actions on other species, many unrelated to *Mercenaria*, a number of exceptions to this theoretical pharmacological outline have been accumulating. In this report, a variety of responses of isolated bivalve ventricles from sundry species will be described. Most of these responses have in common that they are markedly different from those of *Mercenaria mercenaria*. We want to know how the diverse effects are distributed among the species of bivalves, and what the nature of pharmacological diversity might be.

Materials and Methods

The preparation of isolated bivalve ventricles for mechanical recording has often been described, most recently by FLOREY[9]. Some species, oysters in particular, have hearts whose structure is not typical of other bivalves. The preparation of oyster hearts was discussed by GREENBERG[10]. In all cases the perfusion fluid used was either the natural sea water of the location where the experiment was done or artificial sea water[11]. In most experiments, the temperature was controlled at 16 °C. In some cases this was impossible and temperature was ambient (21–29 °C).

Results

Some Aspects of Cockle Heart Pharmacology

Some time ago, the actions on *M. mercenaria* hearts, of epinephrine, norepinephrine and dopamine were compared[12]. Epinephrine and norepinephrine had the usual effects — they increased frequency and tone at relatively high doses (10^{-5} M). Dopamine also had this action but was about 10 times more potent. In addition, at a lower concentration (2–3×10^{-6} M), dopamine depressed the amplitude of beat without increasing tone or frequency. The same difference in the responses to these catecholamines has been observed on the isolated ventricle of *Tresus* ($=Schizothaerus$) *capax* (Fig. 1). Such observations suggested that two receptor sites for catecholamines might exist.

The actions of epinephrine, ACh, and 5-HT, and some related analogs have been tested on hearts of *Clinocardium nuttalli*, as well as other species of cockles. Of interest, in view of the depressor action of threshold doses of dopamine on the *Mercenaria* and *Tresus* hearts, is the finding that epinephrine, itself, acts to diminish the amplitude and frequency of beat of the heart in

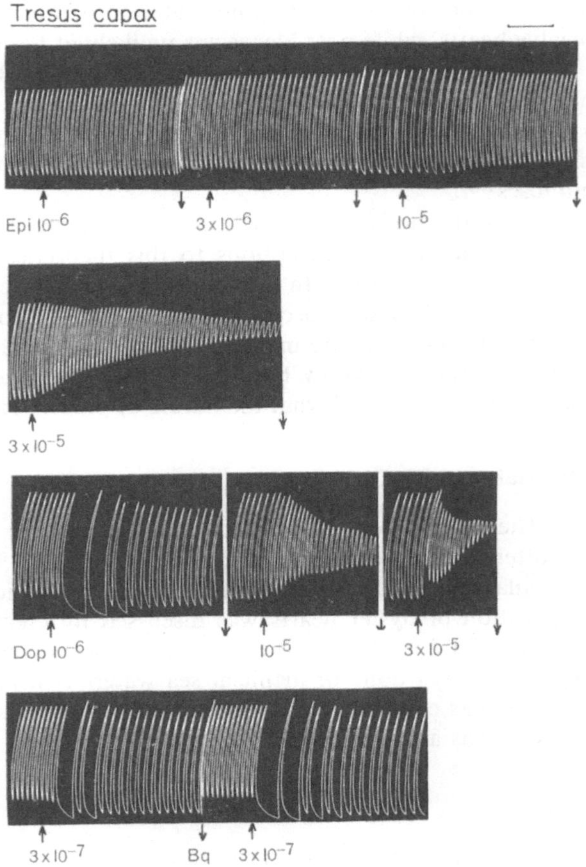

Tresus capax

Fig. 1.
The effect of epinephrine (Epi) and dopamine (Dop) on the isolated heart of *Tresus capax*. The depressor action of Dop is not antagonized by benzoquinonium (Bq), 10^{-5}. Drugs are added at upward-pointing arrows; the bath was washed out at the downward-pointing arrows, while the kymograph drum was stopped for 10 min. Bq was added after the wash and was left in the bath 10 min. before the following dose of Dop, 3×10^{-7}. Dose: moles/liter. Time scale: 1 min. Temp: 15 °C.

Clinocardium. In fact, although a much larger dose is required, the effect of epinephrine (10^{-4} M) is virtually indistinguishable from that of ACh (10^{-7} M) (Fig. 2). However, benzoquinonium chloride blocks the ACh effect, but not that of epinephrine (Fig. 2a), which suggests that the two depressor responses are evoked by the stimulation of different receptors. The depressor effect of dopamine on the heart of *Tresus* is also not antagonized by benzoquinonium (Fig. 1).

Cockle hearts do not respond to LSD in the same way as those of *Mercenaria*. Relatively high doses of LSD (10^{-6} M; 10^{10} times higher than threshold for the *Mercenaria* heart) produce either a small excitation or none at all on isolated ventricles of *Clinocardium* (Fig. 2 and 3). Interestingly, before testing *Mercenaria* hearts, WELSH[13] had noted in 1956 that LSD (3×10^{-6} M) had little excitor effect on the heart of *Arctica* (= *Cyprina*) *islandica* and, in fact, within 2 min of application was an effective antagonist of 5-HT on that preparation.

Fig. 2.
The effects of acetylcholine (ACh) and epinephrine (Epi) on the isolated heart of *Clinocardium nuttalli.*
a) Although the action of ACh is blocked by benzoquinonium (Bq), 10^{-4} M, that of Epi is not. b) Lysergic acid diethylamide (LSD), 10^{-6} M, produces a small increase in amplitude. It partially blocks the ACh effect and completely blocks the action of Epi. Recordings a) and b) are from different preparations. Drugs are added at upward-pointing arrows; downward-pointing arrows indicate washing with the drum stopped for 10 min. Bq and LSD were added to the bath 10 min. before the subsequent dose of ACh or Epi. Dose: moles/liter. Time scale: 1 min. Temp: 15 °C.

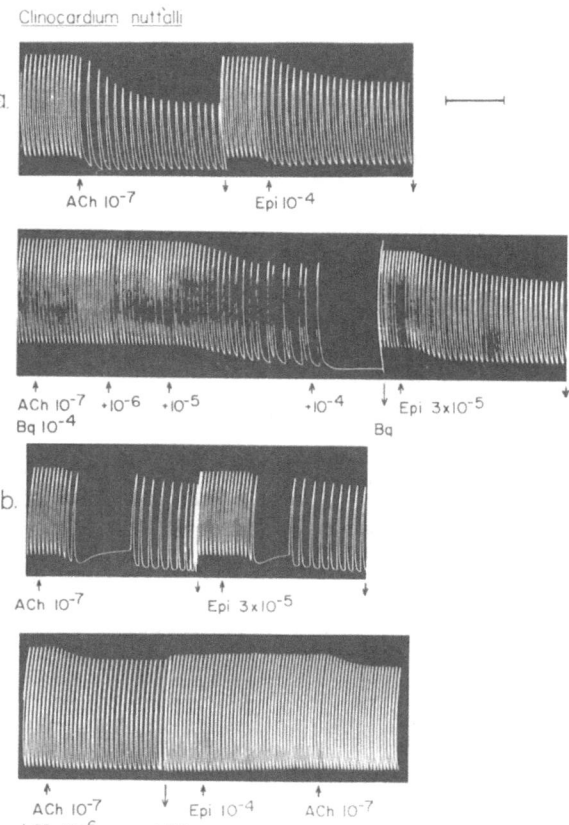

In experiments with *Clinocardium* hearts, LSD partially blocked ACh and completely antagonized the effect of epinephrine (Fig. 2b). Interaction between serotonergic and cholinergic systems was unexpected but had been discussed from time to time[14–16].

The interaction between 5-HT and LSD and its analogs was also tested on *Clinocardium* hearts. First, UML and BOL have virtually no effect on the 5-HT response (Fig. 3a). On the other hand, LSD itself was an effective antagonist of 5-HT, but was even more effective against epinephrine (Fig. 3b). Therefore, it was interesting, especially with regard to the depressor effect of epinephrine, that about 10% of the cockle hearts tested were depressed in tone by 5-HT (Fig. 3b).

The results obtained with *Clinocardium* could, to some extent, be repeated with *Dinocardium robustum* and *Serripes groenlandicus.* However, *Dinocardium* hearts were usually weakly excited by high doses of epinephrine, and hypo-

Fig. 3.
The effect of 5-hydroxy-
tryptamine (5-HT) and
epinephrine (Epi) on
the isolated heart of
Clinocardium nuttalli.
a) The effect of 5-HT
is not blocked by
methysergide (UML)
10^{-6} M. b) Lysergic acid
diethylamide (LSD),
10^{-6} M, produces only
a small increase in
amplitude (compare
with 5-HT 10^{-7} M). The
depressor effects of
5-HT and Epi and the
positive inotropic effect
of 5-HT are antagonized
by LSD. Recordings
a) and b) are from
different preparations.
Drugs are added at the
upward-pointing
arrows; downward-
pointing arrows indicate
washing with the kymo-
graph drum stopped for
10 min. UML and LSD

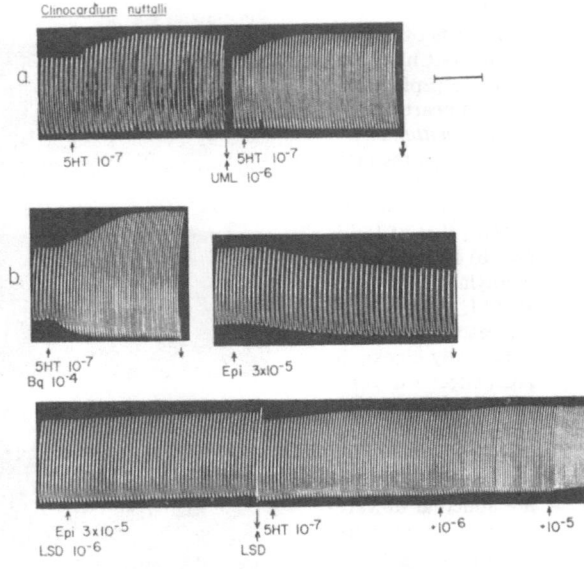

were added to the bath
10 min. before the sub-
sequent dose of 5-HT
or Epi. b) was done in
the presence of benzo-
quinonium (Bq) which

had no effect on the
effects of 5-HT and Epi.
Dose: moles/liter.
Time scale: 1 min.
Temp: 15 °C.

dynamic preparations were excited by LSD (10^{-6}M). Many of the responses
mentioned here were also found by GADDUM and PAASONEN[17] in *Cardium edule*.
Hearts from this species were depressed by both epinephrine and norepinephrine, and were finally stopped in diastole by these drugs at a concentration
of about 2×10^{-6}M. LSD, over a period of 1 or 2 h, tended to depress the
amplitude of the *Cardium edule* heart at rather low doses (3×10^{-9} to 3×10^{-8}M).
The catecholamines and 5-HT were also antagonized by LSD, but the effect
was not so striking as that observed with *Clinocardium*. In general, however,
the cardiac pharmacology in the Family Cardiidae appears to be similar from
species to species.

The Tridacnadae are a family of giant clams in the same superfamily
(Carditacea) as the cockles. The cardiac pharmacology of one of the species of
this family, *Tridacna fossor*, was briefly investigated at the Marine Laboratory
of the Great Barrier Reef Committee on Heron Island, Queensland, Australia.
The hearts of *Tridacna* were inconveniently large; often they could be stuffed
into the muscle bath only with difficulty. Therefore, ventricular strips, consisting
of bundles of parallel muscle trabeculae were prepared. These beat regularly
and responded to added drugs. As with the cockles, LSD 10^{-6}M had almost
no effect on *Tridacna* ventricular muscle, but could, in about half the experi-

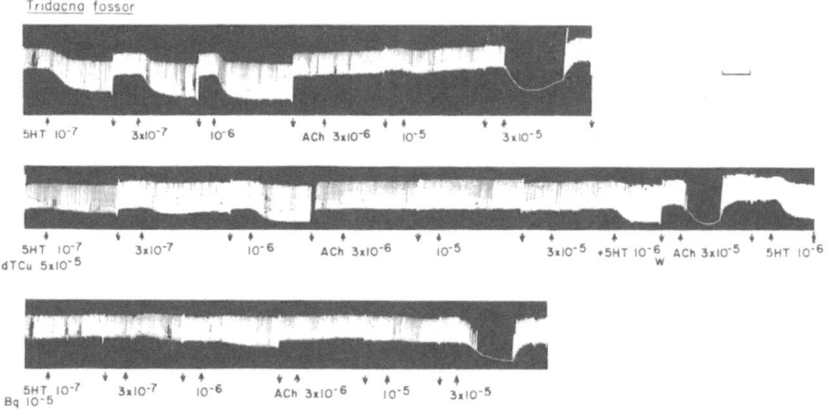

Fig. 4.
The effects of 5-hydroxytryptamine (5-HT) and acetylcholine (ACh) on the isolated heart of the giant clam, *Tridacna fossor*. Drugs were added at upward-pointing arrows; downward-pointing arrows indicate washing with the kymograph drum stopped for 10 min. Top row: Test doses of 5-HT and ACh. Second row: Test doses repeated in the presence of d-tubocurarine (dTCu), added after each wash and 10 min. before the subsequent dose. At W, dTCu was washed out and not replaced; its effects are reversible. Third row: Test doses repeated in the presence of benzo-quinonium (Bq) as above. Dose: moles/liter. Time scale: 1 min. Temp: 27 °C.

ments, block the response to 5-HT. Epinephrine had no effect up to 3×10^{-5} M. Of special interest, however, was the large decrease in diastolic tone produced by 5-HT on nearly all the preparations (Fig. 4). This depression could, in some experiments, be blocked by both *d*-tubocurarine and benzoquinonium although the latter was more effective. On the other hand, ACh depression was more effectively blocked by *d*-tubocurarine. Thus, the 5-HT depression might have been caused, at least in part, by a neurogenic action resulting in a release of ACh the effect of which, in turn, could be blocked by benzoquinonium. Presumably, ACh, added to the perfusion bath and acting directly on muscle, has other sites of action which were more effectively antagonized by *d*-tubocurarine. A similar pattern of responses by the clam rectum has been reported[18].

Tridacna, and the cockles to a lesser extent, are not the only bivalves whose beating is depressed by 5-HT. This experiment can also be repeated with hearts from *Cyrtopleura* (= *Barnea*) *costata*, and the 5-HT depression again can be blocked by benzoquinonium.

Some Notes on Mussel Hearts

If isolated hearts from the Atlantic ribbed mussel, *Modiolus demissus*, are tested with 5-HT, the response is characteristically a profound depression especially of tone. Threshold is relatively low—about 10^{-9} to 10^{-8} M. Of some interest is the fact that other species in the genus *Modiolus*, *M. americanus* (Fig. 5) and *M. modiolus*, do not respond in this way. Depression of the heart by 5-HT remains, therefore, as a taxonomic feature of *M. demissus*.

Fig. 5.
The effect of 5-hydroxy-tryptamine (5-HT) on isolated hearts of *Modiolus demissus* (top and bottom records) and *M. americanus* (middle record). 5-HT added to each preparation, from the top down at the marker. The bath was washed out between doses with the polygraph stopped for 5 min. Dose: moles/liter. Time: 1 min. Temp: 21 °C.

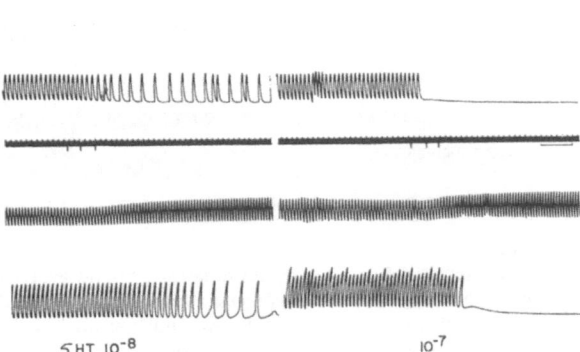

This depression differs from that in *Tridacna* and *Cyrtopleura* in that it is not readily ascribable to the release of ACh. In the first place, the action of ACh is usually solely excitation even when potentiated, as it is to a rather large extent, by eserine sulfate (Fig. 6). In addition, the effect of 5-HT on the *Modiolus* heart is unaffected by either eserine or benzoquinonium.

Of course, 5-HT is known to have a relaxing effect on other molluscan muscles, in particular the anterior byssus retractor muscle (ABRM) of *Mytilus edulis*. HIDAKA et al.[19] showed that 5-HT had no effect on membrane potential and suggested that the relaxing effect of 5-HT on the ABRM is due to a lowering of intracellular free Ca^{++}. In *Modiolus* heart, however, the tension fall is clearly associated with membrane hyperpolarization (Fig. 7).

To complete the story in mussels as far as is possible, it should be mentioned that *Lithophaga bisulcata* hearts are also depressed by 5-HT. In summary, *Mytilus* spp., *Modiolus americanus*, and *M. modiolus* hearts are excited by 5-HT, as most bivalve hearts so far tested seem to be. The *Lithophaga* heart is depressed by 5-HT, but the action is blocked by benzoquinonium. Finally in *M. demissus*, the depression is virtually unaffected by this ACh blocking agent.

The Variable Action of Substance X

Water extracts of the heart and ganglia of the fresh water mussels *Amblema neissleri* and *Elliptio sloatianus* have cardio-excitor activity[7]. If the extracts are filtered through a column of Sephadex G-15, a number of active fractions are obtained. These fractions are usually assayed on the 'standard' *Mercenaria mercenaria* heart preparation. If, however, the effects of the active fractions on

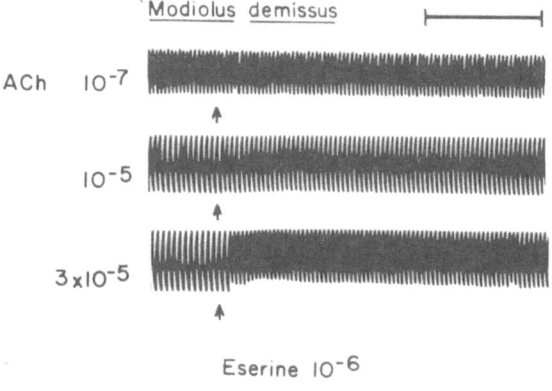

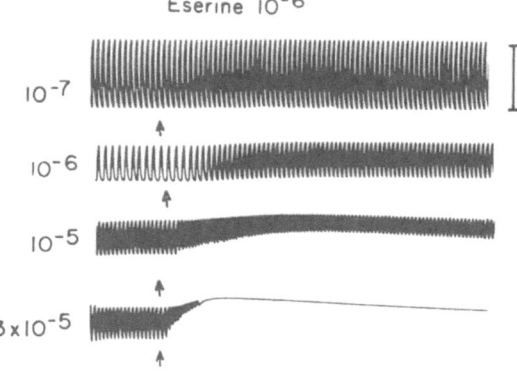

Fig. 6.
The effect of acetyl-
choline (ACh) on the
isolated heart of *Modio-
lus demissus*, before and
after treatment with
eserine. The bath was
washed out for 10 min.
between each dose. In
the lower series re-
sponses, eserine was
added to the bath
immediately after the
wash. Dose: moles/
liter. Time: 1 min.
Tension: 500 mg.
Temp: 21 °C.

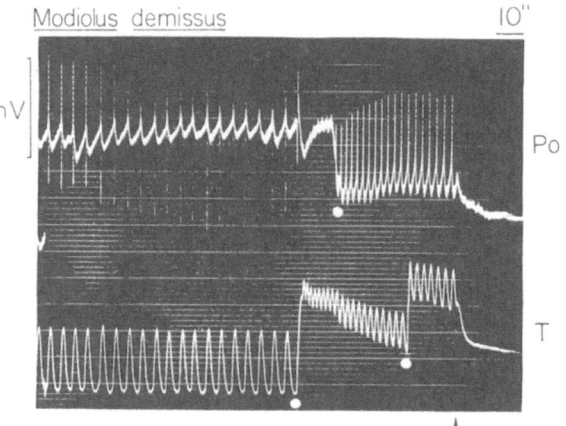

Fig. 7.
The effect of 5-hydroxy-
tryptamine (5-HT) on
the heart of *Modiolus
demissus*. T is tension;
Po is the potential
recording by means of
a suction electrode.
Initially, typical external
diphasic action poten-
tials are recorded. At
the first mark (lower
trace), the electrode is
pressed against the heart,
producing the observed
mechanical disturbance.
At the second mark,
(upper trace), suction
is applied, resulting
in hyperpolarization,
and then recording of
monophasic action po-
tentials. The third mark
(lower trace) indicates
tension balancing. At the
arrow 5-HT (10^{-5} M) was
added to the perfusion
fluid. Time: 10 sec.
Pot: 5 mV.

17 Exper. Suppl. 15

Mercenaria hearts are compared with their effects on *Amblema* hearts, it becomes apparent that the responses of the two hearts to the same fractions are strikingly dissimilar[7] (Fig. 8).

Amblema neisleri

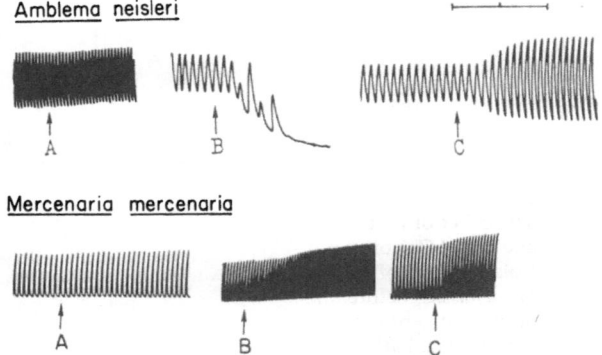

Fig. 8.
The parallel assay of
three fractions of
Substance X from
pooled ganglia of
Amblema neisleri on
isolated hearts of
Amblema and *Merce-
naria mercenaria.* Hearts
were washed after each
dose. Time: 2 min.
Temp: 21 °C.

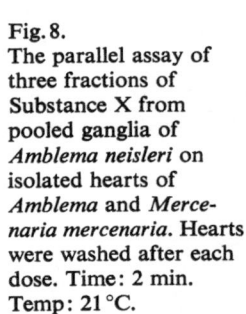

Mercenaria mercenaria

Some Aspects of Oyster Heart Pharmacology

The response of isolated oyster hearts to ACh is distinct from those of venerid and cardiid bivalves in two ways especially. Firstly, although ACh has a depressor action, the threshold is high; usually it is between 10^{-7} and 10^{-6} M, compared with about 10^{-10} M for Veneridae[10]. Secondly, a characteristic ACh effect on many *Crassostrea virginica* or *C. gigas* heart preparations is a marked decrease in frequency and tone accompanied by an increase in the total amplitude of beat (see Fig. 7 in ref.[10]).

ACh has never been proven to be a physiological neurohumor in oysters. Although OKA[20] recorded inhibition of the heart of *Ostrea circumpicta* when the visceral ganglion was electrically stimulated, no evidence relates this inhibition to the release of ACh. Still, VINCENT and JULLIEN[21] found ACh in the heart of *Ostrea edulis*, and depression of isolated oyster hearts by ACh has been reported often (reviewed by GREENBERG[10]).

I have studied the structure-activity relations of ACh on the isolated hearts from a number of oysters, but in particular *Crassostrea nippona*. The structure-activity profile of the cardiac ACh receptors in this species is similar to that in *M. mercenaria* with respect to the same series of *n*-alkytrimethylammonium and ketoamyltrimethylammonium ions (Table)[2, 22, 23]. That is, in spite of quantitative variation, optimal activity among the alkyltrimethylammonium ions is produced by the *n*-amyl member of the series, while 4-ketoamyltrimethylammonium is the most potent of the ketoamyl analogs. Furthermore, not only the depressor receptors of ostreid and venerid hearts, but also all other depressor receptors which have been examined are similar in this regard[24].

An examination of hearts from *C. virginica* and *C. gigas* suggested that the high threshold to ACh, and the inability of high doses to stop the heart permanently, might be due to a high titer of acetylcholinesterase (AChE)[10].

Treatment of *C. virginica* hearts with eserine (10^{-5} g/ml) decreased the threshold by 25–30 times. The effect of eserine on *Mercenaria* hearts is about 10 times smaller[25]. Recently the cholinesterase activity of hearts and ganglia of *C. virginica* and *M. mercenaria* have been compared[26]. While the enzyme activity of the ganglia of the two species was about the same, the activity of the oyster heart was at least 100 times greater than that of the clam heart. The high titer of AChE in the oyster ventricle is probably correlated with a high ACh content; in fact, VINCENT and JULLIEN's measurement[21] of ACh in the heart of *Ostrea circumpicta* was 7 times that found by WELSH[13] for *Mercenaria*. In summary, the high threshold of ACh in its action on oyster hearts, as compared with the 'standard', *Mercenaria*, is not due to any *qualitative* difference in the neurohumor, hydrolyzing enzyme, or receptor. It is explainable in *quantitative* terms, and particularly by the high AChE activity in oyster hearts.

The second characteristic of oyster hearts—the increase in amplitude accompanying a fall in frequency and tone—has been observed in a number of species from diverse families (see Table 2 in GREENBERG[10]). However, the response appears to be much more marked, and occurs with greater frequency, in oysters. A possible explanation of this phenomenon comes from the investigations of IRISAWA and his co-workers, who have been studying the electrophysiology of oyster, as well as mussel, hearts. They have found that ACh doses which inhibit the oyster heart hyperpolarize the membrane and increase spike height[27]; these are the expected electrical correlates of the mechanical responses which have been seen. However, the resting potential of *C. gigas* is low (45.5 mV)[28]. At this potential, sodium permeability is low due to inactivation. Thus, hyperpolarization by ACh may increase sodium permeability resulting in an increased spike height and amplitude of beat of the heart[27]. The resting potential of *Mytilus edulis* heart cells is also low (about 44 mV)[29], but in this preparation ACh depolarizes the membrane. Heart cell resting potentials of *Modiolus demissus* and *Amblema neissleri*, measured in this laboratory, were 50–60 mV. The tone and frequency of the *Amblema* heart are decreased by ACh but the amplitude of beat does not increase markedly, as it does in oysters. Thus, it appears that the small membrane potential of oyster hearts might explain their response to ACh hyperpolarization.

Discussion

The pharmacology of *Mercenaria mercenaria* hearts is not representative of all species of bivalves. Although there are more data available now, the evidence that this must be so has, in reality, been accumulating since JULLIEN and VINCENT's investigation[21] on the heart of *Mytilus galloprovincialis*. GADDUM and PAASONEN's[17] survey of the 5-HT responses of a variety of bivalve heart preparations, and especially PILGRIM's[30] comparative study of ACh effects suggested strongly that the number of 'exceptional' preparations must be quite large. Further work, such as that reported earlier[10] and in the present paper, has served to reinforce this notion. Notwithstanding the wide distribution of variation in pharmacological responses, one must next ask whether such responses are ordered in any systematic way. Our present considerable experience with bivalve heart pharmacology should give us as good a chance as is possible to predict the effects of drugs on the heart of the next species of clam we try.

One possibility is that pharmacological responses are correlated with the environment of the animal. That this is not the case is, by now, abundantly clear. The matter has previously been discussed[10], and subsequent data have only served to support the view that an animal's habitat is unrelated to its cardiac pharmacology. A more promising possibility is that species variation in clam pharmacology is correlated with the systematics of the Class Bivalvia.

First of all, the Class may be divided[31] into three subclasses, one of which (Lamellibranchia) includes the vast majority of species, including all those whose cardiac pharmacology has been examined. Further subdivision of the Lamellibranchia into 6 orders is on the basis of relative size of the muscles, and the structure of shell, gills, and mantle. None of these orders is homogeneous with regard to pharmacology. For one example, pinnids, mussels, and oysters are all in the Order Anisomyaria and their hearts all respond differently to ACh. Again, venerids and cockles are both in the Order Heterodonta and their hearts respond differently to serotonergic drugs. On the other hand, hearts from clams of different orders such as *Tridacna fossor* (Heterodonta) and *Cyrtopleura costata* (Adapedonta) may be affected similarly, in this case by 5-HT.

Within the next lower taxonomic level, the family, however, bivalve cardiac pharmacology tends to become connatural. Still, variations occur and the extent of these depends on the family and the drug. For example, while the effects of ACh are similar on 9 species from 3 genera of mytilids, the pharmacology of 5-HT is not at all uniform in this family. On the other hand, the responses of heart preparations among species within such families as Ostreidae, Cardiidae, Veneridae, and Mactridae have been abundantly tested and such variation as exists is quantitative; most dose-effect relations are similar[32]. Nonetheless, even in these families unusual phenomena have been reported[33,34].

Of the compounds which have been extensively investigated, catecholamines appear to produce the most variable responses both interspecifically and from preparation to preparation within the same species.

In summary, the various mechanical responses of bivalve hearts to neuromuscular drugs seem to have arisen independently many times during the history of the Class. Once having appeared, however, they are sufficiently stable so that knowledge of the family of a particular bivalve often facilitates a prediction as to how its heart will react to pharmacological agents. The precision of the prediction is limited, at present, by the particular family and drug under consideration.

The polyphyletic, variable responses described above have been observed on homologous muscle preparations — isolated ventricles — obtained from species in one subclass of molluscs. The results suggest that pharmacological studies which, for example, attempt to compare the effects of drugs on a variety of visceral and skeletal muscles from a scattering of species representing a number of diverse phyla, may be relatively meaningless[35].

The hearts of bivalve molluscs constitute a class of homologous muscles obtained from a relatively homogeneous group of animals. Nevertheless, these muscles respond diversely to pharmacological stimuli. Consequently, bivalve hearts become, in their sum, a tool by means of which we can attack the general question: What is the nature of species variability in chemical transmission and of the pharmacology associated with this process? An approach to this question is to examine the variation within each part of the transmission process. A first consideration might include simply: (1) The array of neurohumors having a physiological role in the heart of each species, and their relative concentrations, (2) the relative activity of the enzymes associated with each neurohumor, (3) the nature of the receptors with which the neurohumors interact, and (4) the mechanism activated by the complexing of the neurohumor and a receptor.

(1) The wide occurrence of ACh, 5-HT, and catecholamines (predominantly dopamine) in the nervous systems of various bivalves is, by now, well established[3,5,6]. This distribution is not surprising since these compounds are common biochemicals as are the enzymes involved in their metabolism. They are widespread in the nervous and non-nervous tissues of animals and in plants as well[36,37]. Substance X, or at least some of its components, are also being identified wherever they are sought[7,8]. The fact that similar active fractions have been obtained from ganglion extracts from Gastropoda as well as Bivalvia[8], and that similar gel filtration fractions from different species give the same differential assays on *Amblema* and *Mercenaria* hearts[7], suggest that these substances may also have a universal distribution. They may also turn out to be components of various unidentified cardioactive substances from nervous tissue extracts of other molluscs[38,39].

That the amount of ACh present may vary from species to species has already been mentioned. The concentration of 5-HT in bivalve pooled ganglia also varies, though surprisingly little, with species[40,41]. While the principal catecholamine in bivalves is usually dopamine, in *Sphaerium sulcatum* norepinephrine and dopamine are present in comparable concentrations[6]. SWEENEY also found[6] that no monoaminergic motor innervation runs to the heart in *Sphaerium*, suggesting that particular neurohumors may actually be absent from particular tissues in some species.

Of course, mere occurrence of a chemical substance does not necessitate its function as a transmitter agent. Rather, the distribution of these substances suggests that they meet other physiological ends. The proofs that an active drug is, in fact, a neurohumor are well known. However, only ACh and 5-HT have been shown on strong evidence, obtained from only a few species, to play a physiological role in regulating the hearts of bivalves[3,5].

(2) The enzymes involved in the synthesis and oxidation of 5-HT have been identified in various molluscan tissues for some time, as has acetylcholinester-

ase[5]. Only recently have we been able to show, however, that species differences in acetylcholinesterase activity in the hearts of oysters and venus clams could account for observed differences in the responses of these organs to ACh.

(3) Receptors are defined by pharmacological operation, that is, by the drugs and their antagonists which act on a particular preparation, and by their relative potencies. The definition does not imply that the pharmacological receptor is a site of action for a physiological transmitter, although it might be. In this operational sense, two types of ACh receptors have been defined in bivalve hearts, and evidence that they are both physiological has been presented[22,42]. The receptors are identified by challenging the heart with a series of ketoamyltrimethylammonium ions. 'Depressor' receptors respond optimally to 4-ketoamyltrimethylammonium. *Mercenaria*[2] and oyster hearts have only this type of receptor. 'Excitor' receptors respond optimally to the 3-keto compound and predominate in *Mytilus* hearts[24]. *Tresus capax* and *Spisula solidissima* have both types of receptor[24].

Multiple binding sites with which a neurohumor interacts, presumably in different conformations, have been called isoreceptors[43]. The term is naturally applied to the muscarinic and nicotinic receptors in vertebrates. In bivalve hearts both sites are nicotinic[2,24]. Their location in the heart, presynaptic or postsynaptic, has not been established.

The concept of isoreceptors implies the possibility of biological economy, in that a few small, flexible molecules can have a variety of effects even in the same tissue, providing only that sites are available to which they can bind in different conformations. If there are isoreceptors with opposing actions and different thresholds on the same postsynaptic membrane, then a neuron releasing a single transmitter can supply its own feedback. Such a situation has been described for ACh in *Aplysia* abdominal ganglion[44].

If isoreceptors are at different postsynaptic membranes, then each neurone, releasing the same neurohumor, might transmit opposing effects from separate sites. As a purely speculative example, the heart of *Clinocardium nuttalli* (Cardiidae) receives excitatory impulses from the cerebral as well as the visceral ganglion[45]. The heart of *Tapes watlingi* (Veneridae)[33,34] receives cholinergic depressor neurones and serotonergic excitor neurones from the visceral ganglion. Isolated *Tapes* hearts, in the presence of benzoquinonium, are excited by ACh, but this excitation is not mimicking the cardiac excitation produced by stimulation of the visceral ganglion[31,32]. It might be that there are, in bivalves, excitatory postsynaptic cholinergic receptors of motor neurones from the cerebral ganglion.

(4) Upon adsorption of the neurohumor to its receptor, the resulting complex, perhaps by changing the molecular conformation of the membrane, alters the membrane permeability more or less selectively. For example, IRISAWA and SHIGETO[46] have shown that the excitatory action of ACh on the *Mytilus* heart is due to an increased permeability to Na^+ ion. On the other hand, the depressor effect of ACh on the heart of the oyster is due to an increased per-

Table. Relative activities of some acetylcholine analogs on the hearts of *Crassostrea nippona* and *Mercenaria mercenaria*.

Analog	Structure	Relative potency*		(Ref.)
		Crassostrea	*Mercenaria*	
A. *n-ketoamyltrimethylammonium ions:*				
4-ketoamyltrimethylammonium	$CH_3C\ CH_2CH_2CH_2N(CH_3)_3$ ($=O$)	1	1	[21]
3-ketoamyltrimethylammonium	$CH_3CH_2C\ CH_2CH_2N(CH_3)_3$ ($=O$)	30	13	[21]
2-ketoamyltrimethylammonium	$CH_3CH_2CH_2C\ CH_2N(CH_3)_3$ ($=O$)	>30	52	[21]
B. *n-alkyltrimethylammonium ions:*				
n-amyltrimethylammonium	$CH_3CH_2CH_2CH_2CH_2N(CH_3)_3$	1	4	[2]
n-butyltrimethylammonium	$CH_3CH_2CH_2CH_2N(CH_3)_3$	5	19	[2]
n-hexyltrimethylammonium	$CH_3CH_2CH_2CH_2CH_2CH_2N(CH_3)_3$	>10	29	[2]
n-propyltrimethylammonium	$CH_3CH_2CH_2N(CH_3)_3$	50	28	[2]

*) Potency is expressed as the equiactive molar ratio with the effect of 4-ketoamyltrimethylammonium ion equal to 1.

meability to Cl⁻ ion. Since the ACh excitor receptor of *Mytilus* and the depressor receptor of *Crassostrea* are different, as described above, one might suggest that the groupings: (excitation; 3-keto optimum; increased Na⁺ permeability) and (depression; 4-keto optimum; increased Cl⁻ permeability) were necessary associations. However, the ionic mechanisms seem to be widespread in Mollusca[47], and therefore 5-HT, in *Modiolus demissus*, might also act by a Cl⁻ mechanism. Also, while the vertebrate catecholamine α-receptor usually has a pressor action, it has a depressor action in the heart.

All biological phenomena have two aspects: On the one hand, they are general in that analogs can usually be found in most living organisms. On the other hand, the details, even in closely related species, are notoriously variable.

The generality in the transmitter processes considered here, no doubt, resides in the conservative nature of the components involved. Adsorption of common biochemicals, such as acetylcholine, to membrane surfaces undoubtedly evolved early as a mechanism for changing the conformation, and hence the permeability, of the membrane to any one, or a number, of the ions in the external environment. Thus, one might hypothesize that all molluscs have, and have had from the beginning, the same array of neurotransmitters and their associated enzyme systems. Furthermore, to the extent that the relative ion concentrations in the extracellular fluid are similar in animals, the receptor mechanisms would also be similar.

Species variation would seem to rest, first, with quantitative differences in the concentration of the neurohumors, enzyme systems, and receptor sites, and secondly, with the variability of the receptors, and of the particular association they make from among the available ionic channels[48].

References

1   See p. 250.
2   J. H. WELSH and R. TAUB, J. Pharmac. exp. Ther. *99*, 334–342 (1950).
3   G. A. COTTRELL and M. S. LAVERACK, Ann. Rev. Pharmac. *8*, 273–298 (1968).
4   A. M. WRIGHT, M. MOORHEAD and J. H. WELSH, Br. J. Pharmac. *18*, 440–450 (1962).
5   R. B. HILL and J. H. WELSH, *Physiology of Mollusca* (Ed. K. WILBUR and C. M. YONGE; Academic Press, New York 1964), vol. 2, p. 126–174.
6   D. C. SWEENEY, Comp. Biochem. Physiol. *25*, 601–613 (1968).
7   R. A. AGARWAL and M. J. GREENBERG, Proc. XXIVth Intern. Cong. Physiol. Sci., Washington *7*, 4 (1968).
8   N. FRONTALI, L. WILLIAMS and J. H. WELSH, Comp. Biochem. Physiol. *22*, 833–841 (1967).
9   E. FLOREY, Comp. Biochem. Physiol. *20*, 365–377 (1967).
10 M. J. GREENBERG, Comp. Biochem. Physiol. *14*, 513–539 (1965).
11 C. L. PROSSER, *Comparative Animal Physiology* (Saunders, Philadelphia 1950), p. 95.
12 M. J. GREENBERG, Br. J. Pharmac. *15*, 365–374 (1960).
13 J. H. WELSH, J. Mar. Biol. Ass. U.K. *35*, 193–201 (1956).
14 M. H. APRISON, Recent Advance Biol. Psychiat. *4*, 133–146 (1962).
15 N. J. GIARMAN and D. X. FREEDMAN, Pharm. Rev. *17*, 1–25 (1965).
16 D. A. BOROFF and U. FLECK, J. Pharmac. exp. Ther. *157*, 427–431 (1967).
17 J. H. GADDUM and M. K. PAASONEN, Br. J. Pharmac. *10*, 474–483 (1955).

[18] M. J. GREENBERG and T. C. JEGLA, Comp. Biochem. Physiol. *9*, 275–290 (1963).

[19] T. HIDAKA, T. OSA and B. M. TWAROG, J. Physiol., London *192*, 869–877 (1967).

[20] K. OKA, Sci. Rep. Tohoku Imp. Univ. 4th series, Biol. *7*, 133–143 (1932).

[21] D. VINCENT and A. JULLIEN, C.r. Soc. Biol. *127*, 334 (1938).

[22] J. H. WELSH and R. TAUB, Science *112*, 467–469 (1950).

[23] J. H. WELSH and R. TAUB, J. Pharmac. exp. Ther. *103*, 62–73 (1951).

[24] M. J. GREENBERG, Comp. Biochem. Physiol. (1969), in press.

[25] J. H. WELSH and R. TAUB, Biol. Bull. *95*, 346–353 (1948).

[26] T. ROOP and M. J. GREENBERG, Am. Zool. *7*, 737–738 (1967).

[27] H. IRISAWA, A. IRISAWA and N. SHIGETO, this volume.

[28] H. IRISAWA, A. NOMA and R. UEDA, Jap. J. Physiol. *18*, 157–168 (1968).

[29] H. IRISAWA, N. SHIGETO and M. OTANI, Comp. Biochem. Physiol. *23*, 199–212 (1967).

[30] R. L. C. PILGRIM, J. Physiol., London *125*, 208–214 (1954).

[31] J. E. MORTON and C. M. YONGE, *Physiology of Mollusca* (Ed. K. M. WILBUR and C. M. YONGE; Academic Press, New York 1964), vol. 1, p. 4–5.

[32] G. C. CHONG and J. W. PHILLIS, Br. J. Pharmac. *25*, 481–496 (1965).

[33] J. W. PHILLIS, Comp. Biochem. Physiol. *17*, 719–739 (1966).

[34] P. R. CARROLL, G. B. CHESHER, D. F. H. DOUGAN, Comp. Biochem. Physiol. *25*, 913–920 (1968).

[35] N. V. KHROMOV-BORISOV and M. J. MICHELSON, Pharm. Rev. *18*, 1051–1090 (1966).

[36] V. P. WHITTAKER, *Handbuch der experimentellen Pharmakologie* (Subed. G. B. KOELLE; Springer-Verlag, Berlin 1963), vol. 15, p. 1–5.

[37] V. ERSPAMER, *Handbuch der experimentellen Pharmakologie* (Subed. V. ERSPAMER; Springer-Verlag, Berlin 1966), vol. 19, p. 132–165.

[38] M. J. GREENBERG, Am. Zool. *2*, 526 (1962).

[39] C. P. JAEGER, Comp. Biochem. Physiol. *17*, 409–415 (1966).

[40] J. H. WELSH and M. MOORHEAD, Science *129*, 1491–1492 (1959).

[41] J. WELSH and M. MOORHEAD, J. Neurochem. *6*, 146–169 (1960).

[42] M. J. GREENBERG and C. F. McMURRICH, Proc. XXIIIth Intern. Comp. Physiol. Sci. Tokyo 154 (1965).

[43] H. C. MAUTNER, Pharm. Rev. *19*, 107–144 (1967).

[44] H. WACHTEL and E. R. KANDEL, Science *158*, 1206–1208 (1967).

[45] G. E. SILVEY, Comp. Biochem. Physiol. *25*, 257–269 (1968).

[46] H. IRISAWA and N. SHIGETO, personal communication (1968).

[47] M. SATO, G. AUSTIN, H. YAI and J. MARUHASHI, J. gen. Physiol. *51*, 321–346 (1968).

[48] Research supported by grants from the U.S.P.H.S. (HE-09283; National Heart Institute); N.S.F. (Senior Postdoctoral Fellowship); and the U.S.A.E.C. (Div. of Biology and Medicine). I wish to thank the staffs of the marine laboratories at Friday Harbor (University of Washington), Alligator Harbor (Florida State University), Heron Island (Great Barrier Reef Committee), and Misaki (Tokyo University) where much of this work was done. I am in debt to Dr. R. BIRCHER (dec.) (Sandoz Pharmaceuticals) for supplying UML-491 and BOL-148, and to Dr. A. M. LANDS (Sterling-Winthrop Research Institute) for Mytolon chloride (benzoquinonium).

Conclusion and Summary of the Proceedings

by G. A. KERKUT

It is my pleasure and privilege to provide the concluding paper of this meeting. I do not intend to give a detailed account of the factual material presented by the various authors since the authors are more capable of this than I am and the reader will have the relevant material in the preceding pages. Instead I should like to give a personal view concerning some of the features and generalizations that have come from this meeting. Another person would come to different conclusions and this, in a way, is a point in favour of this type of symposium where experts in different but associated subjects can come and talk to each other; what may be an old idea to one person can be a valuable new idea to another and it is this symbiosis of persons and ideas that has made a success of the symposium.

Multiple Techniques

We have had at this meeting experts in different techniques. One striking point has been that, without exception, the speakers have all been concerned with the combination of two or more techniques to their studies. Thus the electron-microscopist has been concerned with setting up experimental conditions so that he can see how the heart muscle cell connections change under different osmotic or pressure changes. The biophysicists have been concerned with the relationship of the physical measurements with the biochemical data. The electrophysiologists have been concerned with the morphological and electron microscope data of their preparations so that they know exactly into what they are sticking their microelectrodes. Thus biochemists, biophysicists, electrophysiologists, embryologists, morphologists have all metamorphosed into people who have lost these special labels and have been more concerned with understanding the various problems of how heart muscle functions. The real concern is the problem itself and one is no longer bothered with a personal identification such as 'I am a Biochemist'. Such an identification is no more complete or helpful than the statement 'I smoke Lucky Camel cigarettes'. The various methods are not ends in themselves, instead they are just means to solve a problem and an investigator can be a pharmacologist one month and a biochemist the next month (providing he has the necessary training and experience), according to the demands of the problem. One often comes across a strange dichotomy in people who feel that because their interest is in biochemical methods of solving problems they must work in a Department with the word Biochemistry over the door. To some extent we are all prisoners of our own education and only feel safe within the walls of its confines. Perhaps one could go to the opposite position and say that the best developments will come from biochemists who are not working in Biochemistry departments but instead in

Physiology, Pharmacology or Biophysics departments. The study of living systems demands an integrated approach in which the various techniques are combined to make the attack. It is the living system that is being studied and not an abstract concept such as Biochemistry, Physiology or Molecular Biology.

Symposia such as the present one, where the living system reigns supreme and where we have to use all our ingenuity, provide the challenge for us to loose our identifying labels and to have a good look instead at the problems of cardiac muscle.

Simple Versus Complex

There are many views concerning the definition of Comparative Physiology. One is that the simpler animals will have simpler systems and thus be easier, to analyse. I do not hold this view. All animals have to compete for survival and reproduction and they do not exist to illustrate the Principles of Evolution. It is, however, possible with the considerable range of animal species to find some animals whose anatomy or physiology allows a more critical, or easier, investigation of a given system. Thus the giant axon of the squid allowed considerable advances to be made in the study of peripheral nerve. Note also that the squids are among the most complex of the invertebrates and that similar studies have not been possible on the simplest invertebrates—the Coelenterates or the Flatworms. Similarly, the cardiac ganglion of *Panulirus* with its 9 neurones has allowed the neuronal interaction and coordination in control of the heart to be studied in very great detail. The differences between neurogenic and myogenic hearts are well shown by both comparative and also embryological studies.

The heart of the hagfish *Myxine* is free from any innervation. Yet the electron micrographs show the presence of secretory granules in cells that otherwise would be called neurosecretory cells. This raises the interesting problem of the endocrine control by the heart. It has recently been suggested that the mammalian heart can produce a steroid that will help control the fluid volume and also the uptake of sodium ions, this steroid being produced by the atrial tissue (LOCKETT 1967, J. Physiol. *193*, 661).

We have been particularly lucky in this symposium in that many of the people who have worked on the Ascidian heart have been present (Drs. FLOREY, JONES, KRIEBEL and ANDERSON), and it was very interesting to hear their discussion of the various problems of this strange heart with its reversal of beat and its economic structure. Quite often during the discussion about the heart one wondered just what the heart was doing in these animals. It is possible that in some animals that have a low blood pressure and an open haemocoel the heart could have a very restricted role. This role could be to provide blood directly to the brain. In the snails the brain is bathed in a well-developed sinus system enclosed by blood vessels and this system receives a good blood supply from the aorta. It is likely that the job of the heart in the snails is to see that the nervous system is well supplied. A similar system most likely holds for the lobsters where the brain quickly fails if its oxygenated supply is cut off. Our ability to differentiate between clever and stupid lobsters or snails is limited to some extent, and so we cannot easily tell if the lobster has part of the brain destroyed unless that part is directly concerned with motor control.

Even in the mammals we know that the brain and heart have the first call on the oxygen capacity of the system and in man the brain, though only 10% the body weight, takes over 25% of the body's oxygen. The studies on the physiology of diving animals have thrown considerable light on the normal functions of the heart and circulatory system even though the diving animals are quite specialized and not necessarily simple. One generalization that can be made about the comparative studies is that for each series of experiments there is usually some animal that will allow precise experimentation to be carried out and which will provide unequivocal results. It may require the services of the ship 'Alpha Helix' or 'Apollo X' to find the animals, but they can be found!

Typical Records

Periodically throughout this meeting we have heard about typical records and atypical records. Some members have had feelings of guilt in that their atypical records, for example, do not show all the features of the EKG. I should like to stand up and support the atypical records. Of course the investigator must have sufficiently good technique to obtain typical records in the 'standard' preparations but the atypical records are often the key to future developments.

To some extent we all try to reproduce the unknown in terms of the known, instead of using our controlled imagination to see what the atypical could indicate. It is possible that the most interesting results are discarded or filed away because they do not immediately fit in with the current views. Perhaps a specific example would help. For some years the membrane potentials in nerve and muscle have been explained in terms of Nernst type equilibria. More recently, however, evidence has accumulated to show that the metabolic systems of some cells can have a direct contribution to the membrane potential and when the electrogenic sodium pump is suddenly switched off (by addition of ouabain or cyanide) there is a rapid fall in the membrane potential by 5–20 mV. In some nerve cells as much as 30% of the resting potential is due to the electrogenic sodium pump. The action of ouabain on heart muscle can be partially explained in terms of this electrogenic pump being stopped; the membrane comes nearer to the depolarization level and this allows the easier production of the muscle action potential. It is likely that the electrogenic pump was noticed for some years but it seemed to be atypical.

Similarly, one should not be too depressed by the strange ionic balances across the insect heart muscle fibres, or by the fact that the isolated chick heart muscles require a reduced oxygen tension before the muscle fibres show a reasonable membrane potential. The atypical results of today can sometimes become the typical results of tomorrow.

Future Developments – Cell Sociology

Though the meeting has been concerned mainly with the work that has been done or that is being performed at present, it is always interesting to speculate as to what may be done in the near future.

One of the present-day limitations to our understanding of the functioning of living systems has been the topic of 'Cell Biology'. Though it has been a useful unifying concept, linking together workers with a common interest in 'The Cell', it is often forgotten that 'The Cell' as such does not usually exist and that we are more likely to meet tissues, organs and animals. The real subject of the present and future is not Cell Biology but Cell Sociology where one is concerned with the way in which the various cells live and interact to form a working and coordinated whole. Interest is being focussed on this point by studies such as the measurements of the resistance between adjacent cells and the discovery that there are some very low-resistance pathways between cells that could allow the passage of material from one cell to another. Investigations by electron microscopy have shown the presence of nexuses and plexuses as well as interconnections between cells. Such links could explain the manner in which the metabolism of the cells could have a common control system. We are still limited in our knowledge concerning the basis of cell-cell interaction. Some evidence is available concerning the types of interaction that take place between nerve and muscle, but with the development of radioactive pulse-labelling techniques and better analytical methods we should soon have some insight into the biochemical language between cells and the factors that keep the cells as continuing functional units within the tissues and organs.

One major problem often overlooked is that our biochemical vocabulary is still quite small and it is likely that there are many new groups of chemicals similar, say, to the prostaglandins that remain to be discovered. It may be that the coordinating chemicals in our cell society are of types as yet not known. We are limited by the types of pure chemicals that are obtainable in bottles on the shelves and often the advances depend on the supply of such chemicals being easily available. The rapid advance in the study of serotonin depended to some extent in the ready availability of pure 5-HT from the chemical suppliers. Nevertheless, it is still possible with the present-day labelling and analytical techniques to show that quite complex substances pass between cells and help maintain cell contact and tissue organization.

Similarly, there is some evidence of interaction and control between organs of the body and some evidence that the heart is able to interact with the kidney and help in the control of the body fluid volume. This type of control is probably a chemical control system. Advances in our knowledge can only be made by the cooperation of chemists, biochemists, physiologists, and pharmacologists, all helping by using their various techniques to solve the specific problems.

Finally, it has been a great pleasure to have attended this successful conference. People with different interests and views have met and talked amicably and freely. All have joined in the free exchange of views and comments and have gained by the process. We have re-met old friends and have also made new friends. To a large extent the success of the conference has been due to the manner and skill with which the conference was organized by Dr. FRANCES McCANN and I should like, on your behalf, to thank her and her colleagues for the care that they have taken of us whilst we have been here, for the excellent projection facilities during the lectures, and for the great deal of careful planning that has gone into the organization of the conference. I hope that we can all meet again in similar surroundings in a few years time to see how many new and exciting advances have been made in the Comparative Physiology of the Heart.